AF595357

Evolutionary Computation & Swarm Intelligence

Evolutionary Computation & Swarm Intelligence

Editors

Fabio Caraffini
Valentino Santucci
Alfredo Milani

MDPI • Basel • Beijing • Wuhan • Barcelona • Belgrade • Manchester • Tokyo • Cluj • Tianjin

Editors
Fabio Caraffini
De Montfort University
UK

Valentino Santucci
University for Foreigners of Perugia
Italy

Alfredo Milani
University of Perugia
Italy

Editorial Office
MDPI
St. Alban-Anlage 66
4052 Basel, Switzerland

This is a reprint of articles from the Special Issue published online in the open access journal *Mathematics* (ISSN 2227-7390) (available at: https://www.mdpi.com/journal/mathematics/special_issues/Evolutionary_Computation_Swarm_Intelligence).

For citation purposes, cite each article independently as indicated on the article page online and as indicated below:

LastName, A.A.; LastName, B.B.; LastName, C.C. Article Title. *Journal Name* **Year**, *Article Number*, Page Range.

ISBN 978-3-03943-454-1 (Hbk)
ISBN 978-3-03943-455-8 (PDF)

Contents

About the Editors

Fabio Caraffini (Ph.D.) received his B.Sc. degree in Electronics Engineering and M.Sc. degree in Telecommunications Engineering from the University of Perugia (Italy) in 2008 and 2011, respectively. He holds a Ph.D. degree in Computer Science, awarded in 2014 by De Montfort University (Leicester, UK), and a Ph.D. degree in Computing and Mathematical Sciences awarded in 2016 by University of Jyväskylä (Finland). Currently, Dr Caraffini is Associate Professor—Research & Innovation at De Montfort University (Leicester, UK) within the School of Computer Science and Informatics and the research Institute of Artificial Intelligence. His research interests include theoretical and applied computational intelligence with a strong emphasis on metaheuristics for optimisation.

Valentino Santucci (Ph.D.) is Assistant Professor in Computer Science and Engineering at the University for Foreigners of Perugia, Department of Humanities and Social Science. In 2012, he received his Ph.D. in Computer Science and Mathematics from the University of Perugia. His main research interests involve the broad areas of Artificial Intelligence and Computational Intelligence. In particular, in the field of Evolutionary Computation, his research focuses on algebraic frameworks for studying combinatorial search spaces and the dynamics of evolutionary algorithms. Other areas of interests include Natural Language Processing and Machine Learning applications to both e-learning and sustainability-related problems. He authored over forty scientific publications, organized special sessions and workshops in international conferences, and served as a guest editor for special issues in top journals.

Alfredo Milani is Associate Professor at the Department of Mathematics and Computer Science, University of Perugia, Italy. He received the title of Doctor in Information Science from University of Pisa, Italy. His research interests include several areas of Artificial Intelligence with a focus on evolutionary algorithms and applications to planning, user interfaces, e-learning, and web-based adaptive systems. He is the author of numerous international journal papers and chair of international conferences and workshops. He is the scientific leader of the KitLab research lab at the University of Perugia.

Preface to "Evolutionary Computation & Swarm Intelligence"

Stochastic optimisation is a broad discipline dealing with problems that cannot be addressed with exact methods due to their complexity, time constraints, and lack of analytical formulation and hypotheses. When a practitioner is asked to solve such problems, the most promising "tools" are heuristic approaches from the fields of Evolutionary Computation and Swarm Intelligence. These are "intelligent" approaches that, unlike exhaustive search methods, explore the search space by following nature-inspired logics, thus being capable of focussing the search on favourable areas and return a near-optimal solution within a reasonable amount of time. Even though the first intelligent optimisation algorithms where already envisaged by Alan Turing in the late '40s and early '50s, their first implementation came decades later thanks to the technological growth in the fields of Electronics, which made it possible for the realisation of personal computers and Computer Science in terms of programming languages. Nowadays, these algorithms are popular and highly employed in a variety of fields including Engineering, Robotics, and Finance, and are lately also being applied in the Health and Care sector. However, their use comes with challenges in which the research community in heuristic optimisation is trying to overcome. Amongst the most important, great efforts are being made to unveil the internal dynamics of heuristic optimisation to provide practitioners with clear indications on how to tune their control parameters to face real problems efficiently and avoid underside behaviours such as premature convergence, high generation of infeasible solutions, etc. The 10 articles forming this book reflect the current state-of-art in heuristic optimisation by showing recent advances in the application of Evolutionary Computation and Swarm Intelligence methods to real-world problems, e.g., related to Robotics, dynamic data clustering, and large-scale optimisation tasks, but also by addressing issues related to algorithmic design and algorithm benchmarking and tuning.

Fabio Caraffini, Valentino Santucci, Alfredo Milani
Editors

Article

A Clustering System for Dynamic Data Streams Based on Metaheuristic Optimisation

Jia Ming Yeoh [1], Fabio Caraffini [1,*], Elmina Homapour [1], Valentino Santucci [2] and Alfredo Milani [3]

1 Institute of Artificial Intelligence, School of Computer Science and Informatics, De Montfort University, Leicester LE1 9BH, UK; jiamingyeoh@gmail.com (J.M.Y.); elmina.homapour@dmu.ac.uk (E.H.)
2 Department of Humanities and Social Sciences, University for Foreigners of Perugia, piazza G. Spitella 3, 06123 Perugia, Italy; valentino.santucci@unistrapg.it
3 Department of Mathematics and Computer Science, University of Perugia, via Vanvitelli 1, 06123 Perugia, Italy; alfredo.milani@unipg.it
* Correspondence: fabio.caraffini@dmu.ac.uk

Received: 31 October 2019; Accepted: 10 December 2019; Published: 12 December 2019

Abstract: This article presents the Optimised Stream clustering algorithm (OpStream), a novel approach to cluster dynamic data streams. The proposed system displays desirable features, such as a low number of parameters and good scalability capabilities to both high-dimensional data and numbers of clusters in the dataset, and it is based on a hybrid structure using deterministic clustering methods and stochastic optimisation approaches to optimally centre the clusters. Similar to other state-of-the-art methods available in the literature, it uses "microclusters" and other established techniques, such as density based clustering. Unlike other methods, it makes use of metaheuristic optimisation to maximise performances during the initialisation phase, which precedes the classic online phase. Experimental results show that OpStream outperforms the state-of-the-art methods in several cases, and it is always competitive against other comparison algorithms regardless of the chosen optimisation method. Three variants of OpStream, each coming with a different optimisation algorithm, are presented in this study. A thorough sensitive analysis is performed by using the best variant to point out OpStream's robustness to noise and resiliency to parameter changes.

Keywords: dynamic stream clustering; online clustering; metaheuristics; optimisation; population based algorithms; density based clustering; k-means centroid; concept drift; concept evolution

1. Introduction

Clustering is the process of grouping homogeneous objects based on the correlation among similar attributes. This is useful in several common applications that require the discovery of hidden patterns among the collective data to assist decision making, e.g., bank transaction fraud detection [1], market trend prediction [2,3] and a network intrusion detection system [4]. Most traditional clustering algorithms developed rely on multiple iterations of evaluation on a fixed set of data to generate the clusters. However, in practical applications, these detection systems are operating daily, whereby millions of input data points are continuously streamed indefinitely, hence imposing speed and memory constraints. In such dynamic data stream environments, keeping track of every historical data would be highly memory expensive and, even if possible, would not solve the problem of analysing big data within the real-time requirements. Hence, a method of analysing and storing the essential information of the historical data in a single pass is mandatory for clustering data streams.

In addition, the dynamic data clustering algorithm needs to address two special characteristics that often occur in data streams, which are known as *"concept drift"* and *"concept evolution"* [5]. Concept drift refers to the change of underlying concepts in the stream as time progresses, i.e., the change in the relationship between the attributes of the object within the individual clusters. For example, customer behaviour in purchasing trending products always changes in between seasonal sales. Meanwhile, concept evolution occurs when a new class definition has evolved in the data streams, i.e., the number of clusters has changed due to the creation of new clusters or the deprecation of old clusters. This phenomenon often occurs in the detection system whereby an anomaly has emerged in the data traffic. An ideal data stream clustering algorithm should address these two main considerations to detect and adapt effectively to changes in the dynamic data environment.

Based on recent literature, metaheuristics for black-box optimisation have been greatly adopted in traditional static data clustering [6]. These algorithms have a general purpose application domain and often display self-adaptive capabilities, thus being able to tackle the problem at hand, regardless of its nature and formulation, and return near-optimal solutions. For clustering purposes, the so-called "population based" metaheuristic algorithms have been discovered to be able to achieve better global optimisation results than their "single solution" counterparts [7]. Amongst the most commonly used optimisation paradigms of this kind, it is worth mentioning the established Differential Evolution (DE) framework [8–10], as well as more recent nature inspired algorithms from the Swarm Intelligence (SI) field, such as the Whale Optimisation Algorithm (WOA) [11] and the Bat-inspired algorithm in [12], here referred to as BAT. Although the literature is replete with examples of data clustering strategies based on DE, WOA and BAT for the static domain, as, e.g., those presented in [13–16], little has been done for the dynamic environment due to the difficulties in handling data streams. The current state of dynamic clustering is therefore unsatisfactory as it mainly relies on algorithms based on techniques such as density microclustering and density grid based clustering, which require the tuning of several parameters to work effectively [17].

This paper presents a methodology for integrating metaheuristic optimisation into data stream clustering, thus maximising the performance of the classification process. The proposed model does not require specifically tailored optimisation algorithms to function, but it is a rather general framework to use when highly dynamic streams of data have to be clustered. Unlike similar methods, we do not optimise the parameters of a clustering algorithm, but use metaheuristic optimisation in its initialisation phase, in which the first clusters are created, by finding the optimal position of their centroids. This is a key step as the grouped points are subsequently processed with the method in [18] to form compact, but informative, microclusters. Hence, by creating the optimal initial environment for the clustering method, we make sure that the dynamic nature of the problem will not deteriorate its performances. It must be noted that microclusters are lighter representations of the original scenario, which are stored to preserve the "memory" of the past classifications. These play a major role since they aid subsequent clustering processes when new data streams are received. Thus, a non-optimal microclusters store in memory can have catastrophic consequences in terms of classification results. In this light, our original use of the metaheuristic algorithm finds its purpose, and results confirm the validity of our idea. The proposed clustering scheme efficiently tracks changes and spots patterns accordingly.

The remainder of this paper has the following structure:

- Section 2 discusses the recent literature and briefly explains the logic behind the leading data stream clustering algorithms;
- Section 3 establishes the motivations and objectives of this research and presents the used metaheuristic optimisation methods, the employed performance metrics and the considered datasets for producing numerical results;
- Section 4 gives a detailed description of each step involved in the proposed clustering system, clarifies its working mechanism and shows methodologies for its implementation;
- Section 5 describes the performance metrics used to evaluate the system and provides experimental details to reproduce the presented results;
- Section 6 presents and comments on the produced results, including a comparison among different variants of the proposed system, over several evaluation metrics;

- Section 7 outlines a thorough analysis of the impact of the parameter setting for the optimisation algorithm on the overall performance of the clustering system;
- Section 8 summarises the conclusions of this research.

2. Background

There are two fundamentals aspects to take into consideration in data stream clustering, namely concept drift and concept evolution. The first aspect refers to the phenomenon when the data in the stream undergo changes in the statistical properties of the clusters with respect to the time [19,20] while the second to the event when there is an unseen novel cluster appearing in the stream [5,21].

Time window models are deployed to handle concept drift in data streams. These are usually embedded into clustering algorithms to control the quantity of historical information used in analysing dynamic patterns. Currently, there are four predominant window models in the literature [22]:

- the "damped time window" model, where historical data weights are dynamically adjusted by fixing a rate of decay according to the number of observations assigned to it [23];
- the "sliding time window" model, where only the most recent past data observations are considered with a simple First-In-First-Out (FIFO) mechanism. as in [24];
- the "landmark time window" model, where the data stream is analysed in batches by accumulating data in a fixed width buffer before being processed;
- the "tilted time window" model, where the granularity level of weights gradually decreases as data points get older.

As for concept evolution, most of the existing data stream clustering algorithms are designed following a two phase approach, i.e., consisting of an online clustering process followed by an offline one, which was first proposed in [25]. In this work, the concept of microclusters was also defined to design the so-called CluStream algorithm. This method forms microclusters having statistical features representing the data stream online. Similar microclusters are then merged into macro-clusters, keeping only information related to the centre of the densest region. This is performed offline, upon user request, as it comes with information losses since merged clusters can no longer be split again to obtain the original ones.

In terms of online microclustering, most algorithms in the literature are distance based [17,22,26], whereby new observations are either merged with existing microclusters or form new microclusters based on a distance threshold. The earliest form of distance based clustering strategy was the process of extracting information about a cluster into the form of a Clustering Feature (CF) vector. Each CF usually consists of three main components: (1) a linear combination of the data points referred to as Linear Sum vector $\overrightarrow{LS}$; (2) a vector $\overrightarrow{SS}$ whose components are the Squared Sums of the corresponding data points' components; (3) the number N of points in a cluster.

As an instance, the popular CluStream algorithm in [25] makes use of CF and the tilted time window model. During the initialisation phase, data points are accumulated to a certain amount before being converted into some microclusters. On the arrival of new streams, new data are merged with the closest microclusters if their distance from the centre of the data point to the centre of the microclusters is within a given radius (i.e., the ϵ-neighbourhood method). If there is no suitable microclusters within this range, a new microclusters is formed. When requested, the CluStream uses the k-means algorithm [27] to generate macroclusters from microclusters in its offline phase. It also implements an ageing mechanism based on timestamps to remove outdated clusters from its online components.

Another state-of-the-art algorithm, i.e. DenStream, was proposed in [18] as an extension of CluStream using the damped time window and a novel clustering strategy named "time-faded CF". DenStream separates the microclusters into two categories: the potential core microclusters (referred to as p-microclusters) and the outlier microclusters (referred to as o-microclusters). Each entry of the CF is subject to a decay function that gradually reduces the weight of each microcluster at a regular evaluation interval period. When the weight falls below a threshold value, the affected p-microclusters are degraded to the o-microclusters, and they are removed from the o-microclusters if the weights deteriorate further. On the other hand,o-microclusters that have their weights improved are promoted to p-microclusters. This concept allows new and old clusters to form online gradually, so addressing the concept evolution issue. In the offline phase, only the p-microclusters are used for generating the final clusters. Similar p-microclusters are merged employing a density based approach based on the ϵ-neighbourhood method. Unlike other commonly used methods, in this case, clusters can assume an arbitrary shape, and no a priori information is needed to fix the number of clusters.

An alternative approach was given in [28], where the proposed STREAM algorithm did not store CF vectors, but directly computed centroids on-the-fly. This was done by solving the "k-means clustering" problem to identify the centroids of K clusters. The problem was structured in a form whereby the distance from data points to the closest cluster had associated costs. Using this framework, the clustering task was defined as a minimisation problem to find the number and position of centroids that yielded the lowest costs. To process an indefinite length of streaming data, the landmark time window was used to divide the streams into n batches of data, and the K-means problem solving was performed on each chunk. Although the solution was plausible, the algorithm was evaluated to be time consuming and memory expensive in processing streaming data.

The OLINDDA method proposed in [29] extends the previously described centroid approach by integrating the ϵ-neighbourhood concept. This was used to detect drifting and new clusters in the data stream, with the assumption that drift changes occurred within the existing cluster region, whilst new clusters formed outside the existing cluster region. The downside of the centroid approach was that the number of K centroids needed to be known a priori, which is problematic in a dynamic data environment.

There is one shortcoming for the two phase approach, i.e., the ability to track changes in the behaviour of the clusters is linearly proportional to the frequency of requests for the offline component [30]. In other words, the higher the sensitivity to changes, the higher the computational cost. To mitigate these issues, an alternative approach has been explored by researchers to merge these two phases into a single online phase. FlockStream [31] deploys data points into a virtual mapping of a two-dimensional grid, where each point is represented as an agent. Each agent navigates around the virtual space according to a model mimicking the behaviour of flocking birds, as done in the most popular SI algorithms, e.g., those in [32–34]. The agent behaviour was designed in a way such that similar (according to a given metric) birds would move in the same direction as the closest neighbours, forming different groups of the flock. These groups can be seen as clusters, thus eliminating the need for a subsequent offline phase.

MDSC [35] is another single phase method exploiting the SI paradigm inspired by the density based approached introduced in DenStream. In this method, the Ant Colony Optimisation (ACO) algorithm [36] is used to group similar microclusters optimally during the online phase. In MDSC, a customised ϵ-neighbourhood value is assigned to each cluster to enable "multi-density" clusters to be discovered.

Finally, it is worth mentioning the ISDI algorithm in [37], which is equipped with a windowing routine to analyse and stream data from multiple sources, a timing alignment method and a deduplication algorithm. This algorithm was designed to deal with data streams coming from different sources in the Internet of Things (IoT) systems and can transform multiple data streams, having different attributes, into cleaner datasets suitable for clustering. Thus, it represents a powerful tool allowing for the use of streams classifiers, as, e.g., the one proposed in this study, in IoT environments.

3. Motivations, Objectives and Methods

Clustering data streams is still an open problem with room for improvement [38]. Increasing the classification efficiency in this dynamic environment has a great potential in several application fields, from intrusion detection [39] to abnormality detection in patients' physiological data streams [40]. In this light, the proposed methodology draws its inspiration from key features of the successful methods listed in Section 2, with the final goal of improving upon the current state-of-the-art.

A hybrid algorithm is then designed by employing, along with standard methods, as, e.g., CF vectors and the landmark time windows model, modern heuristic optimisation algorithms. Unlike similar approaches available in the literature [36,41,42], the optimisation algorithm is here used during the online phase to create optimal conditions for the offline phase. This novel approach is described in detail in Section 4.

To select the most appropriate optimisation paradigm, three widely used algorithms, i.e., WOA, BAT and DE, were selected from the literature and compared between them. We want to clarify that the choice of using three metaheuristic methods, rather than other exact or iterative techniques, was made to be able to deal with the challenging characteristics of the optimisation problem at hand, e.g., the dimensionality of the problem can vary according to the dataset, and the objective functions is highly non-linear and not differentiable, which make them not applicable or time inefficient.

A brief introduction of the three selected algorithms is given below in Section 3.1. Regardless of the specific population based algorithm used for performing the optimisation step, each candidate solution must be encoded as an n-dimensional real valued vector representing the K cluster centres for initialising the following density based clustering method.

Two state-of-the-art deterministic data stream clustering algorithms, namely DenStream and CluStream, are also included in the comparative analysis to further validate the effectiveness of the proposed framework.

The evaluation methodology employed in this work consisted of running classification experiments over the datasets in Section 3.2 and measuring the obtained performances through the metrics defined in Section 3.3.

3.1. Metaheuristic Optimisation Methods

This section gives details on the implementation of the three optimisation methods used to test the proposed system.

3.1.1. The Whale Optimization Algorithm

The WOA algorithm is a swarm based stochastic metaheuristic algorithm inspired by the hunting behaviour of humpback whales [11]. It is based on a mathematical model updated by iterating the three search mechanisms described below:

- the "shrinking encircling prey" mechanism is exploitative and consists of moving candidate solutions (i.e., the whales) in a neighbourhood of a the current best solution in the swarm (i.e., the prey solution) by implementing the following equation:

$$\vec{x}(t+1) = \vec{x}_{\text{best}}(t) - \vec{A} * \vec{D}_{\text{best}} \quad \text{with} \quad \begin{cases} \vec{A} = 2\vec{a} * \vec{r} - \vec{a} \\ \vec{D}_{\text{best}} = 2\vec{r} * \overrightarrow{x_{\text{best}}}(t) - \vec{x}(t) \end{cases} \tag{1}$$

  where: (1) $\vec{a}$ is linearly decreased from two to zero as iterations increase (to represent shrinking, as explained in [7]); (2) $\vec{r}$ is a vector whose components are randomly sampled from $[0, 1]$ (t is the iteration counter); (3) the "$*$" notation indicates the pairwise products between two vectors.
- the "spiral updating position" mechanism is also exploitative and mimics the swimming pattern of humpback whales towards prey in a helix shaped form through Equations (2) and (3):

$$\vec{x}(t+1) = e^{bl} * \cos(2\pi l) * \left|\vec{d}\right| + \vec{x}_{\text{best}}(t) \tag{2}$$

  with:

$$\vec{d} = \left|\vec{x}_{\text{best}}(t) - \vec{x}(t)\right| \tag{3}$$

where b is a constant value for defining the shape of the logarithmic spiral; l is a random vector in $[-1,1]$; the "$|\ldots|$" symbol indicates the absolute value of each component of the vector;

- the "search for prey" mechanism is exploratory and uses a randomly selected solution $\vec{x}_{\text{rand}}$ as an "attractor" to move candidate solutions towards unexplored areas of the search space and possibly away from local optima, according to Equations (4) to (5):

$$\vec{x}(t+1) = \vec{x}_{\text{rand}}(t) - \vec{A} * \vec{D} + \text{rand} \tag{4}$$

with:

$$\vec{D}_{\text{rand}} = \left| 2\vec{a} * \vec{r} * \overrightarrow{x_{\text{rand}}}(t) - \vec{x} \right|. \tag{5}$$

The reported equations implemented a search mechanism that mimics movements made by whales. Mathematically, it is easier to understand that some of them refer to explorations moves across the search space, while others are exploitation moves to refine solutions within their neighbourhood. To have more information on the metaphor inspiring these equations, their formulations and their role in driving the research within the algorithm framework, one can see the survey article in [6]. A detailed scheme describing the coordination logic of the three previously described search mechanism is reported in Algorithm 1.

Algorithm 1 WOA pseudocode.

```
Generate initial whale positions x_i, where i = 1,2,3,...,NP
Compute the fitness of each whale solution, and identify x_best
while t < max iterations do
    for i = 1,2,...,NP do
        Update a, A, C, l, p
        if p < 0.5 then
            if |A| < 1 then
                Update the position of current whale x_i using Equation (1)
            else if |A| ≥ 1 then
                x_rand ← random whale agent
                Update the position of current whale x_i with Equation (4)
            end if
        else if p ≥ 0.5 then
            Update the position of current whale x_i with Equation (2)
        end if
    end for
    Calculate new fitness values
    Update X_best
    t = t + 1
end while
Return x_best
```

With reference to Algorithm 1, the initial swarm is generated by randomly sampling solutions in the search; the best solution is kept up to date by replacing it only when an improvement on the fitness value occurs; the optimisation process lasts for a prefixed number of iterations, here indicated with *max budget*; the probability of using the shrinking encircling rather than the spiral updating mechanism was fixed at 0.5.

3.1.2. The BAT Algorithm

The BAT algorithm was a swarm based searching algorithm inspired by the echolocation abilities of bats [12]. Bats use sound wave emissions to generate an echo that measures the distance of its prey based on the loudness and time difference of the echo and sound wave. To reproduce this system and exploit it for optimisation purposes, the following perturbation strategy must be implemented:

$$f_i = f_{\min} + (f_{\max} - f_{\min}) \cdot \beta \tag{6}$$

$$v_i(t+1) = v_i(t) + (x_i(t) - x_{\text{best}}) \cdot f_i \tag{7}$$

$$x_i(t+1) = x_i(t) + v_i(t) \tag{8}$$

where x_i is the position of the candidate solution in the search space (i.e., the bat), v_i is its velocity, f_i is referred to as the "wave frequency" factor and β is a random vector in $[0,1]^n$ (where n is the dimensionality of the problem). f_{min} and f_{max} represent the lower and upper bounds of the frequency, respectively. Typical values are within 0 and 100. When the bat is close to the prey (i.e., current best solution), it gradually reduces the loudness of its sound wave while increasing the pulse rate. The pseudocode depicted in Algorithm 2 shows the the working mechanism of the BAT algorithm.

Algorithm 2 BAT pseudocode.

Generate initial bats X_i ($i = 1, 2, 3, \ldots, NP$) and their velocity vectors v_i
Compute the fitness values, and find x_{best}
Initialise pulse frequency f_i at x_i
Initialise pulse rate r_i and loudness A_i
while $t <$ *max iterations* **do**
    for $i = 1, 2, 3, \ldots, NP$ **do**
        $x_{\text{new}} \leftarrow$ move x_i to a new position with Equations (6)–(8)
    end for
    for $i = 1, 2, 3, \ldots, NP$ **do**
        if rand() $> r_i$ **then**
            $x_{\text{new}} \leftarrow x_{\text{best}}$ added with a random
        end if
        if rand() $< A_i$ **and** $f(x_{\text{new}})$ improved **then**
            $X_i \leftarrow x_{\text{new}}$
            Increase r_i and decrease A_i
        end if
    end for
    Update x_{best}
    $t = t + 1$
end while
Return x_{best}

To have more detailed information on the equations used to perturb the solutions within the search space in the BAT algorithm, we suggest reading [43].

3.1.3. The Differential Evolution

The Differential Evolution (DE) algorithms are efficient metaheuristics for global optimisation based on a simple and solid framework, first introduced in [8], which only requires the tuning of three parameters, namely the scale factor $F \in [0,2]$, the crossover ratio $CR \in [0,1]$ and the population size NP. As shown in Algorithm 3, despite using crossover and mutation operators, which are typical of evolutionary algorithms, it does not require any selection mechanism as solutions are perturbed one at a time by means of the one-to-one spawning mechanising from the SI field. Several DE variants can be obtained by using different combinations of crossover and mutation operators [44]. The so-called "DE/best/1/bin" scheme was adopted in this study, which employs the best mutation strategy and the binomial crossover approach. The pseudocode and other details regarding these operators are available in [10].

Algorithm 3 DE pseudocode.

```
Generate initial population x_i with i = 1,2,3,...,NP
Compute the fitness of each individual, and identify x_best
while t < max iterations do
    for i = 1,2,3,...,NP do
        X_m ← mutation                                        ▷ "best/1" as explained in [10]
        x_off ← crossover(X_i, X_m)                           ▷ "bin" as explained in [10]
        Store the best individual between x_off and x_i in the i^th position of a new population
    end for
    end
    Replace the current population with the newly generated population
    Update x_best
end while
Return x_best
```

3.2. Datasets

Four synthetic datasets were generated using the built-in stream data generator of the "Massive Online Analysis" (MOA) software [45]. Each synthetic dataset represents different data streaming scenarios with varying dimensions, clusters numbers, drift speed and frequency of concept evolution. These datasets are:

- the *5C5C* dataset, which contains low-dimensional data with a low rate of data changes;
- the 5C10C dataset, which contains low-dimensional data with a high rate of data changes;
- the 10D5C dataset, which is a 5C5C variant containing high-dimensional data;
- the 10D10C dataset, which is a *5C10C* variant containing high-dimensional data.

Moreover, the KDD-99 dataset [46], containing real network intrusion information, was also considered in this study. It must be highlighted that the original KDD-99 dataset contained 494,021 data entries, representing network connections generated in military network simulations. However, only 10% of the entries were randomly selected for this study. Each data entry contained 41 features and one output column to distinguish the attack connection from the normal network connection. The attacks can be further classified into 22 attack types. Streams are obtained by reading each entry of the dataset sequentially.

Details on the five employed datasets are given in Table 1.

Table 1. Name and description of the synthetic datasets and real dataset.

Name	Dimension	Cluster No.	Samples	Drift Speed	Event Frequency	Type
5D5C	5	3–5	100,000	1000	10,000	Synthetic
5D10C	5	6–10	100,000	5000	10,000	Synthetic
10D5C	10	3–5	100,000	1000	10,000	Synthetic
10D10C	10	6–10	100,000	5000	10,000	Synthetic
KDD–99	41	2–23	494,000	Not Known	Not Known	Real

3.3. Performance Metrics

To perform an informative comparative analysis, three metrics were cherry picked from the data stream analysis literature [41,42]. These are referred to as the F-measure, purity and Rand index [47].

Mathematically, these metrics are expressed with the following equations:

$$\text{F-Measure} = \frac{1}{k}\sum_{i=1}^{k} \text{Score}_{C_i} \tag{9}$$

$$\text{Purity} = \frac{1}{k}\sum_{i=1}^{k} \text{Precision}_{C_i} \tag{10}$$

$$\text{Rand Index} = \frac{\text{True Positive} + \text{True Negative}}{\text{All Data Instances}} \tag{11}$$

where:

$$\text{Precision}_{C_i} = \frac{V_{i\text{sum}}}{n_{C_i}} \tag{12}$$

$$\text{Score}_{C_i} = 2 \cdot \frac{\text{Precision}_{C_i} \cdot \text{Recall}_{C_i}}{\text{Precision}_{C_i} + \text{Recall}_{C_i}} \tag{13}$$

$$\text{Recall}_{C_i} = \frac{V_{i\text{sum}}}{V_{i\text{total}}} \tag{14}$$

and:

- C is the solution returned by the clustering algorithm (i.e., the number of clusters k);
- C_i is the the ith cluster ($i = \{1, 2, \ldots, k\}$);
- V_i is the class label with the highest frequency in C_i;
- $V_{i\text{sum}}$ is the number of instances labelled with V_i in C_i;
- $V_{i\text{total}}$ is the total number of V_i instances identified in the totality of clusters returned by the algorithm.

The F-measure represents the harmonic mean of the precision and recall scores, where the best value of one indicates ideal precision and recall, while zero is the worst scenario.

Purity is used to measure the homogeneity of the clusters. Maximum purity is achieved by the solution when each cluster only contains a single class.

The Rand index computes the accuracy of the clustering solution from the actual solution, based on the ratio of correctly identified instances among all the instances.

4. The Proposed System

This article proposes "OpStream", an Optimised Stream clustering algorithm. This clustering framework consisted of two main parts: the initialisation phase and the online phase.

During the initialisation phase, a number λ of data points are accumulated through a landmark time window, and the unclassified points are initialised into groups of clusters via the centroid approach, i.e., generating K centroids of clusters among the points.

In the initialisation phase, the landmark time window is used to collect data points, which are subsequently grouped into clusters by generating K centroid. The latter are generated by solving K-centroid cost optimisation problems with a fast and reliable metaheuristic for optimisation. Hence, their position is optimal and leads to high quality predictions.

Next, during the online phase, the clusters are maintained and updated using the density based approach, whereby incoming data points with similar attributes (i.e., according to the ϵ-neighbourhood method) form dense microclusters in between two data buffers, namely p-microclusters and o-microclusters. These are converted into microclusters with CF information to store a "light" version of previous scenarios in this dynamic environment.

In this light, the proposed framework is similar to advanced single phase methods. However, it requires a preliminary optimisation process to boost its classification performances.

Three variants of OpStream were tested by using the three metaheuristic optimisers described in Section 3. These stochastic algorithms (as the optimisation process is stochastic) are compared against the two DenStream and CluStream state-of-the-art deterministic stream clustering algorithms.

The following sections describe each step of the OpStream algorithm.

4.1. The Initialisation Phase

This step can be formulated as a real valued global optimisation search problem and addressed with the metaheuristic of black-box optimisation. To achieve this goal, a cost function must be designed to allow for the individualisation of the optimal position of the centroid of a cluster. These processes have to be iterated K times to then form K clusters by grouping data according to their distance from the optimal centroids.

The formulation of the cost function plays a key part. In this research, the "Cluster Fitness" (CF) function from [48] was chosen as its maximisation leads to a high intra-cluster distance, which is desirable. Its mathematical formulation, for the κth ($\kappa = 1, 2, 3, \ldots K$) cluster, is given below:

$$CF_\kappa = \frac{1}{K} \sum_{\kappa=1}^{K} S_\kappa \tag{15}$$

from where it can be observed that it is computed by averaging the K clusters' silhouettes "S_κ". These represent the average dissimilarity of all the points in the cluster and are calculated as follows

$$S_\kappa = \frac{1}{n_k} \sum_{i \in C_\kappa} \frac{\beta_i - \alpha_i}{\max\{\alpha_i, \beta_i\}} \tag{16}$$

where α_i and β_i are the "inner dissimilarity" and the "outer dissimilarity", respectively.

The former value measures the average dissimilarity between a data point i and other data points in its own cluster C_{κ^*}. Mathematically, this is expressed as:

$$\alpha_i = \frac{1}{(n_{k^*} - 1)} \sum_{\substack{j \in C_{\kappa^*} \\ j \neq i}} \text{dist}(i, j) \tag{17}$$

with $\text{dist}(i, j)$ being the Euclidean distance between the two points and n_{k^*} is the total number of points in cluster C_{κ^*}. The lower the value, the better the clustering accuracy.

The latter value measures the minimum distance between a data point i to the centre of all clusters, excluding its own cluster C_{κ^*}. Mathematically, this is expressed as:

$$\beta_i = \min_{\substack{\kappa=1,\ldots,K \\ k \neq \kappa^*}} \left(\frac{1}{n_k} \sum_{\substack{j \in C_\kappa \\ k \neq \kappa^*}} \text{dist}(i, j) \right) \tag{18}$$

where n_{k^*} is the number of points in cluster C_{κ^*}. The higher the value, the better the clustering.

These two values are contained in $[-1, 1]$, whereby one indicates the ideal case and -1 the most undesired one.

A similar observation can be done for the fitness function CF_κ [48]. Hence, the selected metaheuristics have to be set up for a maximisation problem. This is not an issue since every real valued problem of this kind can be easily maximised with an algorithm designed for minimisation purposes by simply timing the fitness function by -1, and vice versa.

Regardless of the dimensionality of the problem n, which depends on the dataset (as shown in Table 1), all input data were normalised within $[0, 1]$. Thus, the search space for all the optimisation process was the hyper-cube defined as $[0, 1]^n$.

4.2. The Online Phase

Once the initial clusters have been generated, by optimising the cost function formulated in Section 4.1, clustered data points must be converted into microclusters. This step requires the extraction of CF vectors. Subsequently, a density based approach was used to cluster the data stream online.

4.2.1. Microclusters' Structure

In OpStream, each CF must contain four components, i.e., CF $= [\text{N}, \overrightarrow{\text{LS}}, \overrightarrow{\text{SS}}, \text{timestamp}]$, where

- N$\in \mathbb{N}$ is the number of data points in the microclusters;

- $\overrightarrow{\text{LS}} \in \mathbb{R}^n$ is the Linear Sum of the data points in the microcluster, i.e.,

$$\overrightarrow{\text{LS}} = \sum_{i=1}^{N} \overrightarrow{x_i};$$

- $\overrightarrow{\text{SS}} \in \mathbb{R}^n$ is the squared sum of the data points in the microclusters, i.e.,

$$\overrightarrow{\text{SS}}[j] = \sum_{i=1}^{N} \left(\overrightarrow{x_i}[j] \right)^2; \qquad j = 1, 2, 3, \ldots, n$$

- timestamp indicates when the microclusters was last updated, and it is needed to implement the ageing mechanism, used to remove outdated microclusters while new data accumulated in the time window are available, defined via the following equation:

$$\text{age} = T - \text{timestamp} \tag{19}$$

where T is the current timestamp in the stream and a threshold, referred to as β, is used to discriminate between suitable and outdated data points.

From CF, the centre c and radius r of a microclusters are computed as follows:

$$c = \frac{\overrightarrow{\text{LS}}}{\text{N}} \tag{20}$$

$$r = \sqrt{\frac{\overrightarrow{\text{SS}}}{\text{N}} - \left(\frac{\overrightarrow{\text{LS}}}{\text{N}} \right)^2} \tag{21}$$

as indicated in [18,42].

The obtained r value is used to initialise the ϵ-neighbourhood approach (i.e., $r = \epsilon$), leading to the formation of microclusters as explained in Section 2. This microclusters, which are derived from a cluster formed in the initialisation phase, is now stored in the p-microclusters buffer.

4.2.2. Handling Incoming Data Points

In OpStream, for each new time window, a data point p is first converted into a "degenerative" microcluster m_p containing a single point and having the following initial CF properties:

$$\begin{aligned} m_p.\text{N} &= 1 \\ m_p.\overrightarrow{\text{LS}}_i &= p_i \qquad i = 1, 2, 3 \ldots, n \\ m_p.\overrightarrow{\text{SS}}_i &= p_i^2 \qquad i = 1, 2, 3 \ldots, n \\ m_p.\text{timestamp} &= T \end{aligned}$$

Subsequently, initial microclusters have to be merged. This task can efficiently be addressed by considering pairs of microclusters, say, e.g., m_i and m_j, and computing their Euclidean distance $\text{dist}(c_{m_i}, c_{m_j})$. If m_i is the cluster to be merged, its radius r must be worked out as shown in Section 4.2.1 and then be merged with m_i if:

$$\text{dist}(c_{m_i}, c_{m_j}) \leq \epsilon \qquad (\epsilon = r). \tag{22}$$

Two microclusters satisfying the condition expressed with Equation (22) are said to be "density reachable". The process described above is repeated until there are no longer density reachable microclusters. Every time two microclusters are merged, e.g., m_i and m_j, the CF properties of the newly generated microclusters, e.g., m_k, are assigned as follows:

$$\begin{aligned} m_k.\mathrm{N} &= m_i.\mathrm{N} + m_j.\mathrm{N} \\ m_k.\overrightarrow{\mathrm{LS}} &= m_i.\overrightarrow{\mathrm{LS}} + m_j.\overrightarrow{\mathrm{LS}} \\ m_k.\overrightarrow{\mathrm{SS}} &= m_i.\overrightarrow{\mathrm{SS}} + m_j.\overrightarrow{\mathrm{SS}} \\ m_k.\mathrm{timestamp} &= T \end{aligned}$$

where T is the time at which the two microclusters were merged.

When the condition in Equation (22) is no longer met by a microcluster, this is moved to the p-microclusters buffer. If the newly added microclusters and other clusters in the p-microclusters buffer are density reachable, then they are merged. Otherwise, a new independent cluster is stored in this buffer.

This mechanism is performed by a software agent, referred to as the "Incoming Data Handler" (IDH), whose pseudocode is reported in Algorithm 4 to further clarify this process and allow for its implementation.

Algorithm 4 IDH pseudocode.

```
Input: Data point p
Convert p into microcluster m_p
Initialise merged = false
for mc in p-microclusters do
    if merged is false then
        if m_p is density reachable to mc then
            if new radius ≤ ε_mc then
                Merge m_p with mc
            else
                Add m_p to p-microclusters
            end if
            merged = true
        end if
    end if
end for
if merged is false then
    for each mc in o-microclusters do
        if merged is false then
            if m_p is density reachable to mc then
                if new radius ≤ ε_mc then
                    Merge m_p with mc
                    merged = true
                end if
            end if
        end if
    end for
end if
if merged is false then
    Add m_p to o-microclusters
end if
end
return
```

4.2.3. Detecting and Forming New Clusters

Once microclusters in the o-microclusters buffer are all merged, as explained in Section 4.2.2, only the minimum possible number of microclusters with the highest density exists. The microclusters with the highest number of points N is then moved to an empty set C to initialise a new cluster. After calculating its centre c, with Equation (20), and radius r, with Equation (21), the ϵ-neighbourhood method is again used to find density reachable microclusters. Among them, a process is undertaken to detect the so-called border microclusters [35] inside C, which obviously are not present during the first iteration as C initially contains only one microcluster. Border microclusters are defined as density reachable microclusters that have a density level that is below the density threshold of the first microclusters present in C. Having a threshold that is too high, cluster C will not expand, whilst having a value that is too low, cluster C will contain dissimilar microclusters. Based on the experimental data from the original paper [35], a 10% threshold yields good performance.

Once the border microclusters are identified, only surrounding microclusters that are density reachable to the non-border microclusters are moved to form part of C, according to the process indicated in Section 4.2.2. Figure 1 graphically depicts C. The microclusters marked in red colour do not form as part of C because it is density reachable only to border microclusters of C.

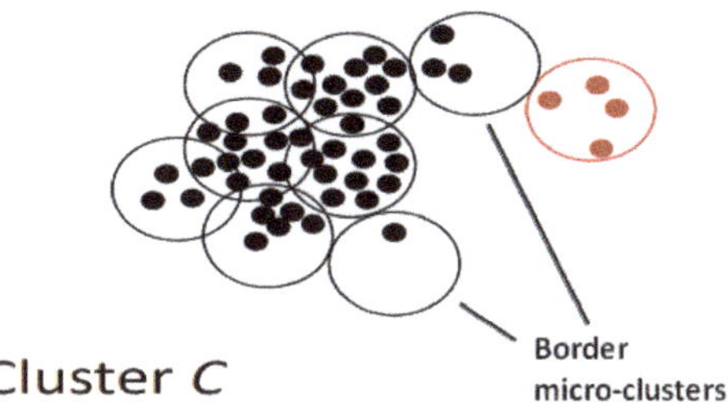

Figure 1. A graphical representation of the "border microclusters" concept [35].

This process is iterated as shown in Algorithm 5. The final version of C is finally moved to the most appropriate buffer according to its size, i.e., if "C.N ≥ minClusterSize", all its microclusters are merged together, and the newly generated cluster C is moved to the p-microclusters buffer. If this does not occur, the cluster C is not generated by merging its microclusters, but they are simply left in the o-microclusters buffer. The recommended method to fix the minClusterSize parameter is:

$$\text{minClusterSize} = \begin{cases} 2 & \text{if } 10\% \text{ of } \lambda \leq 2 \\ 10\% \text{ of } \lambda & \text{otherwise} \end{cases} \tag{23}$$

These tasks are performed by the New Cluster Generator (NCG) software agent, whose pseudocode is shown in Algorithm 5.

Algorithm 5 New cluster generation pseudocode.

```
Input: o-microclusters
while o-microclusters is not empty do
    Initialise cluster C using mc with highest N
    addedMc = true
    while addedMc is true do
        addedMc = false
        for mc in o-microclusters do
            if mc is density reachable to any non-border mc in C then
                Add mc into C
                Remove mc from o-microclusters
                addedMc = true
            end if
        end for
    end while
    if the size of C is ≥ minClusterSize then    ▷ minClusterSize is initialised with Equation (23)
        Merge microclusters mc in C
        Add C into p-microclusters
    end if
end while
```

4.3. OpStream General Scheme

The proposed OpStream method involves the use of several techniques, such as metaheuristic optimisation algorithms, density based and k-means clustering, etc., and requires a software infrastructure coordinating activities as those performed by the IDH and NCG agents. Its architecture is outlined with the pseudocode in Algorithm 6.

Algorithm 6 OpStream pseudocode

```
Launch AS                                                      ▷ initialised with Equation (24)
initialisedFlag = false
while stream do
    Add data point p into window
    if initialisedFlag is true then
        Handle incoming data streams with IDH,                 ▷ i.e., Algorithm 4
    end if
    if window is full then
        if initialisedFlag is false then
            Optimise centres' positions, and initialise clusters   ▷ e.g., with Algorithm 1, 2 or 3
            initialisedFlag = true
        else
            Look for and generate new clusters with NCG,       ▷ i.e., Algorithm 5
        end if
    end if
end while
```

It must be added that an Ageing System (AS) is constantly run to remove outdated clusters. Despite its simplicity, its presence is crucial in dynamic environments. An integer parameter β (equal to four in this study) is used to compute the age threshold as shown below:

$$age\ threshold = \beta \cdot \lambda \tag{24}$$

so that if a microcluster has not been updated in four consecutive windows, it will be removed from the respective buffer.

5. Experimental Setup

As discussed in Section 3, OpStream's performances were evaluated across four synthetic datasets and one real dataset using three popular performance metrics. Two deterministic state-of-the-art stream clustering algorithms, i.e., DenStream [18] and CluStream [25], were also run with the suggested parameter settings available in their original articles for caparison purposes.

The WOA algorithm was initially picked to implement the OpStream framework, as this framework is currently being intensively exploited for classification purposes, but two more variants employing BAT and DE (as described in Section 3) were also run to: (1) show the flexibility of OpStream in the use of different optimisation methods; (2) display its robustness and superiority to deterministic approaches regardless of the optimiser used; (3) establish the preferred optimisation method over the specific datasets considered in this study. For the sake of clarity, these three variants are referred to as WOAS-OpStream, BAT-OpStream and DE-OpStream to represent the respective metaheuristic optimiser used. To reproduce the results presented in this article, the employed parameter setting of each metaheuristic, as well as other algorithmic details are reported below:

- WOA: swarm Size = 20;
- BAT: swarm size = 20, $\alpha = 0.53$, $\gamma = 4.42$, $r_i = 0.42$, $A_i = 0.50$ ($i = 1, 2, 3, \ldots, n$);
- DE: population Size = 20, $F = 0.5$, $CR = 0.5$;
- the "*max Iterations*" value is set to 10 for all three algorithms to ensure a fair comparison (the computational budget was purposely kept low due to the real-time nature of the problem);

- the three optimisation algorithms were equipped with the "toroidal" correction mechanism to handle infeasible solutions, i.e., solutions generated outside of the search space (a detailed description of this operator is available in [10]).

Furthermore, the following parameter values were also required to run the OpStream framework:

- $\lambda = 1000, \epsilon = 0.1, \beta = 4$;

Section 7 explains the role played by these parameters and how their suggested values were determined.

Thus, a total of five clustering algorithms was considered in the experimentation phase. These were executed, with the aid of the MOA platform [45], 30 times over each dataset (the instances' order was randomly changed for each repetition) to produce, for each evaluation metric, average $\pm$ standard deviation values. To further validate our conclusions statistically, the outcome of the Wilcoxon rank-sum test [49] (with the confidence level equal to 0.05) is also reported in all tables with the compact notation obtained from [50], where (1) a "$+$" symbol next to an algorithm indicates that it was outperformed by the reference algorithm (i.e., WOA-OpStream); (2) a "$-$" symbol indicates that the reference algorithm was outperformed; (3) a "$=$" symbol shows that the two stochastic optimisation processes were statistically equivalent.

6. Results and Discussion

A table was prepared for each evaluation metric, each one displaying the average value, standard deviation and the outcome of the Wilcoxon rank-sum test (W) over the 30 performed runs. The best performance on each dataset is highlighted in boldface.

Table 2 reports the results in terms of the F-measure. According to this metric, the three OpStream variants generally outperformed the deterministic algorithms. The only exception is registered over the *KDDC–99* dataset, where DenStream displayed the best performance. From the statistical point of view, WOA-OpStream was significantly better than CluStream (with five "+" out of five cases), clearly preferable to DenStream (with four "+" out of five cases), equivalent to the DE-OpStream variant and approximately equivalent to BAT-OpStream.

Table 2. Average F-measure value $\pm$ standard deviation and Wilcoxon rank-sum test (reference = Whale Optimisation Algorithm (WOA)-OpStream) for WOA-OpStream against BAT-OpStream, Differential Evolution (DE)-OpStream, DenStreamand CluStreamon each dataset. W, Wilcoxon rank-sum test.

Dataset	WOA-OpStream	BAT-OpStream	W	DE-OpStream	W	DenStream	W	CluStream	W
5D5C	**0.924 ± 0.042**	0.907 ± 0.041	+	0.923 ± 0.040	=	0.645 ± 0.016	+	0.584 ± 0.033	+
5D10C	0.868 ± 0.042	0.873 ± 0.048	=	**0.879 ± 0.036**	=	0.551 ± 0.019	+	0.602 ± 0.008	+
10D5C	0.903 ± 0.031	0.899 ± 0.028	=	**0.904 ± 0.030**	=	0.619 ± 0.021	+	0.398 ± 0.006	+
10D10C	0.873 ± 0.035	**0.878 ± 0.028**	=	0.876 ± 0.027	=	0.543 ± 0.020	+	0.380 ± 0.004	+
KDDC–99	0.460 ± 0.000	0.460 ± 0.000	=	0.460 ± 0.000	=	**0.650 ± 0.000**	-	0.140 ± 0.000	+

Similarly, regarding Table 3, WOA-OpStream showed a slightly better statistical behaviour than BAT-OpStream, and it was statistically equivalent to DE-OpStream, also in terms of purity. However, according to this metric, the stochastic classifiers did not outperform the deterministic ones, but had quite similar performances. In terms of the average value over the 30 repetitions, DE-OpStream and DenStream had the highest purity.

Finally, the same conclusions obtained with the F-measure were drawn by interpreting the results in Table 4, where the Rand index metric was used to evaluate the classification performances. Indeed, all three OpStream variants statistically outperformed the deterministic methods. This goes to show that the proposed method was performing very well regardless of the optimisation strategy, and it was always better or competitive with state-of-the-art algorithms. Unlike the case in Table 2, the best performances were in terms of the average value or those obtained with DE rather than WOA.

However, the difference between the two variants was minimal, and the Wilcoxon rank-sum test did not detect differences between the two variants.

Table 3. Average purity ± standard deviation and Wilcoxon rank-sum test (reference = WOA-OpStream) for WOA-OpStream against BAT-OpStream, DE-OpStream, DenStream and CluStream on each dataset.

Dataset	WOA-OpStream	BAT-OpStream	W	DE-OpStream	W	DenStream	W	CluStream	W
5D5C	0.998 ± 0.006	0.996 ± 0.007	+	0.998 ± 0.006	=	**1.000 ± 0.000**	=	0.998 ± 0.004	=
5D10C	0.992 ± 0.016	0.984 ± 0.022	=	0.987 ± 0.022	=	**1.000 ± 0.000**	-	0.998 ± 0.004	-
10D5C	**1.000 ± 0.000**	0.998 ± 0.004	+	**1.000 ± 0.000**	=	**1.000 ± 0.000**	=	**1.000 ± 0.000**	=
10D10C	0.999 ± 0.002	**1.000 ± 0.002**	=	**1.000 ± 0.002**	=	**1.000 ± 0.000**	=	**1.000 ± 0.000**	=
KDDC-99	**1.000 ± 0.000**	**1.000 ± 0.000**	=	**1.000 ± 0.000**	=	**1.000 ± 0.000**	=	0.420 ± 0.000	+

Table 4. Average Rand index ± standard deviation and Wilcoxon rank-sum Test (reference = WOA-OpStream) for WOA-OpStream against BAT-OpStream, DE-OpStream, DenStream and CluStream on each dataset.

Dataset	WOA-OpStream	BAT-OpStream	W	DE-OpStream	W	DenStream	W	CluStream	W
5D5C	**0.951 ± 0.018**	0.945 ± 0.019	+	0.951 ± 0.017	=	0.825 ± 0.005	+	0.596 ± 0.041	+
5D10C	0.944 ± 0.020	0.947 ± 0.020	=	**0.949 ± 0.016**	=	0.753 ± 0.013	+	0.625 ± 0.018	+
10D5C	0.934 ± 0.017	0.932 ± 0.016	=	**0.935 ± 0.017**	=	0.814 ± 0.007	+	0.432 ± 0.033	+
10D10C	0.939 ± 0.020	0.941 ± 0.018	=	**0.942 ± 0.017**	=	0.746 ± 0.016	+	0.400 ± 0.020	+
KDDC-99	0.620 ± 0.000	0.620 ± 0.000	=	0.620 ± 0.000	=	0.820 ± 0.000	-	**0.940 ± 0.000**	-

Summarising, OpStream displayed the best global performance, with WOA-OpStream and DE-OpStream being the most preferable variants. Statistically, WOA-OpStream and DE-OpStream had equivalent performances over different datasets and according to three different evaluation metrics. In this light, the WOA variant was preferred as it required the tuning of only two parameters, against the three required in DE, to function optimally.

A final observation can be done by separating the results from the synthetic datasets and *KDDC–99*. If in the first case, the supremacy of OpStream was evident; a deterioration of the performances can be noted when the later dataset was used. In this light, one can understand that the proposed method presented room for improvement of handling data streams with an uneven distribution of class instances as those presented in *KDDC–99* [51].

7. Further Analyses

In the light of what was observed in Section 6, the WOA algorithm was preferred over DE and BAT to perform the optimisation phase. Hence, it was reasonable to consider the WOA-OpStream variant as the default OpStream algorithm implementation.

This section concludes this research with a thorough analysis of this variant in terms of sensitivity, scalability, robustness and flexibility to handle overlapping multi-density clusters.

7.1. Scalability Analysis

A scalability analysis was performed to test how OpStream behaved, in terms of execution time (seconds) needed to process 100,000 data points per dataset, over datasets having increasing dimension values or an increasing number of clusters. Datasets suitable for this purpose are easily generated with the MOA platform, as previously done for the comparative analysis in Section 6.

This experimentation was performed on a personal computer equipped with an AMD Ryzen 5 2500 U Quad-Core (2.0 GHz) CPU Processor and 8 GB RAM. OpStream was run with the following parameter settings: $\lambda = 1000$, $\epsilon = 0.1$, $\beta = 4$, WOA swarm size equal to 20 and maximum number of allowed iterations equal to 10.

Execution time is plotted over increasing dimension values (for the the data points) in Figure 2.

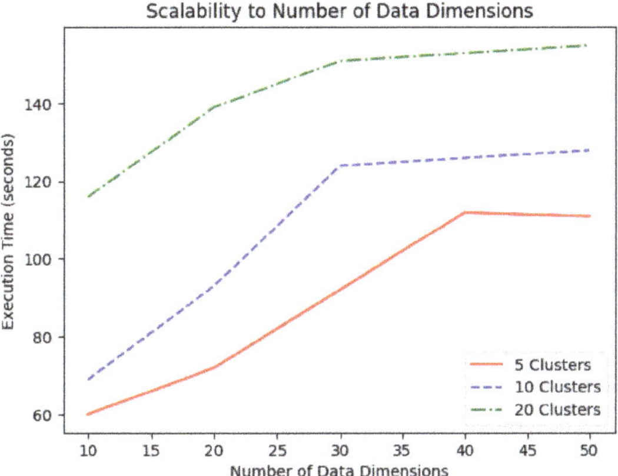

Figure 2. Scalability to the number of data dimensions (data dimension value).

Execution time is plotted over increasing number of clusters (in the datasets) in Figure 3.

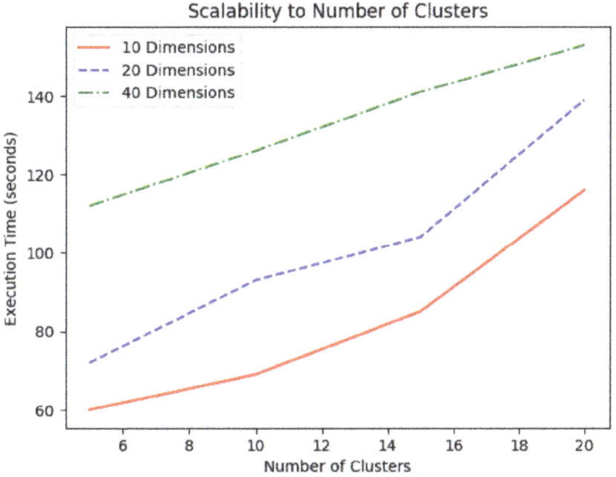

Figure 3. Scalability (number of clusters).

Regardless of the number of clusters, the execution time seemed to grow linearly with the dimensionality of the data points, for low dimension values, to then saturate when the dimensionality was high. The lower the number of clusters, the later the saturation phenomenon took place. With five clusters, this occurred at approximately 40 dimension values. In the case of 20 clusters, saturation occurred earlier at approximately 25 dimension values. This was one of the strengths of the proposed method, as its time complexity did not require polynomial times.

Conversely, no saturation took place when the execution time was measured by increasing the number of clusters. In this case as well, the time complexity seemed to grow linearly with the number of clusters.

7.2. Noise Robustness Analysis

The MOA platform allowed for the injection of increasing noise levels into the datasets *5D10C* and *10D5C*.

The five noise levels indicated in Figures 5 and 6 were used, and the OpStream algorithm was run 30 times for each one of the 10 classification problems (i.e., five noise levels $\times$ 2 datasets) with the same parameter setting used in Section 7.1. Results were collected to display the average F-measure, purity and Rand index relative to *5D10C*, i.e., Table 5, and *10D5C*, i.e., Table 6.

Table 5. Average OpStream performances over *5D10C* at multiple noise levels.

Noise Level	F-Measure	Purity	Rand Index
0%	0.846	0.986	0.934
3%	0.798	0.988	0.909
5%	0.808	0.983	0.892
8%	0.768	0.993	0.881
10%	0.774	0.972	0.880

Table 6. Average OpStream performances over or *10D5C* at multiple noise levels.

Noise Level	F-Measure	Purity	Rand Index
0%	0.902	1.000	0.936
3%	0.856	1.000	0.899
5%	0.854	1.000	0.889
8%	0.848	1.000	0.884
10%	0.835	1.000	0.865

From these results, it is clear that OpStream was able to retain approximately 95% of its original performance as long as the level did not exceed the 5% level. Then, performances slightly decreased. OpStream seemed to be robust to noise, in particular when classifying datasets with high-dimensional data points and a low number of clusters.

7.3. Sensitivity Analysis

Five parameters must be tuned before using OpStream for clustering dynamic data streams. In this section, the impact of each parameter on the classification performance is analysed in terms of the Rand index value.

To perform a thorough sensitivity analysis

- the size λ of the landmark time window model was examined in the range $[100, 5000] \in \mathbb{N}$;
- the ϵ value for the ϵ-neighbourhood method was examined within $[0, 1] \in \mathbb{R}$;
- the effect of the age threshold was examined by tuning β in the interval $[1, 10] \in \mathbb{N}$;
- the WOA swarm sizes under analysis were obtained by adding 5 candidate solutions per experiment, from an initial value of 5 candidate solutions to a maximum of 30 candidate solutions;
- the computational budget for the optimisation process, expressed in terms of "max iterations"number, was increased by 5 iterations per experiment starting with 5 up to a maximum of 30 iterations.

OpStream was run on three datasets for this sensitivity analysis, namely *5D10C*, *10D5C* and *KDDC-99*, and the results are graphically shown in the figures reported below.

Figure 4 shows that too high window sizes were not beneficial, and the best performances were obtained in the rage of $[500, 2000]$ data points. In particular, a peak was obtained with a size of 1000 for the two artificially prepared datasets. Conversely, slightly inferior sizes might be preferred for the *KDDC-99* dataset. In general, there was no need to use more than 2000 data points, as the performance would remain constant or slightly deteriorate.

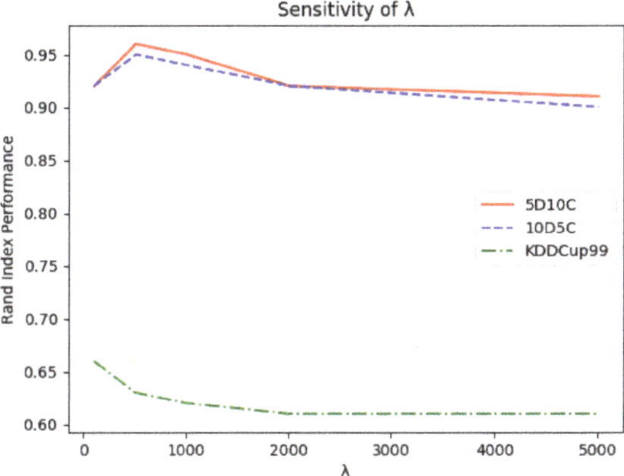

Figure 4. Sensitivity to the windows size parameter λ.

With reference to Figure 5, it was evident that ϵ did not require fine tuning in a wide range as the best performances were obtained within $[0.1, 0.2]$ and then linearly decreased over the remaining admissible values. This can be easily explained as too low values would prevent microclusters from merging while too high values would force OpStream to merge dissimilar clusters. In both cases, the outcome would be a very poor classification. This observation facilitated the tuning process as it meant that it was worth trying values for ϵ of 0.1 and 0.2 and perhaps one or two intermediary values.

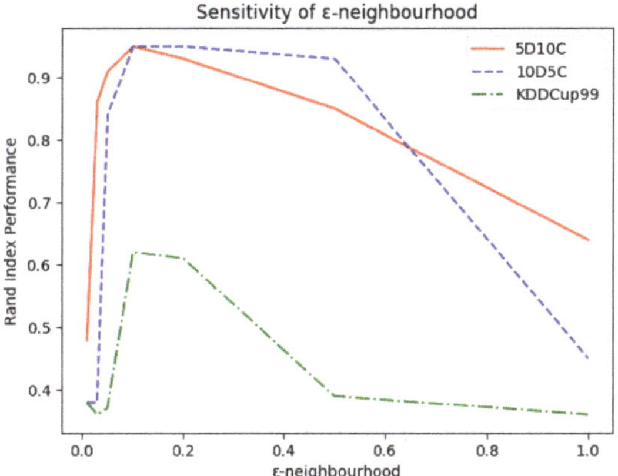

Figure 5. Sensitivity to the ϵ-neighboured parameter ϵ.

As for β, the curves in Figure 6 show that OpStream was not sensitive to the value chosen for removing outdated clusters as long as $\beta \geq 2$. This meant that clusters could be technically left in the buffers for a long time without affecting the performance of the classifier. From a more practical point of view, for memory issues, it was preferable to free buffers from unnecessary microclusters in a timely manner. A sensible choice is $\beta = 4$, as too low values might prevent similar clusters from being merged due to the lack of time required for performing such a process.

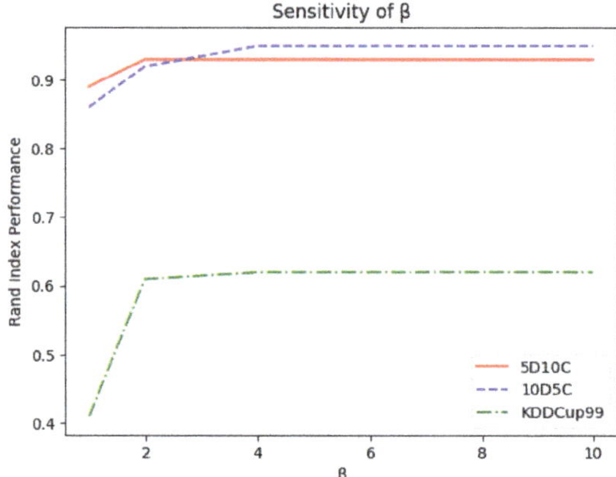

Figure 6. Sensitivity to the Ageing System (AS) parameter β.

It can be noted that a small number of candidate solutions was used for the optimisation phase. This choice was made for multiple reasons. First, it has been recently shown that a high number of solutions can increase the structural biases of the algorithm [52], which is not wanted as the algorithm has to be "general-purpose" to handle all possible scenarios obtained in the dynamic domain. Second, due to the time limitations related to the nature of this application domain, a high number of candidate solutions was to be avoided as it would slow down the converging process. This is not admissible in the real-time domain where also the computational budget is kept very low. Third, as shown in Figure 7, the WOA method used in OpStream seemed to work efficiently regardless of the employed number of candidate solutions, as long as it was greater than 20.

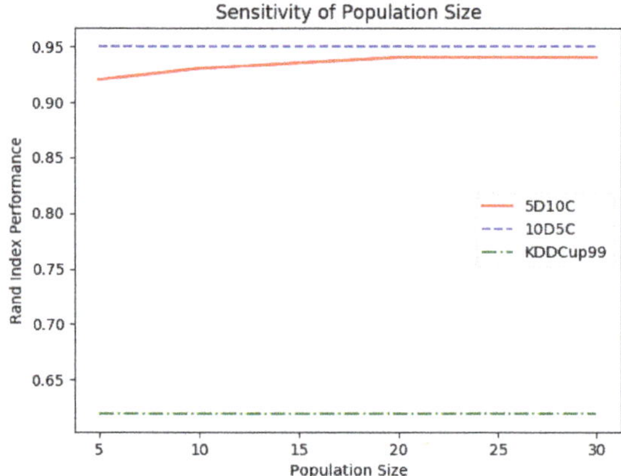

Figure 7. WOA sensitivity to the swarm size.

Similar conclusions can be made for the computational budget. According to Figure 8, it was not necessary to prolong the duration of the WOA optimisation process for more than 10 iterations. This makes sense in dynamic domains, where the problem changes very frequently, thus making the exploitation phase less important.

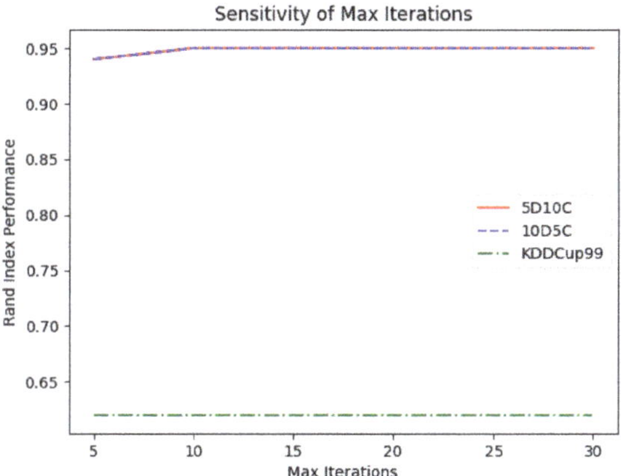

Figure 8. Sensitivity to $maxIterations$.

7.4. Comparison with Past Studies on Intrusion Detection

One last comparison was performed to complete this study. This was performed on a specific application domain, i.e., network intrusion detection, by means of the "KDD–cup 99" database [53]. The comparison algorithms employed in this work, i.e., DenStream and CluStream, were both tested on this dataset in their original papers [18,25], respectively. Despite the fact that OpStream is not meant for datasets with overlapping multi-density clusters, as in KDD–cup 99, we executed it over such a dataset to test its versatility. Results are displayed in Table 7 where the last column indicates the average performance of the clustering method by computing:

$$\text{AVG} = \frac{\text{F-Measure} + \text{Purity} + \text{Rand Index}}{3}. \tag{25}$$

Table 7. Results obtained with the KDD–cup 99 [53] dataset for intrusion detection.

Algorithm	F-Measure	Purity	Rand Index	AVG
OpStream	0.46	1.00	0.62	0.69
DenStream	0.65	1.00	0.82	0.82
ClusStream	0.14	0.42	0.94	0.50

Surprisingly, OpStream had an AVG better performance than CluStream, due to the fact that it significantly outperformed it in terms of F-measure and purity and displayed a state-of-the-art behaviour in terms of the purity value. As expected, DenStream provided the best performance, thus being preferable in this application domain unless a fast real-time response is required. In the latter case, its high computational cost could prevent DenStream from being successfully used [18].

8. Conclusions and Future Work

Experimental numerical results showed that the proposed OpStream algorithm was a promising tool for clustering dynamic data streams as it was competitive and outperformed the state-of-the-art on several occasions. This approach could then be applied in several challenging application domains where satisfactory results are difficult to obtain with clustering methods. Thanks to its optimisation driven initialisation phase, OpStream displays high accuracy, robustness to noise in the dataset and versatility. In particular, we found out that its WOA implementation was efficient, scalable (both in term of dataset dimensionality and number of clusters) and resilient to parameters' variations. Moreover, due to a low number of parameters to be tuned in WOA, this optimisation algorithm was preferred over other approaches returning similar accuracy values as DE and BAT. Finally, this study clearly showed that hybrid clustering methods are promising and more suitable than classic approaches to address challenging scenarios.

Possible improvements can be done to address some of the aspects arising during the experimental section. First, the deterioration of the performance over unevenly distributed datasets, as *KDDC-99*, will be investigated. A simple solution to this problem is to embed non-density based clustering algorithms into the OpStream framework. Second, since the proposed methods do not benefit from preceding optimisation processes (as shown in Figure 8), probably because of the dynamic nature of the problem, the optimisation algorithm employing "restart" mechanisms will be implemented and tested. These algorithms usually work on a very short computational budget and handle dynamic domains better than others by simply re-sampling the initial point where a local search routine is applied, as, e.g., [54], or by also adding to it information from the previous past solution with the "inheritance" method [55–57].

It is also worthwhile to extend OpStream to handle overlapping multi-density clusters in dynamic data streams, as these cases are not currently addressable and are common in some real-world scenarios, such as network intrusion detection [51] and Landsat satellite image discovery [58].

Author Contributions: All authors contributed to the draft of the manuscript and read and approved the final manuscript. J.M.Y. made a major contribution by implementing the proposed system and the code to run all experiments. F.C. made a major contribution in writing the manuscript with J.M.Y. and in the implementation and correct use of the optimisation algorithms. E.H., V.S. and A.M. contributed in writing the manuscript, checking the validity of the proposed evaluation methods and their appropriateness.

Funding: This research received no external funding.

Conflicts of Interest: The authors declare no conflict of interest.

References

1. Modi, K.; Dayma, R. Review on fraud detection methods in credit card transactions. In Proceedings of the 2017 International Conference on Intelligent Computing and Control (I2C2), Coimbatore, India, 23–24 June 2017.
2. Moodley, R.; Chiclana, F.; Caraffini, F.; Carter, J. Application of uninorms to market basket analysis. *Int. J. Intell. Syst.* **2019**, *34*, 39–49. [CrossRef]
3. Moodley, R.; Chiclana, F.; Caraffini, F.; Carter, J. A product-centric data mining algorithm for targeted promotions. *J. Retail. Consum. Serv.* **2019**. [CrossRef]
4. Zarpelão, B.B.; Miani, R.S.; Kawakani, C.T.; de Alvarenga, S.C. A survey of intrusion detection in Internet of Things. *J. Netw. Comput. Appl.* **2017**, *84*, 25–37. [CrossRef]
5. Masud, M.M.; Chen, Q.; Khan, L.; Aggarwal, C.; Gao, J.; Han, J.; Thuraisingham, B. Addressing Concept-Evolution in Concept-Drifting Data Streams. In Proceedings of the 2010 IEEE International Conference on Data Mining, Sydney, Australia, 14–17 December 2010; pp. 929–934.
6. Gharehchopogh, F.S.; Gholizadeh, H. A comprehensive survey: Whale Optimization Algorithm and its applications. *Swarm Evol. Comput.* **2019**, *48*, 1–24. [CrossRef]
7. Hardi, M.; Mohammed, S.U.U.; Rashid, T.A. A Systematic and Meta-Analysis Survey of Whale Optimization Algorithm. *Comput. Intell. Neurosci.* **2019**, *2019*, 25. [CrossRef]
8. Storn, R.; Price, K. Differential Evolution—A Simple and Efficient Heuristic for global Optimization over Continuous Spaces. *J. Glob. Optim.* **1997**, *11*, 341–359. [CrossRef]
9. Caraffini, F.; Kononova, A.V. Structural bias in differential evolution: A preliminary study. *AIP Conf. Proc.* **2019**, *2070*, 020005. [CrossRef]
10. Caraffini, F.; Kononova, A.V.; Corne, D. Infeasibility and structural bias in Differential Evolution. *Inf. Sci.* **2019**, *496*, 161–179. [CrossRef]
11. Mirjalili, S.; Lewis, A. The Whale Optimization Algorithm. *Adv. Eng. Softw.* **2016**, *95*, 51–67. [CrossRef]
12. Yang, X.S. A New Metaheuristic Bat-Inspired Algorithm. *Nat. Inspired Coop. Strateg. Optim.* **2010**, *284*, 65–74. [CrossRef]
13. Chen, G.; Luo, W.; Zhu, T. Evolutionary clustering with differential evolution. In Proceedings of the 2014 IEEE Congress on Evolutionary Computation (CEC), Beijing, China, 6–11 July 2014; pp. 1382–1389.

14. Carnein, M.; Trautmann, H. evoStream—Evolutionary Stream Clustering Utilizing Idle Times. *Big Data Res.* **2018**, *14*, 101–111. [CrossRef]
15. Nasiri, J.; Khiyabani, F. A Whale Optimization Algorithm (WOA) approach for Clustering. *Cogent Math. Stat.* **2018**, *5*. [CrossRef]
16. Nandy, S.; Sarkar, P. Chapter 8–Bat algorithm–based automatic clustering method and its application in image processing. In *Bio-Inspired Computation and Applications in Image Processing*; Academic Press: Cambridge, MA, USA, 2016; pp. 157–185.
17. Kokate, U.; Deshpande, A.; Mahalle, P.; Patil, P. Data Stream Clustering Techniques, Applications, and Models: Comparative Analysis and Discussion. *Big Data Cogn. Comput.* **2018**, *2*, 32. [CrossRef]
18. Cao, F.; Ester, M.; Qian, W.; Zhou, A. Density based Clustering over an Evolving Data Stream with Noise. In Proceedings of the 2006 SIAM Conference on Data Mining, Bethesda, MD, USA, 20–22 April 2006; Volume 2006, pp. 328–339.
19. Sun, J.; Fujita, H.; Chen, P.; Li, H. Dynamic financial distress prediction with concept drift based ontime weighting combined with Adaboost support vector machine ensemble. *Knowl. Based Syst.* **2017**, *120*, 4–14. [CrossRef]
20. Brzezinski, D.; Stefanowski, J. Prequential AUC: Properties of the area under the ROC curve for data streams with concept drift. *Knowl. Inf. Syst.* **2017**, *52*, 531–562. [CrossRef]
21. ZareMoodi, P.; Kamali Siahroudi, S.; Beigy, H. Concept-evolution detection in non-stationary data streams: A fuzzy clustering approach. *Knowl. Inf. Syst.* **2019**, *60*, 1329–1352. [CrossRef]
22. Carnein, M.; Trautmann, H. Optimizing Data Stream Representation: An Extensive Survey on Stream Clustering Algorithms. *Bus. Inf. Syst. Eng. Int. J. Wirtsch.* **2019**, *61*, 277–297. [CrossRef]
23. Gao, X.; Ferrara, E.; Qiu, J. Parallel clustering of high-dimensional social media data streams. In Proceedings of the 2015 15th IEEE/ACM International Symposium on Cluster, Cloud and Grid Computing, Shenzhen, China, 4–7 May 2015; pp. 323–332.
24. Gao, L.; Jiang, Z.Y.; Min, F. First-Arrival Travel Times Picking through Sliding Windows and Fuzzy C-Means. *Mathematics* **2019**, *7*, 221. [CrossRef]
25. Aggarwal, C.C.; Yu, P.S.; Han, J.; Wang, J. A Framework for Clustering Evolving Data Streams. In Proceedings of the 2003 VLDB Conference, Berlin, Germany, 9–12 September 2003; Volume 29, pp. 81–92.
26. Madhulatha, T.S. Overview of streaming-data algorithms. *arXiv* **2012**, arXiv:1203.2000.
27. Hartigan, J.A.; Wong, M.A. Algorithm AS 136: A K-Means Clustering Algorithm. *Appl. Stat.* **1979**, *28*, 100–108. [CrossRef]
28. O'Callaghan, L.; Mishra, N.; Meyerson, A.; Guha, S.; Motwani, R. Streaming-data algorithms for high-quality clustering. In Proceedings of the 18th International Conference on Data Engineering, San Jose, CA, USA, 26 February–1 March 2002; pp. 685–694.
29. Spinosa, E.J.; de Leon, F.; de Carvalho, A.P.; Gama, J.A. OLINDDA: A Cluster based Approach for Detecting Novelty and Concept Drift in Data Streams. In Proceedings of the 2007 ACM Symposium on Applied Computing, Seoul, Korea, 11–15 March 2007; pp. 448–452.
30. Forestiero, A.; Pizzuti, C.; Spezzano, G. A single pass algorithm for clustering evolving data streams based on swarm intelligence. *Data Min. Knowl. Discov.* **2013**, *26*, 1–26. [CrossRef]
31. Forestiero, A.; Pizzuti, C.; Spezzano, G. FlockStream: A Bio-Inspired Algorithm for Clustering Evolving Data Streams. In Proceedings of the 2009 21st IEEE International Conference on Tools with Artificial Intelligence, Newark, NJ, USA, 2–4 November 2009.
32. Alswaitti, M.; Albughdadi, M.; Isa, N.A.M. Density based particle swarm optimization algorithm for data clustering. *Expert Syst. Appl.* **2018**, *91*, 170–186. [CrossRef]
33. Shamshirband, S.; Hadipoor, M.; Baghban, A.; Mosavi, A.; Bukor, J.; Varkonyi-Koczy, A.R. Developing ANFIS-PSO Model to Predict Mercury Emissions in Combustion Flue Gases. *Mathematics* **2019**. [CrossRef]
34. Kong, F.; Jiang, J.; Huang, Y. An Adaptive Multi-Swarm Competition Particle Swarm Optimizer for Large-Scale Optimization. *Mathematics* **2019**, *7*, 521. [CrossRef]
35. Fahy, C.; Yang, S. Finding and Tracking Multi-Density Clusters in Online Dynamic Data Streams. *IEEE Trans. Big Data* **2019**. [CrossRef]
36. Dorigo, M.; Di Caro, G. Ant colony optimization: a new meta-heuristic. In Proceedings of the 1999 Congress on Evolutionary Computation-CEC99 (Cat. No. 99TH8406), Washington, DC, USA, 6–9 July 1999; Volume 2, pp. 1470–1477.

37. Tu, D.Q.; Kayes, A.S.M.; Rahayu, W.; Nguyen, K. ISDI: A New Window based Framework for Integrating IoT Streaming Data from Multiple Sources. In Proceedings of the 33rd International Conference on Advanced Information Networking and Applications, AINA 2019, Matsue, Japan, 27–29 March 2019; pp. 498–511.
38. Krempl, G.; Žliobaite, I.; Brzeziński, D.; Hüllermeier, E.; Last, M.; Lemaire, V.; Noack, T.; Shaker, A.; Sievi, S.; Spiliopoulou, M.; et al. Open Challenges for Data Stream Mining Research. *SIGKDD Explor. Newsl.* **2014**, *16*, 1–10. [CrossRef]
39. Yin, C.; Xia, L.; Wang, J. Data Stream Clustering Algorithm Based on Bucket Density for Intrusion Detection. In *Advances in Computer Science and Ubiquitous Computing*; Park, J.J., Loia, V., Yi, G., Sung, Y., Eds.; Springer: Singapore, 2018; pp. 846–850.
40. Huang, G.; Zhang, Y.; Cao, J.; Steyn, M.; Taraporewalla, K. Online mining abnormal period patterns from multiple medical sensor data streams. *World Wide Web* **2014**, *17*, 569–587. [CrossRef]
41. Fahy, C.; Yang, S.; Gongora, M. Finding Multi-Density Clusters in non-stationary data streams using an Ant Colony with adaptive parameters. In Proceedings of the 2017 IEEE Congress on Evolutionary Computation (CEC), San Sebastián, Spain, 5–8 June 2017; pp. 673–680.
42. Fahy, C.; Yang, S.; Gongora, M. Ant Colony Stream Clustering: A Fast Density Clustering Algorithm for Dynamic Data Streams. *IEEE Trans. Cybern.* **2019**, *49*, 2215–2228. [CrossRef]
43. Yang, X.S.; Hossein Gandomi, A. Bat algorithm: A novel approach for global engineering optimization. *Eng. Comput.* **2012**, *29*, 464–483. [CrossRef]
44. Opara, K.R.; Arabas, J. Differential Evolution: A survey of theoretical analyses. *Swarm Evol. Comput.* **2019**, *44*, 546–558. [CrossRef]
45. Bifet, A.; Holmes, G.; Kirkby, R.; Pfahringer, B. MOA: Massive Online Analysis. *J. Mach. Learn. Res.* **2010**, *11*, 1601–1604.
46. University of California. *KDD Cup 1999*; University of California: Irvine, CA, USA, 2007.
47. Rand, W.M. Objective Criteria for the Evaluation of Clustering Methods. *J. Am. Stat. Assoc.* **1971**, *66*, 846–850. [CrossRef]
48. Hedar, A.R.; Ibrahim, A.M.M.; Abdel-Hakim, A.E.; Sewisy, A.A. K-Means Cloning: Adaptive Spherical K-Means Clustering. *Algorithms* **2018**, *11*, doi:10.3390/a11100151. [CrossRef]
49. Wilcoxon, F. Individual comparisons by ranking methods. *Biom. Bull.* **1945**, *1*, 80–83. [CrossRef]
50. Caraffini, F. The Stochastic Optimisation Software (SOS) Platform. Available online: https://doi.org/10.5281/zenodo.3237024 (accessed on 1 December 2019).
51. Tavallaee, M.; Bagheri, E.; Lu, W.; Ghorbani, A.A. A detailed analysis of the KDD CUP 99 dataset. In Proceedings of the 2009 IEEE Symposium on Computational Intelligence for Security and Defense Applications, Ottawa, ON, Canada, 8–10 July 2009.
52. Kononova, A.V.; Corne, D.W.; Wilde, P.D.; Shneer, V.; Caraffini, F. Structural bias in population based algorithms. *Inf. Sci.* **2015**, *298*, 468–490. [CrossRef]
53. Rosset, S.; Inger, A. KDD-cup 99: Knowledge Discovery in a Charitable Organization's Donor Database. *SIGKDD Explor. Newsl.* **2000**, *1*, 85–90, doi:10.1145/846183.846204. [CrossRef]
54. Caraffini, F.; Neri, F.; Gongora, M.; Passow, B. Re-sampling Search: A Seriously Simple Memetic Approach with a High Performance. In Proceedings of the IEEE Symposium Series on Computational Intelligence, Workshop on Memetic Computing, Singapore, 16–19 April 2013; pp. 52–59.
55. Iacca, G.; Caraffini, F. Compact Optimization Algorithms with Re-Sampled Inheritance. In *Applications of Evolutionary Computation*; Kaufmann, P., Castillo, P.A., Eds.; Springer: Cham, Switzerland, 2019; pp. 523–534.
56. Caraffini, F.; Iacca, G.; Yaman, A. Improving (1+1) covariance matrix adaptation evolution strategy: A simple yet efficient approach. *AIP Conf. Proc.* **2019**, *2070*, 020004. [CrossRef]
57. Caraffini, F.; Neri, F.; Epitropakis, M. HyperSPAM: A study on hyper-heuristic coordination strategies in the continuous domain. *Inf. Sci.* **2019**, *477*, 186–202. [CrossRef]
58. Li, X.; Ye, Y.; Li, M.J.; Ng, M.K. On cluster tree for nested and multi-density data clustering. *Pattern Recognit.* **2010**, *43*, 3130–3143. [CrossRef]

Article

Application of Differential Evolution Algorithm Based on Mixed Penalty Function Screening Criterion in Imbalanced Data Integration Classification

Yuelin Gao [1,*], Kaiguang Wang [1,*], Chenyang Gao [2], Yulong Shen [2] and Teng Li [2]

1 Ningxia Province Key Laboratory of Intelligent Information and Data Processing, North Minzu University, Yinchuan 750021, China

2 School of Cyber Engineering, Xidian University, Xi'an 710071, China; ghygao2008@gmail.com (C.G.); ylshen@mail.xidian.edu.cn (Y.S.); tengli@xidian.edu.cn (T.L.)

* Correspondence: gaoyuelin@263.net (Y.G.); wkg13759842420@foxmail.com (K.W.); Tel.: +86-139-9510-0900 (Y.G.); +86-183-2997-0138 (K.W.)

Received: 20 November 2019; Accepted: 10 December 2019; Published: 13 December 2019

Abstract: There are some processing problems of imbalanced data such as imbalanced data sets being difficult to integrate efficiently. This paper proposes and constructs a mixed penalty function data integration screening criterion, and proposes Differential Evolution Integration Algorithm Based on Mixed Penalty Function Screening Criteria (DE-MPFSC algorithm). In addition, the theoretical validity and the convergence of the DE-MPFSC algorithm are analyzed and proven by establishing the Markov sequence and Markov evolution process model of the DE-MPFSC algorithm. In this paper, the entanglement degree and enanglement degree error are introduced to analyze the DE-MPFSC algorithm. Finally, the effectiveness and stability of the DE-MPFSC algorithm are verified by UCI machine learning datasets. The test results show that the DE-MPFSC algorithm can effectively improve the effectiveness and application of imbalanced data classification and integration, improve the internal classification of imbalanced data and improve the efficiency of data integration.

Keywords: imbalanced data; screening criteria; DE-MPFSC algorithm; Markov process; entanglement degree; data integration

MSC: 90C59; 54A05

---

1. Introduction

In data classification, the most common methods used are clustering methods based on statistical theory [1] such as naive Bayes [2] and artificial neural networks (ANNs) [3]. These methods can use large-scale imbalanced data as clustering targets [4]. However, with the development of large data networks, electronic data integration, imbalanced data analysis, large text databases, quantum data decoding and encoding analysis, and other data types in recent years, the efficient processing of data has become the foremost subject of current data classification and processing. The imbalanced data classification system studied in this paper belongs to this subject. When the imbalanced data is classified and processed, integration and classification of the data has gradually become the mainstream method for processing imbalanced data but this method usually results in a higher computational cost [5]. In view of the cause, we have identified a meaningful conclusion—the efficient clustering of imbalanced data processing tends to increase the processing cost.

For the processing of imbalanced data, there are some mature algorithms such as sampling methods [6,7], cost-sensitive algorithms [8,9] and one-class classification [8,9]. However, there are some problems in that the processing of imbalanced data is mainly confined to the level of computer algorithm analysis and big data analysis, and lacks the function of efficient integration, which can result in privacy or data breaches, resulting in significant losses to data owners. Distributed and compatible construction of evolutionary algorithm structures is an innovative idea, which was proposed by Santucci [10]. Centroid classification algorithm will provide new ideas for us to explore and solve imbalanced data classification problems. Mikalef proposed a new idea about data processing. Application level and resource level of big data can improve investment efficiency and business performance, which idea was proposed by Patrick Mikalef and it is applied preferentially to business [11]. In addition to the above algorithms, there are other scholars who have performed research on imbalanced data processing. To seek parameter optimization to reduce the cost before training the model, Thai-Nghe proposed a fault-tolerant recognition algorithm that sets the artificial cost rate of data processing to ultrahigh parameters [12]. In the integration of imbalanced data, most experts focus on how to improve the efficient integration and the integration of imbalanced data, such as the boosting data analysis method [13] mentioned by Freund and the bagging data integration analysis algorithm [14] proposed by Breiman. Sun proposed a modular data analysis integration strategy that classifies and integrates multiple imbalanced data and analyzed data using modular strategies [15]. Chawla proposed a data integration strategy combining the (Synthetic Minority Oversampling Technique) SMOTE algorithm and boosting algorithm used in the sampling process [16]. Lachiche [17] proposed a new effective recursive algorithm combining $1BC$ and $2BC$ Bayesian structural systems, improving the difference between computer systems and artificial data in low-dimensional levels.

For generalized imbalanced data processing, the introduction of the above algorithms is limited to the specific algorithm program of data integration and the results of data classification. However, for the number of feature samples and the feature categories about the imbalanced data, the above algorithms may cause higher data classification costs and classification errors. Overall, in the previous research methods, most of them are integrated strategies under a single integrated classifier structure model such as support vector machine, decision tree analysis and the naive Bayesian method [18]. These methods determine the integration effect based on data performance as a classification criterion, which is not conducive to the continuous integration and classification of imbalanced datasets. To solve the problem of high-efficiency classification of imbalanced data, this paper establishes a differential evolution integration algorithm with global convergence ability [19,20] based on the screening criterion of the mixed penalty function (differential evolution integration algorithm based on mixed penalty function screening criteria—the DE-MPFSC algorithm), which provides a useful method for solving the high-efficiency integration and classification of imbalanced data.

A: Source of The Idea

The idea of the DE-MPFSC algorithm is derived from the following three excellent overview papers. The purpose of the first paper [21] is to summarize and organize the comprehensive foundation and the recent proposals on parameter adaptation about differential evolution (DE), which inspired us that the optimal parameters analysis should be carried out before the algorithm design. The second paper [22], presents the basic structure, parameter analysis and the latest applications of evolutionary algorithms. It describes the related principles of various evolutionary algorithms in detail. Comparing the performance of DE algorithm is an effective way to further improve the understanding of evolutionary algorithm. The third paper [23] introduces the various structures and main variants of the DE algorithm. The applications of the DE algorithm in the fields of commerce, agriculture and economics was emphasized in the literature. The theoretical design in this paper [23] is consistent with the ideas about the topological structure of the divided regions [20]. Because the population is so sensitive to the spatial structure of the search area during the evolution process, we carefully considered this idea and proposed a screening criterion based on region segmentation. However, when we reviewed their contributions, we had one common finding: for the variant of DE, when we test the algorithm, we implement it based on a procedural evaluation system of test functions and dimensions. However, the evaluation system ignores the important influence of convergence speed and convergence precision on individual evolution. Based on this finding, we systematically discuss the influence of convergence precision and convergence speed on the evolution process based on the structural indicators of the evolutionary algorithm. Finally, we propose an important idea: judge whether the evolutionary algorithm is better than before based on the degree of entanglement between convergence precision and convergence speed, then, further verify the scientificity of the idea proposed by us through numerical experiments.

In terms of improvements in the DE algorithm, in a multimodal interactive environment, balancing the parameters F and CR of the evolutionary algorithm is an important idea. In multi-objective parameter design and adaptive parameter analysis, the idea of self-adaptation was proposed by Janez Brest, namely (Self-Adapting Parameter Setting in Differential Evolution)JDE [21,24] and applied in the field of computer and machine learning, which can be also applied to large-scale problems [25]. The dynamic parameter design idea of DE-MPFSC algorithm is based on this research and numerical experiments are also analyzed in detail. The Opposition Based Differential Evolution (OBDE), proposed by Rahnamayan et al. (2008) [22,26], mainly emphasizes the impact of the evolution speed on the DE algorithm. The OBDE algorithm can enhance the adaptability of the algorithm by improving the learning efficiency, which idea has a priority role in solving multi-objective problems. Our inspiration for the idea of decision space segmentation in this paper comes from it, which is innovative for improving the decision structure of evolutionary algorithms. The (Differential Evolution with Global and Local neighborhoods) DEGL algorithm [23,27] has made some improvements in the individual optimization space, emphasizing the influence of topological structure and spatial neighborhood on the evolutionary algorithm and highlighting the role of the idea of spatial segmentation, which can guide us to think deeply about the search area of the evolutionary algorithm [28,29]. The new model will be widely used in technological improvement. In Reference [20], we independently proposed the spatial topology concept of the DE algorithm and gave relevant proofs. We independently proposed the spatial topology concept of the DE evolution algorithm and gave relevant proofs. Topology concept is a priority direction in expanding search area. Qin and Suganthan proposed a new version of adaptive differential evolution algorithm [22,30]. Through the adaptation of mutation factor F and cross probability CR, the (Self-Adaptive Differential Evolution) SADE algorithm can adjust the individual search precision and search speed from a micro perspective, which will fundamentally change the search pattern of individuals in the spatial neighborhood and will promote the healthy development of the DE algorithm and provide useful help for the dynamic adjustment strategy of the DE-MPFSC algorithm. In addition, the Modified DE algorithm (MDE) was proposed by Zou et al. [21,31]. The main idea of MDE is to mix the adaptive mechanism of mean and Gaussian distribution to improve the ability of parameters F and CR to update themselves. The Fitness Adaptation Differential Evolution (FADE) was proposed by Ghosh et al. [21,32], the strategy selection mechanism of the main idea is that individual selection behaviors can be randomly generated during the evolution process. When individual populations evolve in spatial neighborhoods, the tentative strategy of selecting individuals is also an effective way to expand the diversity of the population. The spatial structure of individual evolution is not completely random. The evolutionary pattern of individuals will be affected by mutations, resulting in structural deviations of individuals, which is not conducive to large-scale population evolution [33], which was proposed by Caraffini. The structured properties of evolutionary algorithms is one of the important issues we will explore in the future. Especially in terms of the convergence and stability of evolutionary algorithms, this effect is significant. To this end, we should explore new ideas to reduce structural deviations. We roughly reviewed the variants of the DE algorithm and related research work from recent years. Based on the above viewpoints, we have developed an improved model of the DE algorithm and proposed the DE-MPFSC algorithm, which opens up new idea for further research on the DE algorithm and related applications.

B: Technical Route

a. According to the idea of region segmentation, propose the region screening criteria for individuals in the evolution process;

b. Construct the operation mechanism of the DE-MPFSC algorithm, including the introduction of the dynamic mutation factor F and the dynamic crossover probability CR;

c. According to the structural evaluation system, the idea about entanglement degree of the convergence speed and convergence precision of the algorithm is proposed;

d. Compare the four classic DE algorithms JDE, OBDE, DEGL and SADE to further verify the advantages of the DE-MPFSC algorithm.

The first advantage of the DE-MPFSC algorithm is that it can classify and extract the original imbalanced data, generate an imbalanced data point searching area and balanced data point searching area, further redistribute the data searching unit. The second advantage is that it can clarify the progressive boundaries of imbalanced points and balanced points of the classified conditions. When the mixed penalty function is at a progressive boundary, DE algorithms only search inside or outside all data points and cannot cross data boundaries, which is not conducive to decreasing the time for data integration and classification. However, the DE-MPFSC algorithm can optimize the data structure by using the mixed penalty function. The third advantage can improve the accuracy and global optimization of imbalanced data classification and purification. The fourth advantage can analyze many different types of data structures. Because of the DE-MPFSC algorithm having self-adaptability, it can use the global convergence of heuristic algorithms to greatly shorten the search time of the searching area. The DE-MPFSC algorithm expands the searching space, improving the integration efficiency of imbalanced data, which results in wide adaptability. The full-text structure is arranged as follows:

Part one: The first and second chapters introduce the current situation of imbalanced data integration and classification, the advantage analysis of the DE-MPFSC algorithm and the form and normalization of the internal and external penalty functions.

Part two: The third chapter is mainly the processing of the constraint conditions for the mixed penalty function, the formal construction of the DE-MPFSC algorithm and property analysis of the mixed penalty function screening criteria.

Part three: The forth chapter is mainly the theoretical analysis of the DE-MPFSC algorithm. The validity and convergence of the DE-MPFSC algorithm are analyzed from the mathematical point of view.

Part four: In the fifth chapter, we creatively introduce the entanglement degree and entanglement degree error to compare the performance of the DE-MPFSC by numerical experiments.

Part five: The sixth chapter mainly establishes verification indicators of the Classification Accuracy (CA), Adjusted Rand Index (ARI), Normalized Mutual Information (NMI) datasets (due to the overall structure of this paper, we list the full name of the datasets in this section. The relevant definition of the datastes is in the data test chapter and the reference is attached). We analyze and verify the theoretical nature of the DE-MPFSC algorithm through the imbalanced data set (Unified Communications Irvine Machine Learning Repository, namely UCI) UCI machine learning data set;

Part six: Gives relevant conclusions.

2. Prerequisite Knowledge

The general form of the constrained optimization problem will be expressed as follows:

$$\min\ f(x); s.t. \begin{cases} g_i \leq 0, i = 1, 2, \cdots, j \\ h_i = 0, i = j+1, j+2, \cdots, m, \end{cases} \tag{1}$$

where $\mathbf{X} = (x_1, x_2, \cdots, x_n)$ are the decision variables of the objective function $f(x)$, g_i is inequality constraint describing the variable, which role of the inequality constraint is to form the search area in the feasible domain. h_i is equality constraint that forms a boundary value condition in the feasible domain, which role is to control the boundary of the search area. We call the feasible solution containing only equality constraints the positive constraint solution where its constraints are active constraints; otherwise, non-active constraints.

2.1. Basic Steps of the DE Algorithm

The differential evolutionary (DE) algorithm [19,20], proposed by Storn and Price in 1995 to solve Chebyshev inequalities, is an efficient global optimization algorithm, which adopts floating-point vector coding to search in continuous space [20]. There are higher stability, lower volatility and better convergence than others. The specific form of the DE algorithm is as follows.

2.1.1. Population Initialization

Assume that the population individuals of the DE algorithm is $\mathbf{X}(t) = (X_1, (X_2, \cdots, (X_n))^{\mathsf{T}}$ [20,34]. Then, population individuals are following:

$$X_i^t = (x_{i1}^t, x_{i2}^t, \cdots, x_{iD}^t), i = 1, 2, \cdots, NP, \tag{2}$$

where t is the number of iterations and NP is the number of population [20].

Initialization settings: suppose the dimension of the optimization problem is D. The maximum number of iterations is T, then the initialization operation of the DE algorithm is expressed as follows [20]:

$$X_i^0 = (x_{i1}^0, x_{i2}^0, \cdots, x_{iD}^0) \tag{3}$$

$$x_{ij}^0 = a_{ij} + rand(0,1) \cdot (b_{ij} - a_{ij}), i = 1, 2, \cdots, NP; j = 1, 2, \cdots, D, \tag{4}$$

where, $a_{ij}, b_{ij} \in R$.

2.1.2. Individual Mutation

When the individual evolves with the DE algorithm, the mutation sites of the individual originate from the two individuals of the parent $(x_{i_1}^t, x_{i_2}^t)$ [20] in the tth generation parental individuals, where $i_1, i_2 \in NP$. Then, the differential vector is defined as $D_{i_{1,2}} = (x_{i_1}^t - x_{i_2}^t)$. For any vector individual X_i^t, the mutation operation [35,36] is defined as

$$V_i^{t+1} = X_{i_3}^t + F \cdot (X_{i_1}^t - X_{i_2}^t), \tag{5}$$

where $NP \geq 4$, F is the mutation factor and $i_1, i_2, i_3 \in \{1, 2, \cdots, NP\}$ and i_1, i_2, i_3 are not all the same during population evolution [20].

2.1.3. Individual Crossover

Test individuals U_i^{t+1} are generated from mutated individual V_i^{t+1} and individual X_i^t before iteration. At the same time, we introduce random function $rand(0,1)$ and the crossover probability CR to improve the diversity and stability of individual evolution and ensure that one mutation site of the test individual U_i^{t+1} is provided by at least the individual from the last iteration (x_i^t) or V_i^{t+1}. The individual crossover [20,36] is following:

$$(u_{ij}^{t+1}) = \begin{cases} (v_{ij}^{t+1}), if\ rand_j(0,1) \leq CR, \\ (x_{ij}^t), otherwise. \end{cases} \quad i = 1, 2, \cdots, NP; j = 1, 2, \cdots, D, \tag{6}$$

where $CR = 0$ and $CR = 1$ are two extreme cases of individual crossover [20]. The former is conducive to maintaining the global optimization ability of the population and the latter is conducive to increasing the diversity of the population. $CR \in (0,1)$ is conducive to expanding the global search range of population individuals, increasing population diversity and improving the search precision of the population individuals [20].

2.1.4. Individual Selection

The selection operation of the DE algorithm in the evolutionary group is based on the greedy search mechanism. This search mechanism is based on "the individual fitness function value after iteration being smaller than the individual fitness function value" before iteration as a standard [20]. The DE algorithm chooses between test individual X_i^t and mutation individual U_i^{t+1} and mutation individual with small fitness function values will be evolved to the next generation [35,37]. The selection effect of the selection operator [20,35,37] in the population is described by the following equation:

$$X_i^{t+1} = \begin{cases} U_i^{t+1}, if\ f(U_i^{t+1}) \leq f(X_i^t), \\ X_i^t, otherwise. \end{cases} \quad i = 1, 2, \cdots, NP. \tag{7}$$

2.2. Internal Penalty Function

When the internal penalty function [38] solves the nonlinear optimization problem, the most important thing is to limit the iteration points in the nonempty feasible domain, which conducts punishment to the feasible points towards the boundary. In addition, the closer the distance towards the feasible domain boundary, the bigger the penalty probability. The purpose of penalty term is to balance the tendency of the feasible domain to be away from the optimal solution when the feasible domain approaches the boundary and the specific form of the internal penalty function is as follows:

$$\Psi_1(x, r_k) = f(x) + r_k B(x), \tag{8}$$

where $B(x)$ is the penalty term of the internal penalty function, which satisfies the following conditions: $B(x)$ is continuous inside the feasible domain $int(D)$, $B(x)$ is a non-negative function. When $x \to Round(D)$, $B(x) \to +\infty$. r_k is the internal penalty factor, $\{r_k\}$ is a monotone decreasing internal penalty factor sequence, which satisfies the following conditions: $r_k \geq 0$, $\lim_{k\to+\infty} r_k = 0$. Let x_k be an approximate value after the k iteration, then there are $\lim_{k\to+\infty} B(x) = +\infty$ and $\lim_{k\to+\infty} \min \Psi_1(x, r_k) = \lim_{k\to+\infty} \min(f(x) + r_k B(x)) = \min f(x)$.

Normalization of the Internal Penalty Function

Because of the logarithmic function or semi-exponential function converging better in the nonlinear optimization theory, we use the logarithmic function to reduce the interior penalty function term into a formula as follows:

$$B(x) = -\sum_{i=1}^{j} ln(-\frac{g_i}{\{\min + \max\}_{int(D)} g_i}), \tag{9}$$

where $\{\min + \max\}_{int(D)} g_i$ is the average of the maximum and minimum values within the feasible domain $int(D)$, which can balance the searching speed.

2.3. External Penalty Function

The main method of the external penalty function in solving the nonlinear optimization problem is to gradually approach the feasible domain boundary from the outside of the feasible domain [39], which will punish the feasible point violating the constraint conditions but will not punish the minimum point that satisfies the constraint conditions [40]. In addition, its specific form is described as follows:

$$\Psi_2(x,\mu_k) = f(x) + \mu_k P(x), \tag{10}$$

where $P(x)$ is the penalty term of the external penalty function, which satisfies these conditions as follows: $P(x)$ is continuous, $P(x) = 0, \forall x \in int(D); P(x) > 0, \forall x \notin D$. μ_k is the external penalty factor, which is a monotone increasing positive external penalty factor sequence that satisfies these conditions: $\mu_k > 0$, $\lim_{k\to+\infty}\mu_k = +\infty$. Let x_k be an approximate value after the k iteration, there is $\lim_{k\to+\infty} P_k = 0$, then $\lim_{k\to+\infty}\min\Psi_2(x,\mu_k) = \lim_{k\to+\infty}\min(f(x)+\mu_k P(x)) = \min f(x)$.

Normalization of the External Penalty Function

To balance the equality constraint conditions and inequality constraint conditions on the penalty term of the external penalty function, we normalize the specific form of P_k. In addition, its form is described as follows:

$$P(x) \quad = \sum_{i=1}^{j}(\max\{\{\min+\max\}_{int(D)}g_i, 0\})^2 + \sum_{i=j+1}^{m}(\{\min+\max\}_{Round(D)}h_i)^2, \tag{11}$$

where $\{\min+\max\}_{Round(D)}h_i$ is the average of the maximum and minimum values of the feasible domain boundary $Round(D)$, which can balance the searching speed.

3. Differential Evolution Integration Algorithm Based on Mixed Penalty Function Screening Criteria

3.1. Mixed Penalty Function

Simply using internal or external penalty function can transform the constraint problem into an unconstrained problem, which reduces the computational difficulty. However, there are two major defects. First, the position of the effective solution under the constraint conditions often has swing effects when the values of the respective penalty factors are different. In other words, if the optimal point lies on the constraint boundary, the optimal point will not be searched by using the two algorithms, but the iterative sequence generated by the two algorithms will be infinitely close to the optimal point. Second, the higher penalty factors are not to easy to be controlled. For example, for the external penalty function, the penalty factor is too large, which causes the iterative points to rotate around the optimal point and be mistaken for the optimal point, which causes incorrect solutions; if the penalty factor is too small, the condition number of matrix of the penalty function term often becomes too large, so that it is difficult for the iterative sequence to converge to the optimal point. To balance the optimal defects of the two methods, we will establish a mixed penalty function as follows:

$$\begin{aligned}\Psi(x,r_k) = f(x) - Pr_1(k)r_k\sum_{i=1}^{j} ln(-\frac{g_i}{\{\min+\max\}_{int(D)}g_i}) + Pr_2(k)\frac{1}{r_k}(\sum_{i\notin I}(\max\{\{\min+\max\}_{int(D)}g_i,0\})^2 \\ + \sum_{i=j+1}^{m}(\{\min+\max\}_{Round(D)}h_i)^2)\end{aligned} \tag{12}$$

or

$$\Psi(x,r_k) = f(x) + r_k B(x) + \frac{1}{r_k}P(x), \tag{13}$$

where $\mu_k = \frac{1}{r_k}$, $P(x) = \sum_{i=1}^{j}(\max\{\{\min+\max\}_{int(D)}g_i,0\})^2 + \sum_{i=j+1}^{m}(\{\min+\max\}_{Round(D)}h_i)^2 - \sum_{i\notin I} ln(-\frac{g_i}{\{\min+\max\}_{int(D)}g_i})$, $I = \{x|g_i(x) \le 0, i = 1,2,\cdots,j\}$, $\{r_k\}$ is a monotone decreasing penalty factor sequence. $Pr_t, t = 1,2$ is the probability of fuzzy iterative points (that is, when the penalty factor reaches the approximate level of the optimal point by iterating, the closer the optimal point is, the bigger the degree of of the iterative point approximating the optimal point. Then, when $k \to +\infty$, the degree of approximation between x_k and x^* will be in a fuzzy state), which can adjust different kinds of penalty functions.

3.2. The Screening Criteria of the Mixed Penalty Function

When solving nonlinear constrained optimization problems (1), to effectively balance the swing problem generated by the equality penalty and inequality constraints. Let the iterative point be x_k after k iterations, and the accuracy of the iterative solution x_k be a sufficiently small real value $\varepsilon \in (10^{(-4)}, 10^{(-5)})$. According to the brief screening rules of the effective solutions proposed by Deb [41], we establish the screening criteria of the mixed penalty function based on the internal penalty function (this screening criterion was established to relax the spatial search area in order to facilitate cross-regional search of the objective function), the external penalty function and their normalization, as follows:

(1) When the two effective solutions x_{k_1} and x_{k_2} are inside the feasible domain $int(D)$ under the constraint conditions, the effective solution with a smaller internal penalty function value is considered the optimal approximate solution and then $Pr_1 = \frac{x_k}{\sum_{i \in I} x_k}$, x_k satisfies the formula $f(x_k) = \min(f(x_{k_1}), f(x_{k_2}))$;

(2) When the two effective solutions x_{k_1} and x_{k_2} appear in the form of certain solutions and are located at the boundary of the feasible domain $Round(D)$ under the constraint conditions, we take the mean value of the two effective solutions as an optimal approximate solution. Then, $Pr_1 \frac{\min\{x_{k_1}, x_{k_2}\}}{\sum_{i \in I \cup i \notin D} x_k}$, $Pr_2 = \frac{\max\{x_{k_1}, x_{k_2}\}}{\sum_{i \in I \cup i \notin D} x_k}$;

(3) When the two effective solutions x_{k_1} and x_{k_2} appearing in the form of uncertain solutions are located at the boundary of the feasible domain $Round(D)$ under the constraint conditions, let x_{k_1} be an internal iterative solution, then all internal iterative solutions constitute a positive monotonic decreasing sequence $\{x_{k_1}\}$ within the feasible domain. In addition, let x_{k_2} be an external iterative solution, then all external iterative solutions constitute a positive monotonic increasing sequence outside the feasible domain. At the same time, let $x_{k_1} = \inf\{x_{k_1}\}$, $x_{k_2} = \sup\{x_{k_2}\}$, then we take the mean value of the two effective solutions $\frac{\inf\{x_{k_1}\} + \sup\{x_{k_2}\}}{2}$ as an optimal approximate solution. Further, there is the following: $Pr_1 = \frac{\max\{\inf\{x_{k_1}\}, \sup\{x_{k_2}\}\}}{\sum_{i \in int(I) \cup i \notin I} x_k}$, $Pr_2 = \frac{\max\{\inf\{x_{k_1}\}, \sup\{x_{k_2}\}\}}{\sum_{i \in int(I) \cup i \notin I} x_k}$;

(4) When the two effective solutions x_{k_1} and x_{k_2} are outside the feasible domain under constraint conditions, the effective solution with the larger exterior penalty function value will be considered as the optimal approximate solution and then $Pr_2 = \frac{x_k}{\sum_{i \in I} x_k}$, x_k will satisfy the formula $f(x_k) = \max(f(x_{k_1}), f(x_{k_2}))$;

(5) When the solution of the constraint is an invalid solution, in order to ensure the global nature of the algorithm search, we will perform the domain expansion operation based on the probability of the two invalid solutions: Firstly, the invalid solution x_1' is taken as the center and the original space search area is dimension expanded with the preset precision $\varepsilon \in (-1, 1)$ as the radius, so that the constraint condition is relaxed and the invalid solution is validated.

The screening criterion possesses the following properties:

1. The internal penalty function or the external penalty function can only detect the inside of the feasible domain or the outside of the feasible domain; they are unable to detect the feasibility of the area by crossing the regional boundary, which can divide the feasible domain into two parts, interior and exterior. However, by adding the probability condition based on the penalty function, we can balance the approximate optimal solution x_k according to the approximate degree of the iterative points x_k and the global optimal point x^*. In addition, the criterion (2) or (3) plays a balancing role in the optimal conditions, avoiding the phenomenon of the solution that cannot move in the interior and exterior of the feasible domain.

2. When the iterative points x_k completely fall in the interior $int(D)$ or exterior $out(D)$ of the feasible domain, the criteria (1) or (4) has the characteristics of being able to flexibly select the optimal area, which cannot only avoid the loss of the approximate global optimal point x^* caused by a single optimal area but also ensure searching the iterative points in the whole area. Further, it can avoid the occurrence of an approximate solution with less precision due to the singularity of the solution.

3. The criterion (1) or (4) adopts the selection strategy in probability, which has better optimal effect for the strong constraint optimization problem [42]. This is essentially an intensive strategy for elite retention strategies, so it can be applied to stronger constraint optimization problems.

4. The internal penalty function tends to select the iterative points that are far from the feasible domain boundary and the external penalty function tends to select the iterative points that are closer to the feasible domain boundary $Round(D)$, both of which are not conducive to the global convergence of the optimal point. However, the criterion can effectively avoid the dispersion distribution of iterative points and can find the global optimal point faster in the condition of the induction of the differential evolution algorithm.

3.3. The Processing of the Constraint Conditions

In the constraint optimization problems shown by (1), there are two kinds of constraints: equality constraints and inequality constraints. Inequality constraints tend to expand the searching range, which is beneficial to global searching and can reduce the probability of fault tolerance. However, The equality constraints present a one-dimensional linear region, which can narrow the optimal range and is not conducive to global optimization. For this reason, we conduct high-dimension processing on the equality constraints:

$$g_i = |h_i| - \varepsilon_i \leq 0, \tag{14}$$

where $\varepsilon_i \geq 0$ is a sufficiently small real-valued slack variable, which increases the searching range by expanding the dimension, which is beneficial to global searching and global convergence. However, the phenomenon will reduce the feasible domain searching ratio, which is not beneficial to finding the global convergent point x^*. To solve the problem, we adopt the method of the self-adaptive slack variable to deal with the equality constraints [43].

$$If(R_n(k) \leq R_l), Then\ \varepsilon_i(t+1) = \beta_l \varepsilon_i(k) \tag{15}$$

$$If(R_n(k) \geq R_u), Then\ \varepsilon_i(t+1) = \beta_u \varepsilon_i(k), \tag{16}$$

where R_n is the proportion of feasible solutions satisfying the slack variable in the current population, $0 \leq R_l \leq R_u \leq 1, 0 \leq \beta_l \leq 1 \leq \beta_u, \varepsilon_i(0)$ is the maximum value that violates the effective solution in the initial population.

3.4. Implementation of Differential Evolution Integrated Algorithm Based on the Screening Criterion of the Mixed Penalty Function

In this paper, the mixed penalty function screening criterion is applied to the DE algorithm, the validity and the convergence of the algorithm are analyzed by the Markov model. Finally, we obtain the differential evolution integrated algorithm based on mixed function screening criteria (DE-MPFSC algorithm).

In the DE-MPFSC algorithm, let the space dimension be $D = n$, the population individuals are a set of $S_X = \{(x_i^t, \delta_i) | i = 1, 2, \cdots, NP\}$, x_i^t is an n-dim decision variable, δ_i is the n-dim individual iterative step variable. The initial population is evenly distributed in the n-dim space constraints. The mutation operator, crossover operator and selection operator DE algorithm can act on the n-dim space constraints and work on all individuals in the initial population to improve their searching ability, where the searching step calculation formula of the DE-MPFSC algorithm is as follows:

$$\delta_i^{t+1} = \delta_i^t + exp(\tau \cdot N(0,1) + \tau' \cdot N_i(0,1)) \quad (17)$$

$$x_i^{t+1} = x_i^t + \delta_i^{t+1} \cdot N_i(0,1), \quad (18)$$

where t is the number of iterations, and τ and τ' are the self-adaptive learning rate of the population individuals. To ensure the accuracy of the calculation, we calculate according to the method proposed by Schwefel [44]: $\tau = (\sqrt{2\sqrt{n}})^{-1}$ and $\tau' = (\sqrt{2n})^{-1}$. $N(0,1)$ and $N_i(0,1)$ are all real uniform Gaussian distributions with a mean of 0 and a variance of 1. Then, we select individuals in the algorithm according to the mixed penalty function screening criteria and keep the outstanding individuals of ζ parent individuals and ξ progeny individuals to the next generation. The DE-MPFSC algorithm steps are as follows:

STEP 1: Initializing variables: According to the number of the population and the number of individuals, let $t = 0$, then generate the initial population within ψ parents individuals and φ progeny individuals and each of the individuals corresponds with $\{(x_i^t, \delta_i)|i = 1, 2, \cdots, NP\}$.

STEP 2: Calculating the fitness function value of the initial variable: Calculate the fitness function values of ψ parent individuals and record the maximum value and minimum value. At the same time, calculate the fitness function values at each iterative points inside and outside the feasible domain under the condition of the initial conditions and records the maximum value and minimum value.

STEP 3: According to Formulas (5), (17) and (18), generate corresponding ψ mutation individuals.

STEP 4: Calculating the fitness function value of corresponding ψ mutation individuals and calculating the probability of the fitness function value of the iterative points of different regions based on screening criterion of the mixed penalty function.

STEP 5: Conducting selection and crossover operations for corresponding ψ mutation individuals according to Formulas (6) and (7), generating φ progeny individuals and calculating their fitness function values and recording the iterative points that accord with the accuracy of the problem.

STEP 6: Consisting of the ψ mutation individuals and the φ progeny individuals generated by the selection and crossover operations into a new $\psi + \varphi$ individuals. According to the screening criteria, choose ψ individuals of t generation as $t + 1$ generation parents individuals and record the iterative points x_{k+1} that accord with the accuracy of the problem.

STEP 7: Generating a series of iterative points $\{x_{k_1}\}$,$\{x_{k_2}\}$ in two parts inside and outside the feasible domain.

STEP 8: According to the iterative points sequence $\{x_{k_1}\}$,$\{x_{k_2}\}$ generated by step 7, the former are arranged in a monotonic decreasing sequence and the latter is arranged in a monotonic increasing sequence.

STEP 9: Calculating Pr_1, Pr_2 according to the probability formula of the screening criteria.

STEP 10: Substituting the result of step 9 into Formulas (12) or (13), calculating the fitness function values and iterative points and judging whether the accuracy is satisfied.

STEP 11: If the iterative points satisfy the screening criteria and accuracy, then terminate the algorithm flows, otherwise let $t = t + 1$, return step 3.

4. Theoretical Analysis of DE-MPFSC Algorithm

Karmer [45] pointed out through experiments and analysis that, when solving the nonlinear optimization problem with constraints, if the optimal solution is at the feasible domain boundary, using a single objective function as the test function is not conducive to searching the optimal point, which easily falls into local optimization, resulting in an error solution. To effectively avoid the phenomenon, this paper uses a selection variable with adaptive accuracy to modify the equality constraints, expands the feasible domain searching range, establishes the DE-MPFSC algorithm. First, we theoretically analyze the validity of the DE-MPFSC algorithm (judging whether the Markov model conditions of the DE-MPFSC algorithm is true) and the convergence analysis (judging whether the DE-MPFSC algorithm possesses local or global convergence).

4.1. Validity Analysis of DE-MPFSC Algorithm

The validity analysis of the DE-MPFSC algorithm is that whether the Markov model condition of the DE-MPFSC algorithm is true in the constrained feasible domain. It is necessary to judge whether the state transition of the population sequence is a Markov chain. So, we introduce the corresponding definitions and lemmas.

Definition 1 ([46,47]). *Let $\{\hat{X}_n; n \geq 0\}$ be a random variable with a discrete value. The whole discrete values are recorded as S, which is called the state space. If $\forall n \geq 1, i_k \in S(k \leq n+1)$ and $P\{\hat{X}_{n+1} = \frac{i_{n+1}}{\hat{X}_n} = i_n, \cdots, \hat{X}_0 = i_0\} = P\{\hat{X}_{n+1} = \frac{i_{n+1}}{\hat{X}_n} = i_n\}$, then $\{\hat{X}_n; n \geq 0\}$ is called the Markov chain.*

Lemma 1 ([46,47]). *The joint distribution of homogeneous Markov chain $p_{ij}^n = P\{\hat{X}_{n+m} = \frac{j}{\hat{X}_m} = i\}$ is determined by the initial distribution $P\{\hat{X}_0 = i\} = p_i, i \in S$ and the individual transition probability $p_{ij} = P\{\hat{X}_n = \frac{j}{\hat{X}_{n-1}} = i\}, (i, j \in S)$.*

Lemma 2 ([46,47]). *For the homogeneous Markov chain $p_{ij}^n = P\{\hat{X}_{n+m} = \frac{j}{\hat{X}_m} = i\}$ and $\forall n, m \geq 0, i, j \in S$, there are $p_{ij}^{n=m} = \sum_{k \in S} p_{ik}^n p$.*

Definition 2 ([46,47]). *Let $\{\hat{X}_n; n \geq 0\}$ be the finite homogeneous Markov chain, p_{ij}^n is the nth iterative individual transition probability. If $\exists n \geq 1$ makes $p_{ij}^n \geq 0$, which is called a case in which status i can be transferred to state j; otherwise, is called another case in which the status i cannot be transferred to state j.*

Definition 3 ([46,47]). *For Markov chain $\{\hat{X}_n; n \geq 0\}$, the greatest common divisor d_i of the state set $\{i | n \geq 1, p_{ij}^n \geq 0\}$ is called the generalized period of the state set i. If $d_i > 1$, then state set i is periodical. If $d_i = 1$, then state set i is nonperiodic. If d_i is a nonpositive real number, then i cannot be periodical.*

Lemma 3 ([46,47]). *If Markov chain $\{\hat{X}_n; n \geq 0\}$ is irreducible, that is, that all individual states are connected to each other, and $\exists j \leq N$, and $\exists j \leq N$ make $p_{ij} > 0$, then the chain is nonperiodic. Its transfer matrix is the primitive random matrix and there is a stationary distribution on the transfer matrix. Where there is a stationary distribution and a limited distribution $\lim_{n \to \infty} P\{X_n = j\} = o_j$ on the transfer random matrix of the Markov chain being nonperiodic, irreducible and a finite state.*

Now, consider the Markov processing model $\{(x_i^t, \delta_i) | i = 1, 2, \cdots, NP\}$, the progeny x_k^{t+1} and iterative steps δ_i of the DE-MPFSC algorithm are determined by the following formula.

$$\delta_i^{t+1} = \begin{cases} \gamma \delta_i^t, If\ \Psi(x_k^t + \delta_i^t C^t) < \Psi(x_k^t), \\ and\ g_i(x_k^t + \delta_i^t C^t) \leq 0 \\ \gamma^{-1} \delta_i^t, otherwise \end{cases} \tag{19}$$

$$x_k^{t+1} = \begin{cases} x_k^t + \delta_i^t C^t, If\ \Psi(x_k^t + \delta_i^t C^t) < \Psi(x_k^t), \\ and\ g_i(x_k^t + \delta_i^t C^t) \leq 0 \\ x_k^t, otherwise \end{cases}, \tag{20}$$

where $\gamma > 1$ is a compilation parameter, $\{C^t, t \geq 0\}$ is a random vector and independent and identically distributed.

Assume that the new individuals generated by the mutation operation of the DE algorithm are distributed on a circle centered on x_k^t with a radius of δ_i^t. When the mutation individual generated by the DE-MPFSC algorithm is valid, the step increases; otherwise, the step decreases.

Where effective mutation individuals are superior than the parents. Then, each evolution can be performed simultaneously from the inside and outside of the feasible domain.

Markov Processing Model of DE-MPFSC Algorithm

Assume the individuals in the population are expressed as $X_i^t = (x_{i1}^t, x_{i2}^t, \cdots, x_{iD}^t)$, $i = 1, 2, \cdots, NP$ [20], where t is the number of iterations, NP is population size [20], the mutation operator, crossover operator, and selection operator are, respectively, $F, T_c, T_s, T = F \cdot T_c \cdot T_s$. Then, the iterative equation of the DE-MPFSC algorithm is $X(t+1) = T(X(t)) = (F \cdot T_c \cdot T_s)(X(t))$. We obtain the population sequence $\{\hat{X}_t; t \geq 0\}$ after initializing the population.

Theorem 1. *Let the population sequence of the DE-MPFSC algorithm be* $\{\hat{X}_t; n \geq 0\}$*; then, the sequence is a finite, homogeneous, irreducible and nonperiodic Markov chain.*

Proof. Let $X(t)$ be the population of the DE-MPFSC algorithm, the population individuals can be $X_i^t = (x_{i1}^t, x_{i2}^t, \cdots, x_{iD}^t), i = 1, 2, \cdots, NP$. $S = \{\hat{X}_n; n \geq 0\}$ is the state space of the population sequence, which is the finite space. Because of $X(t+1) = T(X(t)) = (F \cdot T_c \cdot T_s)(X(t))$,where there is no connection between operators $F, T_c, T_s, T = F \cdot T_c \cdot T_s$ of the DE-MPFSC algorithm and the iterative times t, and $X(t+1)$ is only related to $X(t)$. In other words, the evolved population individual is only related to the corresponding individual before the evolution, not the number of evolutions. Therefore, $\{\hat{X}_t; t \geq 0\}$ is a finite state Markov chain. Due to

$$\begin{aligned} & P\{T(\hat{X}_t)_k = X_k(t+1)\} \\ & = \sum_{X_k'(t+1) \in S} \sum_{(X_1^k(t), X_2^k(t)) \in S} P\{T_s(X(t))(X_1^k(t), X_2^k(t))\} \cdot P\{T_c(X_1^k(t), X_2^k(t)) X_k'(t+1)\} \cdot P\{F(X_k'(t+1)) \\ & = X_k(t+1)\}, \end{aligned} \tag{21}$$

where $X_k'(t+1)$ is a population of individuals generated by a crossover operation, and $\forall X(t) \in S$. Then, there is the case in which $(X_1^k(t), X_2^k(t))$ and $X_k'(t+1)$ make

$$P\{T_s(X(t)) = (X_1^k(t), X_2^k(t))\} > 0 \tag{22}$$

$$P\{T_c(X_1^k(t), X_2^k(t)) = X_k'(t+1)\} > 0. \tag{23}$$

Further, for $\forall X_k'(t+1)$ and $\forall X_k(t+1)$, there is

$$P\{F(X_k'(t+1)) = X_k(t+1)\} > 0 \tag{24}$$

Then,

$$P\{T(\hat{X}_t)_k = X_k(t+1)\} > 0. \tag{25}$$

Based on formula (22), (23), (24), we know that formula (25) is connected with n, then

$$P\{T(X(t)) = X(t+1)\} = \prod_{i=1}^{NP} P\{T(\hat{X}_t)_k = X_k(t+1)\} > 0 \tag{26}$$

has no connection with n. □

Relying on these proving processes, we know that the population sequence $\{\hat{X}_t; n \geq 0\}$ generated by the DE-MPFSC algorithm is a homogeneous, irreducible and nonperiodic Markov chain. The conclusion is true.

4.2. Convergence Analysis of DE-MPFSC Algorithm

Definition 4 ([48–50]). *$\{\hat{X}_n; n \geq 0\}$ is the set that is strongly convergent to the global optimal solution in probability. If $\lim_{n\to\infty} P\{[\hat{X}_n \subset M]\} = 1$ is true, which is marked as $\hat{X}_n \to M(P.S.)$.*

Definition 5 ([48–50]). *$B \subset S$ is a satisfactory solution set. If $\forall X \in B, Y \notin B$ make $f(X) > f(Y)$.*

Theorem 2. *(Theorem 2 shows that the Markov chain of the DE-MPFSC algorithm is convergent in the distribution and does not depend on the choice of the initial population, which is expressed as $o(Z) = \sum_{Y\in S} P\{X_t = \frac{Z}{X_0} = Y\}$). (Limit Distribution Theorem) There is a limited distribution for the population sequence $\{\hat{X}_n; n \geq 0\}$ of the DE-MPFSC algorithm.*

Proof. From Theorem 1, we know that the sequence $\{\hat{X}_t; n \geq 0\}$ of the DE-MPFSC algorithm is a finite, homogeneous, irreducible and nonperiodic Markov chain. Then, from lemma 3, we obtain

$$\lim_{t\to\infty} P\{T(X_t) = Y\} = \lim_{t\to\infty} \sum_{X_0\in S} P\{T(X_t) = \frac{Y}{X_0}\}P\{X_0\} = o(Y). \tag{27}$$

Then, $o(Y)$ is the distribution on state population S and

$$\sum_{Y\in S} P\{X_t = \frac{Z}{X_0} = Y\} = o(Z), (Z \in S). \tag{28}$$

□

Theorem 3. *(The Strong Convergence Theorem in the Probability of the DE-MPFSC Algorithm) (Theorem 3 shows that the convergence of the Markov sequence $\{X_t\}$, $\{X_t\}$ on the state population S of the DE-MPFSC algorithm is related to parameters $\{\alpha_t^D\}, \{\beta_t^D\}$. α_t^D is the probability that the population individuals in a Markov sequence $\{X_t\}$ will leave the next satisfactory solution sets [50]. However, from Formulas (1), (2) of theorem 3, α_t^D is a negative correlation with β_t^D. So, for the Markov sequence $\{X_t\}$, there is a nonconvergent state. In order to enhance the convergence analysis of the satisfactory solution to the DE-MPFSC algorithm, we need to perform strong convergence analysis around all feasible points that satisfy the limit distribution. The theorem theoretically explains the necessary conditions for the absolute convergence of the DE-MPFSC algorithm.). Let P be the probability distribution on the state population S, $\{X(t)\}$ is the Markov sequence on S, M is the global optimal solution set, D is any satisfactory solution set. If $\{\tilde{\alpha}_t^D\}, \{\tilde{\beta}_t^D\}$ satisfies these following conditions: (1) $\sum_{t=1}^{\infty}(1-\tilde{\beta}_t^D) = \infty$; (2) $\lim_{t\to\infty} \frac{\tilde{\alpha}_t^D}{1-\tilde{\beta}_t^D} = 0$.*

Then, X_t is strongly convergent in probability to satisfactory solution sets D, which is to say $\lim_{t\to\infty} P\{X_t \subset D\} = 1$. Where $\tilde{\alpha}^D = P\{X(t+1) \cap D^c \neq \emptyset\}, \beta_t^D = P\{X(t+1) \cap D^c \neq \emptyset\}\}$.

Proof. Let $P_0 = P\{X_t \cap D = \emptyset\}$, from the *Bayes* formula, we know

$$\begin{aligned} &P_0(t+1) = P\{X(t) \cap D = \emptyset\} \\ &= P\{X(t+1) \cap D \neq \emptyset\} \cdot P\{X(t) \cap D \neq \emptyset\} + P\{X(t+1) \cap D = \emptyset\}\} \cdot P\{X(t) \cap D = \emptyset\} \\ &\leq \tilde{\alpha}_t^D + \tilde{\beta}_t^D P_0(t). \end{aligned} \tag{29}$$

From Formula (4), we know $\forall \varepsilon > 0, \exists N_1$, when $t \geq N_1$, $\frac{\tilde{\alpha}_t^D}{1-\tilde{\beta}_t^D} \leq \frac{\varepsilon}{2}$; thus, $\tilde{\alpha}_t^D \leq \frac{\varepsilon}{2}(1-\tilde{\beta}_t^D)$. From Formula (29), we obtain

$$(P_0(t+1) - \frac{\varepsilon}{2}) - \tilde{\beta}_t^D(P_0(t) - \frac{\varepsilon}{2}) \leq P_0(t+1) - \tilde{\beta}_t^D P_0(t) - \tilde{\alpha}_t^D \leq 0. \tag{30}$$

Further, there is $P_0(t+1) - \frac{\varepsilon}{2} \le \tilde{\beta}_t^D(P_0(t) - \frac{\varepsilon}{2})$, then we know

$$(P_0(t+1) \le \frac{\varepsilon}{2} + \prod_{k=1}^{t} \tilde{\beta}_t^D(P_0(t) - \frac{\varepsilon}{2}) \tag{31}$$

Additionally, $\lim_{t\to\infty} \frac{\tilde{\alpha}_t^D}{1-\tilde{\beta}_t^D} \Leftrightarrow \prod_{k=1}^{t} \tilde{\beta}_t^D = 0$. Then, when $\exists N_2, t \ge N_2$, there is $\prod_{k=1}^{t} \tilde{\beta}_t^D \le \frac{\varepsilon}{2}$. Further, when $t \ge \max\{N_1, N_2\}$, there is $P_0(t+1) \le \varepsilon$. Therefore, we obtain

$$\lim_{t\to\infty} P\{X(t) \cap D = \varnothing\} = \lim_{t\to\infty} P_0(t) = 0. \tag{32}$$

Further, there is $\lim_{t\to\infty} P\{X_t \cap D \ne \varnothing\} = 1$. This shows that DE-MPFSC is strongly convergent. □

5. Entanglement and Numerical Test of DE-MPFSC Algorithm Stability

We focus on the effect of the DE-MPFSC algorithm in imbalanced data integration, rather than focusing on the advantages of this algorithm over other algorithms. To test the advantages of the DE-MPFSC in processing the imbalanced data and to ensure the high property of the DE-MPFSC algorithm in the imbalanced data integration process compared to the traditional DE algorithm, we will conduct test experiments on the stability and effectiveness of the DE-MPFSC algorithm. Therefore, based on the topology structure between the convergence rate and convergence precision of the individuals and the quantum relationship [20], including the entanglement phenomenon of individual convergence speed and convergence precision in the evolution process, we introduce the concept of entanglement degree $\frac{1}{\zeta}$ and the entanglement degree error ζ, analyze the stability of DE-MPFSC algorithm and DE algorithm from system theory.

5.1. Entanglement Degree $\frac{1}{\zeta}$ and Entanglement Degree Error ζ

Lemma 4 ([20]). *Let f_ε be a continuous differentiable function defined in a complete normed linear space and $f_\varepsilon(v_1, \cdots, v_n) \in L_2(\mathbb{R}_n), M \in Sp(2n + P_t.\mathbb{R})$, M^n is a complete normed linear space. $\{S_i^n | i = 1, 2, \cdots, n\}$ is a generalized n dimension complete normed linear subspace and $M^n = S_1^n \cup S_2^n \cup \cdots \cup S_n^n$ forms an open cover of M^n, then $\forall \lambda_i^t \in \{\lambda_i^t | i = 1, 2, \cdots, n\}, \exists S_i^n, \lim_{t\to\infty} \lambda_i^t \sim \lambda_i$ or $\lim_{t\to\infty} \lambda_i^t = (\lambda_\varepsilon)_i, \lambda_i \in S_i^n$. If $\det(B) \ne 0$, then the following formula,*

$$\begin{aligned}\Delta_v^2 \cdot \Delta_{x_\beta^\varepsilon}^2 &\ge (\frac{\sqrt{\lim_{t\to\infty} \lambda_1^t} + \cdots + \sqrt{\lim_{t\to\infty} \lambda_n^t}}{2})^2 \\ &= (\frac{\sqrt{(\lambda_\varepsilon)_1} + \cdots + \sqrt{(\lambda_\varepsilon)_n}}{2})^2.\end{aligned} \tag{33}$$

The lemma is detailed in Reference [20], which theoretically shows that the convergence accuracy and convergence speed of the evolutionary individual in the iterative process represent an uncertain quantum relationship. The lemma is important for analyzing the distribution of the optimal point of the segmentation regions. The feasible points distributed in different regions force the internal and external penalty functions to search in different segmentation regions by the jumping function of probability. Let the region have two local optimal feasible points x_1 and x_2, which are distributed to satisfy $\Delta_v^2 \cdot \Delta_{x_\beta^\varepsilon}^2 \ge (\frac{\sqrt{x_1}+\sqrt{x_2}}{2})^2$ and $\sqrt{\lim_{t\to\infty} x_1^t} = \sqrt{x_1}$, $\sqrt{\lim_{t\to\infty} x_2^t} = \sqrt{x_2}$, where t is the number of individual iterations.

The individual entanglement degree we define is $\frac{1}{\zeta} = \frac{v_{X(t)}}{\iota_{X(t)}}$, which is the numerical ratio between the convergence speed and convergence precision of individuals during population evolution. The higher the entanglement degree, the better the convergence stability of the algorithm or the faster convergence of the algorithm and the higher convergence precision. The entanglement degree error ζ is the correlation of the entanglement effect, which is the efficiency of the entanglement decided by $\frac{1}{\xi}\ln Cov(NP) = \zeta \ln E(NP - X_i)$. The smaller the entanglement degree error, the smaller the fluctuation of the algorithm in the iterative process. Where, $v_{X(t)}$ is the individual convergence speed, $\iota(X(t))$ is the individual convergence precision, $\frac{\ln Cov(NP)}{\ln E(NP - X_i)}$ is the entanglement coefficient, ξ is the discrete degree between entangled individuals. About the entangled relationship between convergence precision and convergence speed, we can refer to the topological geometric relationship described by the author of Reference [20]. The effect of entanglement degree clearly shows the efficiency of algorithm convergence in the geometry.

5.2. Entangled Image and Analysis of DE-MPFSC Algorithm

Since data integration is closely related to data size and data dimension, we first reduces dimensionality of high-dimensional data before performing data experiments. Let the number of population individuals is 1000. The experiment is divided into 5 times on average, and the number of individuals per test is gradually increased from 200. Test runs an average of 5 times. The initial variation factor and initial crossover probability of the algorithm are determined by the literature [19,20]. The initial mutation factor and the initial crossover probability are respectively set to $F = 0.2, CR = 0.5$. In the following experiments, $F = 0.2, CR = 0.5$ are dynamically adjusted according to the DE-MPFSC algorithm. The Figures 1–10 below shows the numerical image of the DE-MPFSC algorithm.

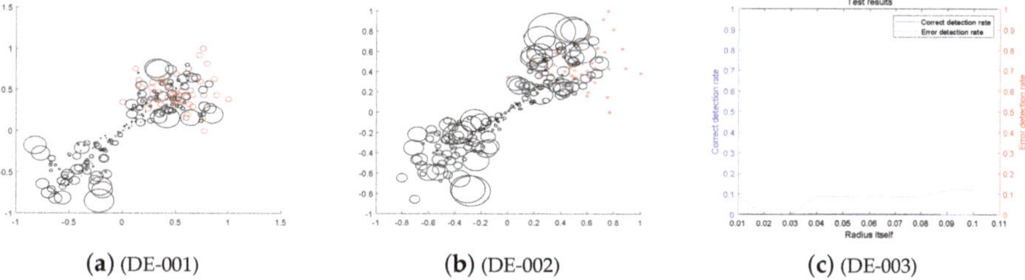

(a) (DE-001) **(b)** (DE-002) **(c)** (DE-003)

Figure 1. The entangled image and error of the differential evolution (DE) algorithm running 200 times.

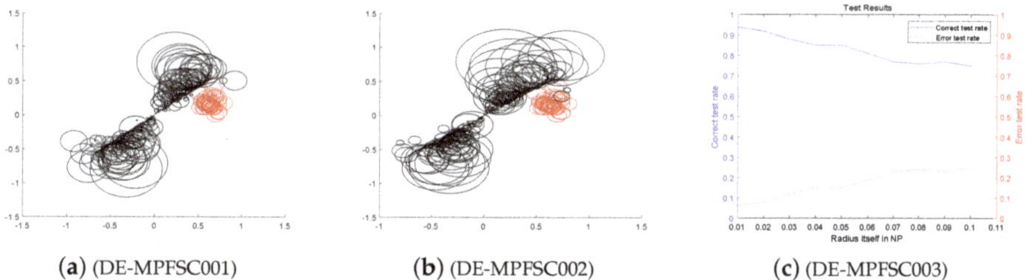

(a) (DE-MPFSC001) **(b)** (DE-MPFSC002) **(c)** (DE-MPFSC003)

Figure 2. The entangled image and error of the Differential Evolution Integration Algorithm Based on Mixed Penalty Function Screening Criteria (DE-MPFSC) algorithm running 200 times.

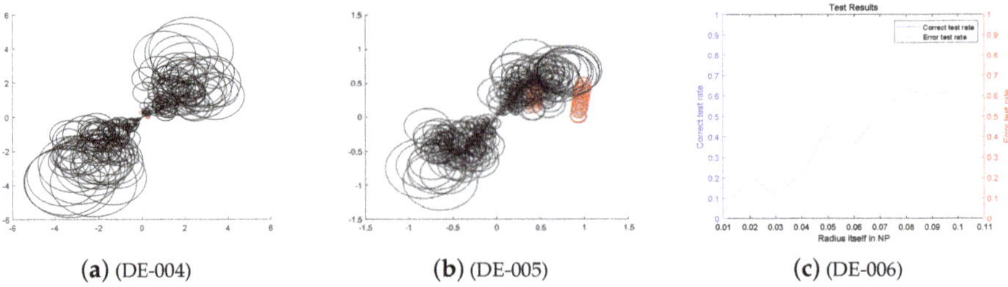

(**a**) (DE-004) (**b**) (DE-005) (**c**) (DE-006)

Figure 3. The entangled image and error of the DE algorithm running 400 times.

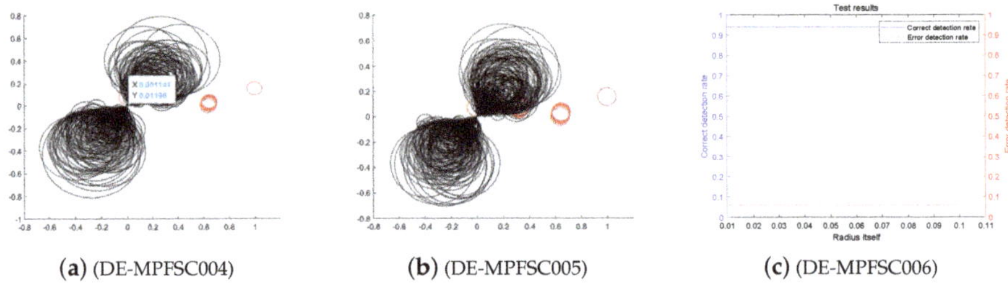

(**a**) (DE-MPFSC004) (**b**) (DE-MPFSC005) (**c**) (DE-MPFSC006)

Figure 4. The entangled image and error of the DE-MPFSC algorithm running 400 times.

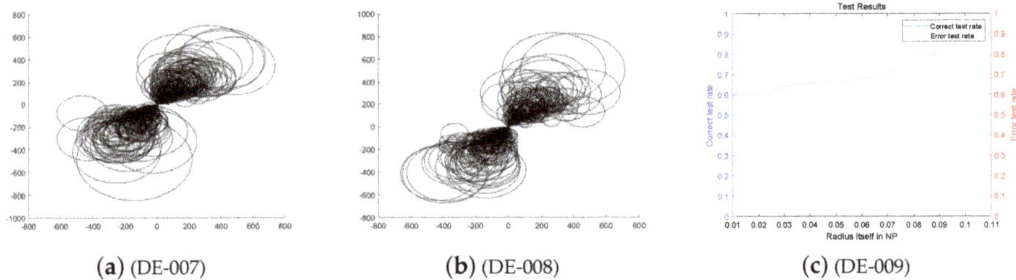

(**a**) (DE-007) (**b**) (DE-008) (**c**) (DE-009)

Figure 5. The entangled image and error of the DE algorithm running 600 times.

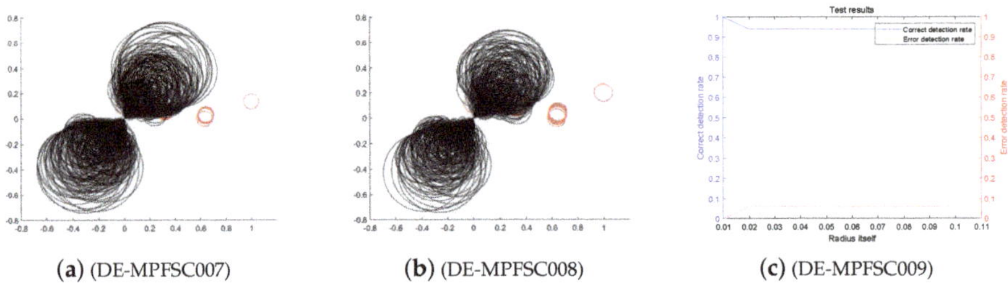

(**a**) (DE-MPFSC007) (**b**) (DE-MPFSC008) (**c**) (DE-MPFSC009)

Figure 6. The entangled image and error of the DE-MPFSC algorithm running 600 times.

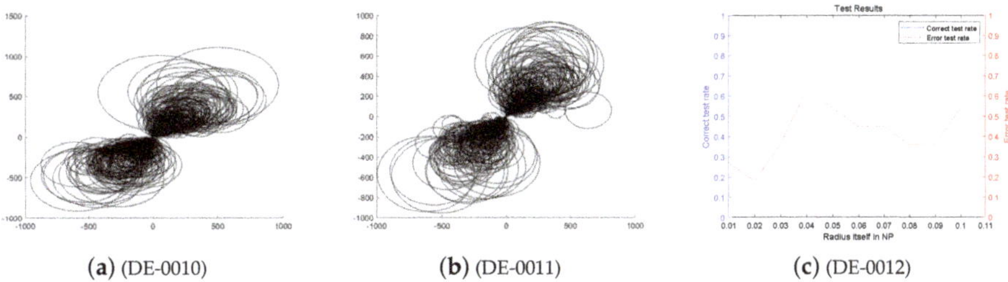

(a) (DE-0010) (b) (DE-0011) (c) (DE-0012)

Figure 7. The entangled image and error of the DE algorithm running 800 times.

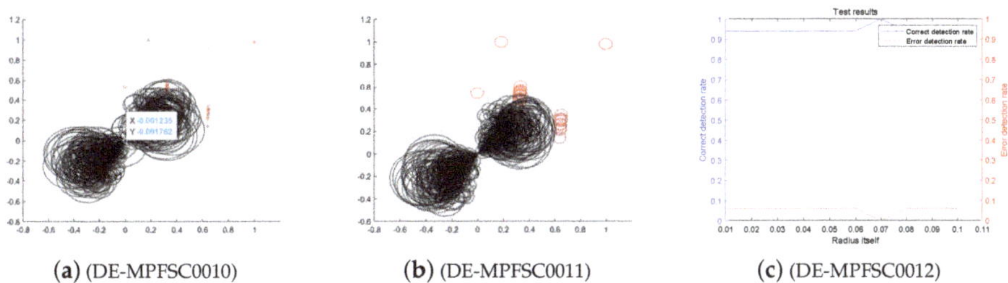

(a) (DE-MPFSC0010) (b) (DE-MPFSC0011) (c) (DE-MPFSC0012)

Figure 8. The entangled image and error of the DE-MPFSC algorithm running 800 times.

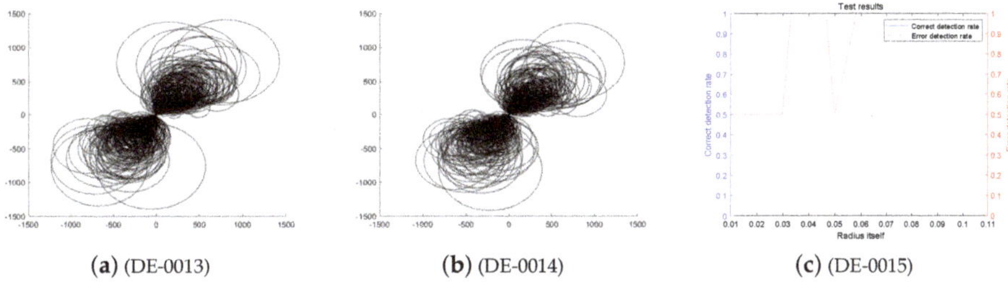

(a) (DE-0013) (b) (DE-0014) (c) (DE-0015)

Figure 9. The entangled image and error of the DE algorithm running 1000 times.

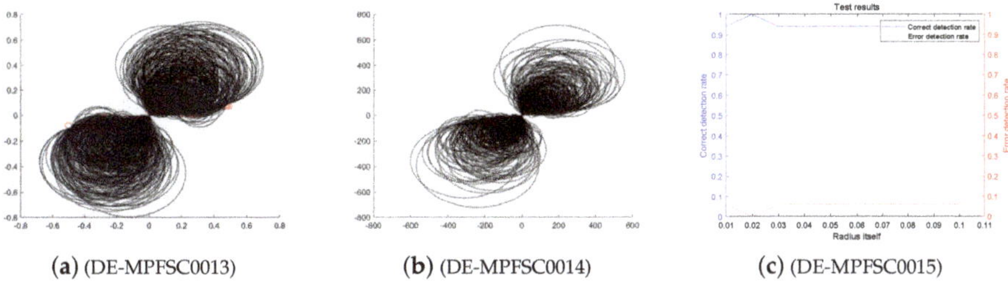

(a) (DE-MPFSC0013) (b) (DE-MPFSC0014) (c) (DE-MPFSC0015)

Figure 10. The entangled image and error of the DE-MPFSC algorithm running 1000 times.

Figures 1–10 above present entanglement degree analysis about the stability and validity between convergence precision and convergence speed of population individuals.

First, we analyze the stability and effectiveness of the DE-MPFSC algorithm from the number of population individuals. When the number of individuals are 200, 400, 600, 800, 1000, the DE algorithm entanglement degree about convergence precision and convergence speed are divided into two branches. However, the convergence speed and convergence precision are higher discrete and they do not converge to the entanglement center point along the convergence curve. The error test rates are 0.1, 0.65, 0.65, 0.55, 1, the correct test rate is 0 and the detection efficiency is 0, which shows that when the population individuals are 200, 400, 600, 800, 1000, the stability of the DE algorithm is not high and it is easy to generate fluctuations. Compared with the DE algorithm, the entanglement degree of the DE-MPFSC algorithm about convergence precision and convergence speed are divided into two branches. The convergence speed and convergence precision are lower discrete and they converge to the entanglement center point along the convergence curve. The error test rates are 0.25, 0.05, 0.05, 0.05, 0.05, the correct test rate are 0.75, 0.95, 0.95, 0.95, 0.95 and the detection efficiency are 3, 19, 19, 19, 19, which shows that when the population individuals are 200, 400, 600, 800, 1000, the stability of the DE-MPFSC algorithm is higher than DE algorithm.

In particular, when the number of individuals is 1000, the convergence speed and convergence precision are slightly lower than that of 800 individuals but the distribution is still scattered. They generate a more dramatic marginal differentiation phenomenon on branches and they do not converge to the entanglement center point along the convergence curve. The error test rate is 1, the correct test rate is 0 and the detection efficiency is 0. The situation must not be used as a basis for effective convergence, which shows that when the population individuals is 1000, the stability of the DE algorithm cannot be accepted and it is easy to generate fluctuations. Compared with the DE algorithm, the entanglement degree of the DE-MPFSC algorithm is divided into two branches. The convergence speed and convergence precision are lower discrete, which can further weak marginal differentiation phenomenon on branches and they converge to the entanglement center point along the convergence curve.

In summary, the stability of the DE-MPFSC algorithm is higher than that of the DE algorithm.

5.3. Numerical Test and Analysis of DE-MPFSC Algorithm

Since the data integration is closely related to the data size and data dimension, we set the population individuals to 1000, which is divided into 5 experiments and the average operation is 5 times for each experiment. According to the recommendations of Professor Storn R and Professor Price in Reference [19], the initial mutation factor and initial crossover probability are set to $F = 0.2, CR = 0.5$. Tables 1–5 below shows the numerical effects of (Differential Evolution Algorithm Based on Mixed Penalty Function Screening Criteria)DE-MPFSC algorithm and (Differential Evolution)DE algorithm.

Table 1. Numerical Analysis of Entanglement Degree $\frac{1}{\xi}$ and Entanglement Degree Error ζ of the Differential Evolution (DE) algorithm and the Mixed Penalty Function Screening Criteria (DE-MPFSC) algorithm 200 times.

NP	DE					$DE-MPFSC$				
	F	CR	ξ	ζ	$\varepsilon\%$	F^*	CR^*	ξ	ζ	$\varepsilon\%$
200	0.1	0.9	0.8	0.5	±0.625	0.1	0.9	0.12	0.05	±0.416
200	/	/	0.73	0.66	±0.904	0.5	0.6	0.20	0.06	±0.3
200	/	/	0.81	0.49	±0.604	0.3	0.2	0.19	0.11	±0.578
200	/	/	0.75	0.54	±0.72	0.2	0.3	0.15	0.03	±0.2
200	/	/	0.8	0.56	±0.7	0.2	0.3	0.15	0.02	±0.13
	DE					$DE-MPFSC$				
Average	0.1	0.9	0.778	0.55	±0.7106	0.26	0.46	0.162	0.054	±0.3248
	DE					$DE-MPFSC$				
Variance	0	0	0.00127	0.0046	±0.01407	0.023	0.083	0.00107	0.00123	±0.03164

Table 2. Numerical Analysis of Entanglement Degree $\frac{1}{\xi}$ and Entanglement Degree Error ζ of the DE algorithm and the DE-MPFSC algorithm 400 times.

NP	DE					DE − MPFSC				
	F	CR	ξ	ζ	$\varepsilon\%$	F^*	CR^*	ξ	ζ	$\varepsilon\%$
400	0.1	0.9	0.77	0.43	±0.558	0.1	0.9	0.15	0.015	±0.1
400	/	/	0.85	0.29	±0.341	0.4	0.5	0.25	0.02	±0.08
400	/	/	0.81	0.55	±0.679	0.2	0.2	0.20	0.05	±0.25
400	/	/	0.80	0.63	±0.787	0.2	0.3	0.30	0.042	±0.14
400	/	/	0.79	0.51	±0.645	0.2	0.3	0.22	0.02	±0.09
	DE					DE − MPFSC				
Average	0.1	0.9	0.804	0.482	±0.602	0.22	0.44	0.224	0.0294	±0.132
	DE					DE − MPFSC				
Variance	0	0	0.00088	0.01672	±0.02802	0.012	0.078	0.00313	0.00024	±0.00487

Table 3. Numerical Analysis of Entanglement Degree $\frac{1}{\xi}$ and Entanglement Degree Error ζ of the DE algorithm and the DE-MPFSC algorithm 600 times.

NP	DE					DE − MPFSC				
	F	CR	ξ	ζ	$\varepsilon\%$	F^*	CR^*	ξ	ζ	$\varepsilon\%$
600	0.1	0.9	0.69	0.38	±0.550	0.1	0.9	0.15	0.015	±0.1
600	/	/	0.70	0.43	±0.614	0.6	0.8	0.33	0.10	±0.303
600	/	/	0.81	0.59	±0.728	0.3	0.4	0.19	0.09	±0.474
600	/	/	0.75	0.41	±0.547	0.2	0.3	0.22	0.011	±0.05
600	/	/	0.79	0.49	±0.620	0.2	0.3	0.27	0.07	±0.259
	DE					DE − MPFSC				
Average	0.1	0.9	0.748	0.46	±0.6118	0.28	0.54	0.232	0.0572	±0.2372
	DE					DE − MPFSC				
Variance	0	0	0.00282	0.0069	±0.0054	0.037	0.083	0.00492	0.00175	±0.02869

Table 4. Numerical Analysis of Entanglement Degree $\frac{1}{\xi}$ and Entanglement Degree Error ζ of the DE algorithm and the DE-MPFSC algorithm 800 times.

NP	DE					DE − MPFSC				
	F	CR	ξ	ζ	$\varepsilon\%$	F^*	CR^*	ξ	ζ	$\varepsilon\%$
800	0.1	0.9	0.63	0.44	±0.698	0.1	0.9	0.15	0.015	±0.1
800	/	/	0.64	0.31	±0.484	0.5	0.4	0.27	0.03	±0.111
800	/	/	0.75	0.50	±0.667	0.3	0.3	0.33	0.06	±0.182
800	/	/	0.70	0.50	±0.714	0.2	0.3	0.30	0.017	±0.057
800	/	/	0.71	0.36	±0.507	0.2	0.3	0.27	0.02	±0.074
	DE					DE − MPFSC				
Average	0.1	0.9	0.686	0.422	±0.614	0.26	0.44	0.264	0.0284	±0.1048
	DE					DE − MPFSC				
Variance	0	0	0.00253	0.00722	±0.01205	0.023	0.068	0.00468	0.00035	±0.00231

The above five data sheets in Tables 1–5 are data analysis of the entanglement degree $\frac{1}{\xi}$, entanglement degree errorζ and error rate ε about convergence precision and convergence speed. We experimented with 200 individuals as the cardinal number, repeating five times per experiment. We record the entanglement degree $\frac{1}{\xi}$, entanglement degree error ζ and error rate ε about DE algorithm and DE-MPFSC algorithm in real time. At the same time, we record the dynamic mutation factor F

and the dynamic crossover probability CR of the DE-MPFSC algorithm and analyze the data tables one by one.

Table 5. Numerical Analysis of Entanglement Degree $\frac{1}{\xi}$ and Entanglement Degree Error ζ of the DE algorithm and the DE-MPFSC algorithm 1000 times.

NP	DE					DE − MPFSC				
	F	CR	ξ	ζ	$\varepsilon\%$	F^*	CR^*	ξ	ζ	$\varepsilon\%$
1000	0.1	0.9	0.69	0.40	±0.579	0.1	0.9	0.15	0.015	±0.1
1000	/	/	0.82	0.66	±0.805	0.3	0.4	0.30	0.04	±0.133
1000	/	/	0.67	0.49	±0.731	0.3	0.3	0.29	0.02	±0.068
1000	/	/	0.77	0.58	±0.753	0.2	0.3	0.27	0.019	±0.070
1000	/	/	0.79	0.44	±0.557	0.2	0.3	0.32	0.018	±0.056
	DE					DE − MPFSC				
Average	0.1	0.9	0.748	0.514	±0.685	0.22	0.44	0.266	0.0224	±0.0854
	DE					DE − MPFSC				
Variance	0	0	0.00422	0.01118	±0.01219	0.007	0.068	0.00453	0.0001	±0.00097

In summary, when the number of individuals is 200, 400, 600, 800 or 1000, the range of entanglement error rate of the DE algorithm and the DE-MPFSC algorithm are respectively $\varepsilon \in (\pm 0.714\%, \pm 0.904\%)$ and $\varepsilon \in (\pm 0.133\%, \pm 0.578\%)$, which shows that the entanglement degree of the DE-MPFSC algorithm is significantly higher than that of the DE algorithm, and the stability of the DE-MPFSC algorithm is better. The range of entanglement degree of the DE algorithm and the DE-MPFSC algorithm are respectively $\frac{1}{\xi} \in (0.65, 0.9)$ and $\frac{1}{\xi} \in (0.15, 0.35)$, which shows that the entanglement degree of the DE-MPFSC algorithm is significantly higher than that of the DE algorithm, and the fluctuation of the DE-MPFSC algorithm is lower. The entanglement error ranges of the DE algorithm and the DE-MPFSC algorithm are respectively $\zeta \in (0.25, 0.7)$ and $\zeta \in (0.015, 0.1)$, which shows that the entanglement degree of the DE-MPFSC algorithm is significantly higher than that of the DE algorithm, and the DE-MPFSC convergence of the algorithm is stronger. In addition, there IS an additional discovery during the experiment that the optimal dynamic mutation factor and the optimal dynamic crossover probability of the DE-MPFSC algorithm are $F = 0.2$ and $CR = 0.3$ respectively.

In summary, the convergence, stability and effectiveness of the DE-MPFSC algorithm are higher than the DE algorithm, which has obvious advantages for the integration of individuals.

5.4. Test and Analysis of DE-MPFSC Algorithm about Several Classic DE Algorithms

5.4.1. Entanglement Degree Test of DE-MPFSC Algorithm about Several Classic DE Algorithms

We choose four classic DE improved algorithms JDE, OBDE, DEGL and SADE for related tests. Through entanglement degree test and data analysis, we can verify the related performance of the DE-MPFSC algorithm and its superiority in data integration. From the previous experiments, we can know that the DE-MPFSC is better than the DE in stability, effectiveness and convergence. At the same time, the optimal dynamic mutation factor $F = 0.2$ and the optimal crossover probability $CR = 0.3$ of the DE-MPFSC algorithm were obtained through data experiments. In the following test experiments, we use them as the initial mutation factor and the initial cross probability to better compare with other classic DE algorithms. Because the function of data integration is closely related to the data size and dimension, the algorithm in this paper sets the number of individuals to 1000 and are divided into 5 experiments. The number of individuals gradually increases in each test and the test runs an average of 4 times. We record relevant experimental indicators: dynamic mutation probability F, dynamic cross probability CR, entanglement degree $\frac{1}{\xi}$, entanglement error ζ, error rate $\varepsilon\%$, average individual loss efficiency. Figures 11–15 below shows a numerical results of the experiment.

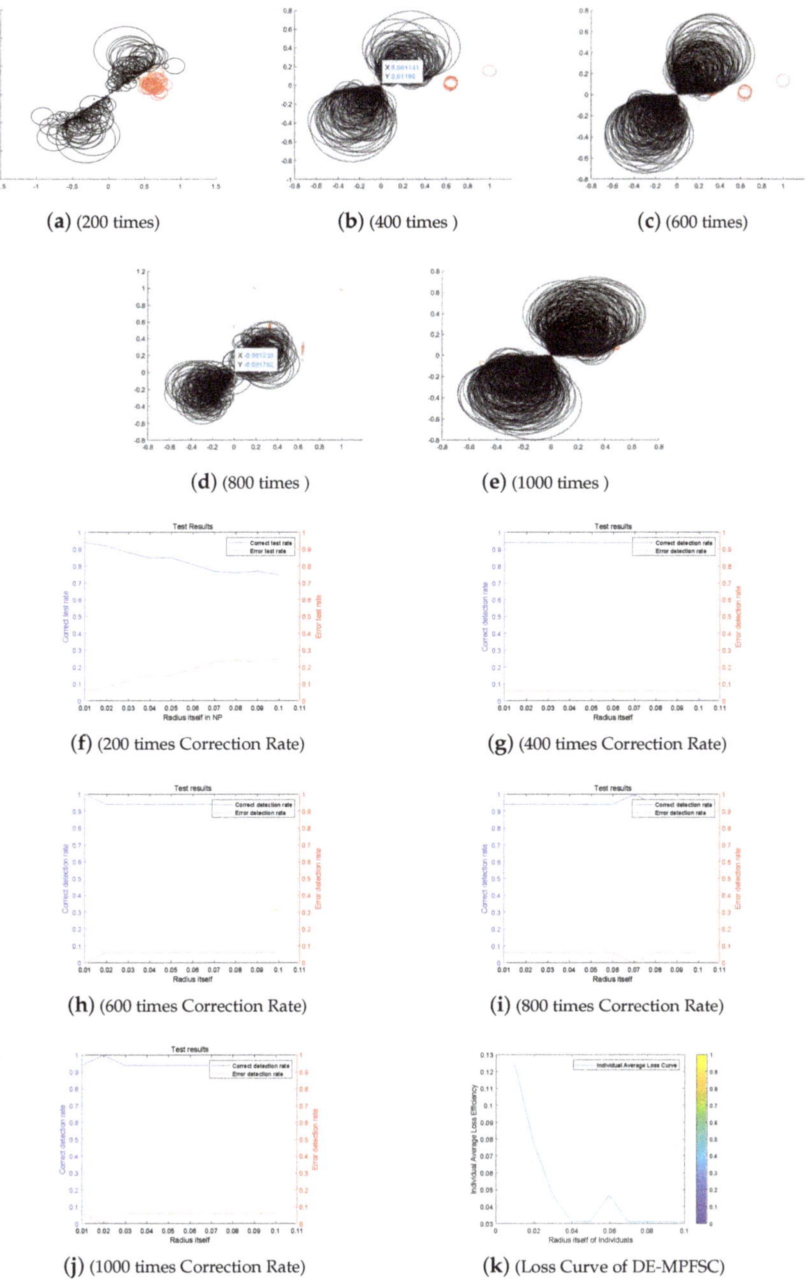

(a) (200 times) **(b)** (400 times) **(c)** (600 times)

(d) (800 times) **(e)** (1000 times)

(f) (200 times Correction Rate) **(g)** (400 times Correction Rate)

(h) (600 times Correction Rate) **(i)** (800 times Correction Rate)

(j) (1000 times Correction Rate) **(k)** (Loss Curve of DE-MPFSC)

Figure 11. The Entanglement Degree, Correction Rate and Loss Curve of DE-MPFSC algorithm.

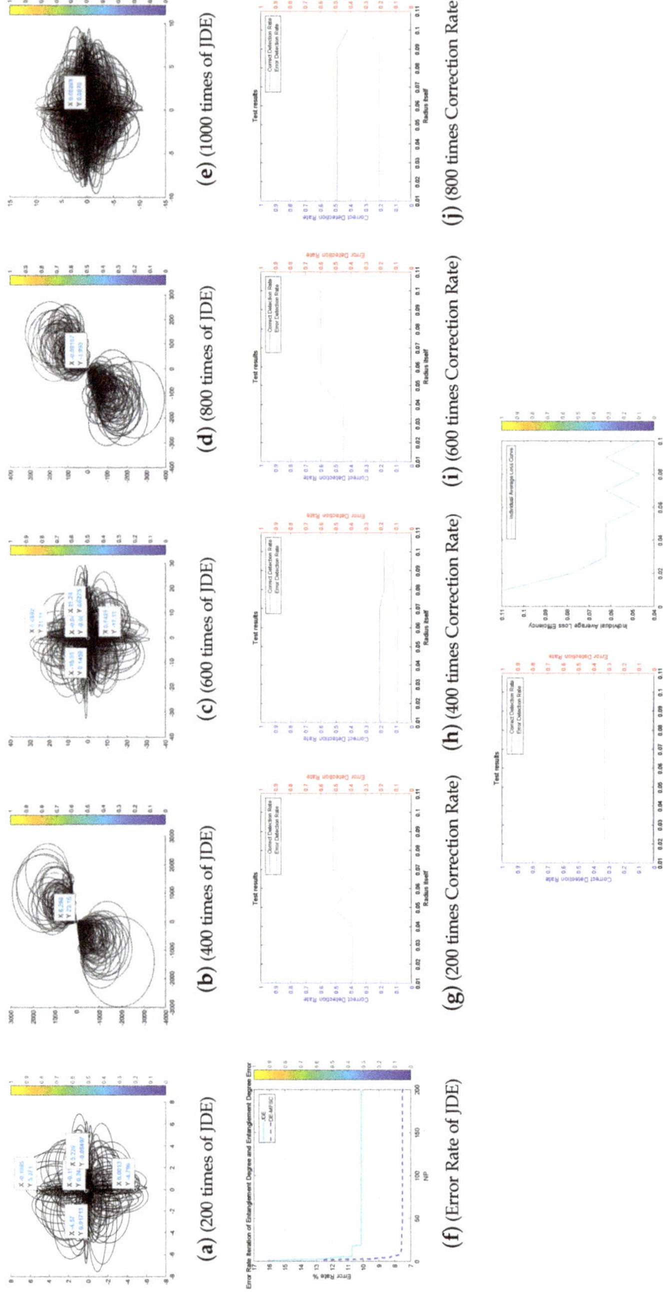

(a) (200 times of JDE) **(b)** (400 times of JDE) **(c)** (600 times of JDE) **(d)** (800 times of JDE) **(e)** (1000 times of JDE)

(f) (Error Rate of JDE) **(g)** (200 times Correction Rate) **(h)** (400 times Correction Rate) **(i)** (600 times Correction Rate) **(j)** (800 times Correction Rate)

(k) (1000 times Correction Rate) **(l)** (Loss Curve of JDE)

Figure 12. The Entanglement Degree, Correction Rate and Loss Curve of the (Self-Adapting Parameter Setting in Differential Evolution) JDE algorithm.

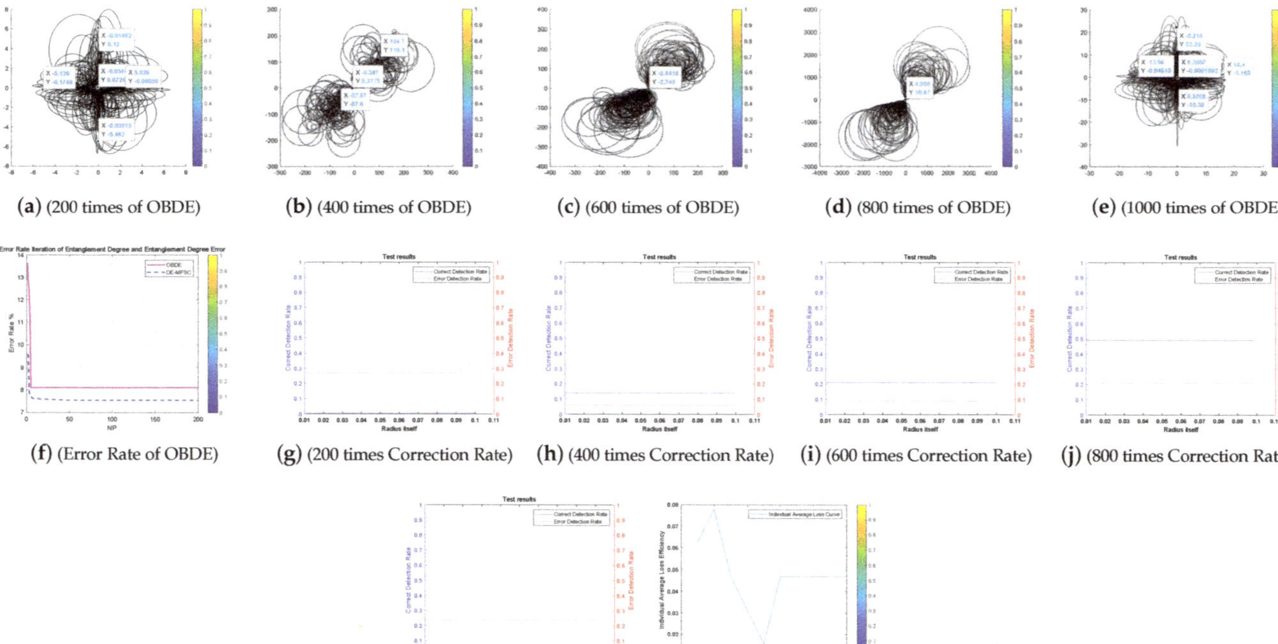

Figure 13. The Entanglement Degree, Correction Rate and Loss Curve of the Opposition Based Differential Evolution (OBDE) algorithm.

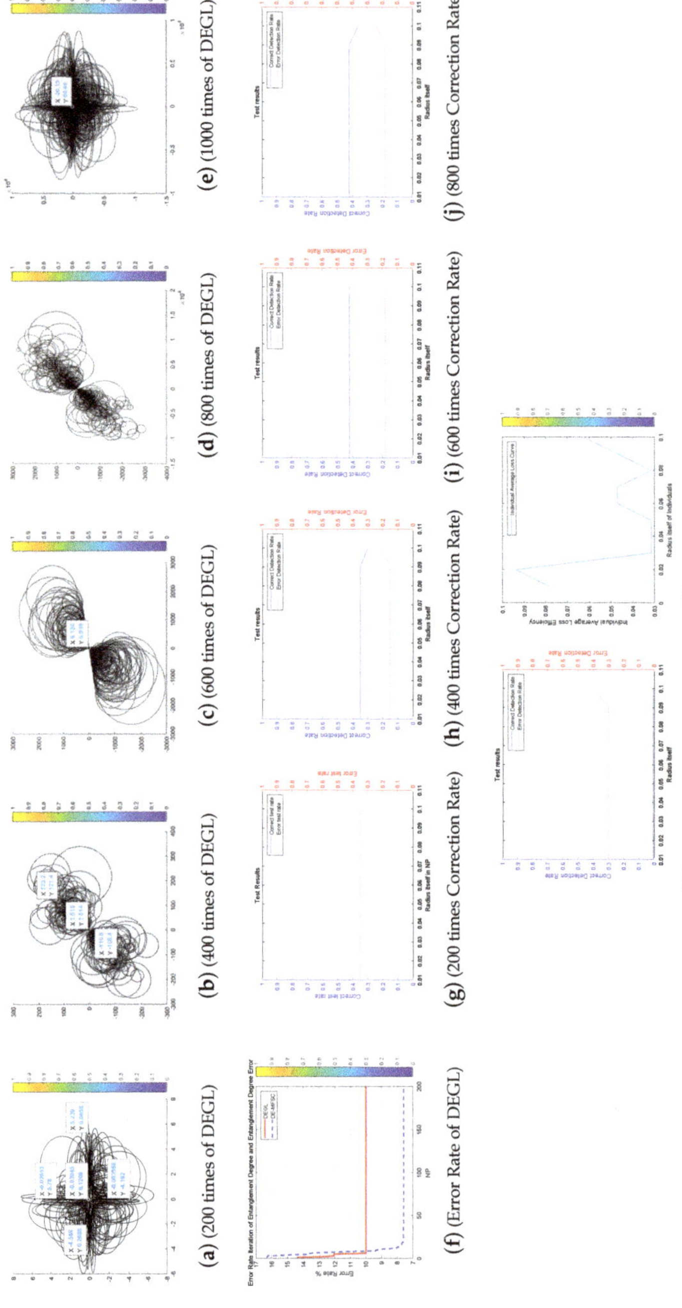

(a) (200 times of DEGL) **(b)** (400 times of DEGL) **(c)** (600 times of DEGL) **(d)** (800 times of DEGL) **(e)** (1000 times of DEGL)

(f) (Error Rate of DEGL) **(g)** (200 times Correction Rate) **(h)** (400 times Correction Rate) **(i)** (600 times Correction Rate) **(j)** (800 times Correction Rate)

(k) (1000 times Correction Rate) **(l)** (Loss Curve of DEGL)

Figure 14. The Entanglement Degree, Correction Rate and Loss Curve of the (Differential Evolution with Global and Local neighborhoods) DEGL algorithm.

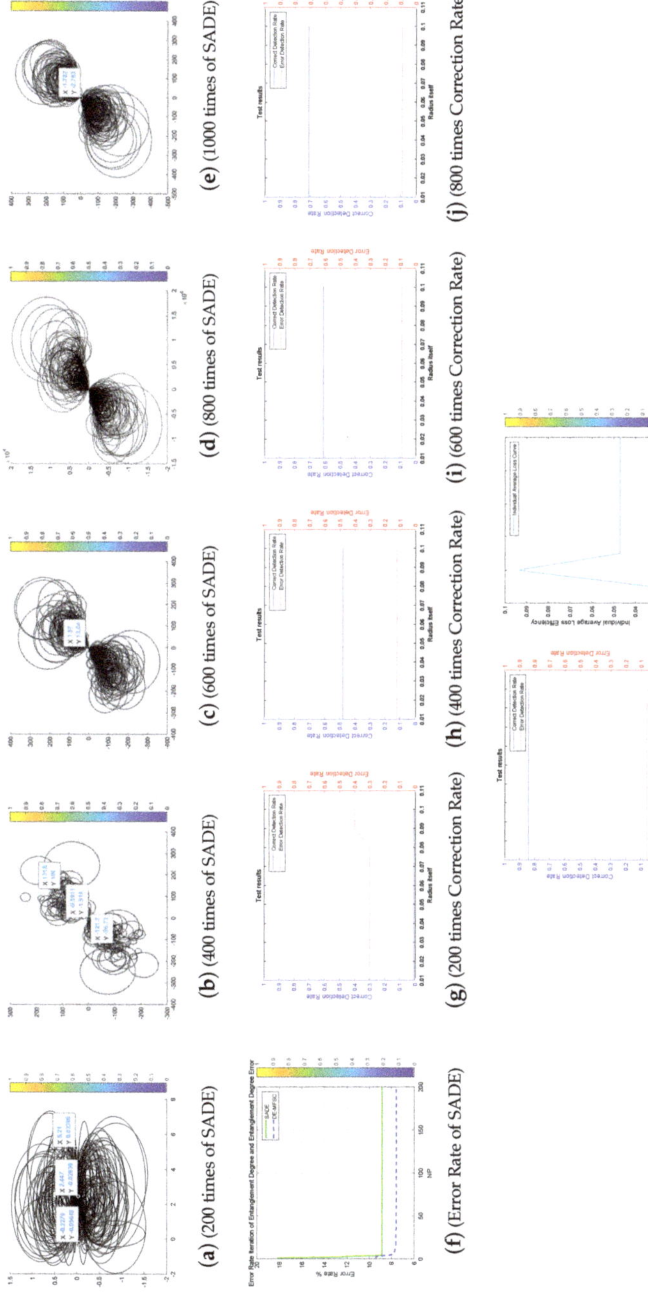

(**a**) (200 times of SADE) (**b**) (400 times of SADE) (**c**) (600 times of SADE) (**d**) (800 times of SADE) (**e**) (1000 times of SADE)

(**f**) (Error Rate of SADE) (**g**) (200 times Correction Rate) (**h**) (400 times Correction Rate) (**i**) (600 times Correction Rate) (**j**) (800 times Correction Rate)

(**k**) (1000 times Correction Rate) (**l**) (Loss Curve of SADE)

Figure 15. The Entanglement Degree, Correction Rate and Loss Curve of the (Self-Adaptive Differential Evolution) SADE algorithm.

We analyze the above Figures 11–15. First we analyze from the entanglement degree. In the process of the DE-MPFSC algorithm, the convergence speed and convergence precision of population individuals gradually stabilized as the population increased. The density and integration of the two entangled images gradually increase and, at the same time, the edge differentiation phenomenon is further weakened on each branch, which make the entangled branch converge to the center of the entanglement along the convergence curve. Compared with the DE-MPFSC algorithm, the entanglement phenomenon of the convergence speed and convergence precision of the JDE algorithm is less obvious. When the number of individuals is 200, 600 and 1000, the entangled image is highly dispersed, which causes a severe edge differentiation on the entangled branches. Although, when the number of individuals is 400 and 600, the state of entanglement is better than that of 200, 600 and 800, the continuity of individual evolution is not reached as a whole, it does not converge to the center of entanglement along the convergence curve. JDE's entangled image shows irregular fluctuations. The OBDE algorithm is slightly better than the JDE algorithm in fluctuation, when the number of individuals is 200 and 1000, the entangled branches are severely differentiated and the dispersed degree is higher than that of JDE, which shows that the OBDE algorithm is not suitable. When the number of individuals is 400, 600 and 800, the dispersed degree of the entangled branches of the OBDE algorithm is lower but the edge dispersion of the entangled branches is more obvious than the DE-MPFSC algorithm. OBDE and JDE algorithms also do not show evolutionary continuity. The performance of the DEGL algorithm on evolutionary fluctuation is weaker than JDE and OBDE. When the number of individuals is 200 and 1000, the entangled branches are highly dispersed and do not converge to the entangled center. The DEGL is similar to the JDE algorithm and the OBDE algorithm. The individual convergence speed and convergence accuracy completely deviate from the entanglement center. The number of entangled branches integrated with 400 and 800 individuals is much higher than that with 200 and 1000 individuals. Although DEGL converges to the center of entanglement along the convergence curve, the edge differentiation of entangled branches is still not resolved. The continuity, volatility and convergence of the SADE algorithm are the closest to the DE-MPFSC algorithm but, unfortunately, a discrete state occurs when the number of individuals is 200, which makes the entanglement center distributed on the fitted line rather than an entanglement center. When the number of individuals is 400, 600, 800 and 1000, the edge dispersion of the entangled branches is weakened but it is not used for large-scale evolution.

Then we analyze from the correct test rate of entangled branches. The entanglement degree of the DE-MPFSC algorithm is divided into two branches about convergence precision and convergence speed. From the image, the convergence speed and convergence precision are lower discrete and they converge to the entanglement center point along the convergence curve. The detection efficiency are 3, 19, 19, 19, 19, which shows that when the population individuals are 200, 400, 600, 800, 1000, the stability of the DE-MPFSC algorithm is higher. Due to the fluctuation of the JDE algorithm, its correct test rate is 0, and its error test rates are 0.4, 0.4, 0.95, 0.6 and 0.95, respectively. JDE's higher error test rate is easy to eliminate optimal individuals. The OBDE algorithm is weaker than the JDE algorithm in fluctuation. Its correct test rate is 0 but its error test rates are 1, 1, 0.4, 0.4, 1, which are higher than that of the JDE algorithm. When the number is 1000, the error test rate suddenly rebounds. The DEGL algorithm has a non-zero in correct test rate, which at least shows that there is a certain ability to adapt to individual mutations. But when the number of individuals is 200 and 1000, the error test rate of entangled branches is 1 and 0.95, which is still too high. When the number of individuals is 400 and 800, the correct test rate of entangled branches declines. The instability of this entanglement behavior is still not conducive to the evolution of mutation individuals. The correct test rate of the SADE algorithm are 0, 0, 0.35, 0.38, 0.35, and the error test rates are 0.4, 0.15, 0.15, 0.2, 0.15, which are much better than the JDE, OBDE and DEGL algorithms. Stability and continuity should be more conducive to the evolutionary behavior of large-scale individuals. However, compared to the DE-MPFSC algorithm, because of the correct test rate of the DEGL algorithm being too low, the detection of mutation individuals is not adaptive.

Finally, from the individual loss curve, we can know that the fluctuation of the JDE is higher than that of DE-MPFSC, OBDE, DEGL and SADE, indicating that during the individual evolution process, the population individual loss degree is higher. Especially for large-scale evolutionary populations, the risk of mutation individuals being lost during evolution will greatly increase. The individual loss degree of OBDE and DEGL rebounded in the later stages of evolution. The loss degree of mutation individuals showed an increasing trend relative to DE-MPFSC, JDE and SADE. The individual loss of SADE and DE-MPFSC is similar but the individual loss of SADE is higher than the latter. Because the individual losses of JDE, OBDE, DEGL and SADE are higher than DE-MPFSC, the advantages of the latter are obvious in stability. By analyzing the error rate curve, we can know that when DE-MPFSC is processing large-scale individuals (based on 200 individuals), the error rate of individual loss is lower than other algorithms.

5.4.2. Numerical Test of DE-MPFSC Algorithm about Several Classic DE Algorithms

We analyze the test results of the above data Table 6. We already know that when the number of individuals is 200, 400, 600, 800, 1000, the entanglement error range of DE-MPFSC is $\varepsilon \in (\pm 0.133\%, \pm 0.578\%)$, The entanglement range of DE-MPFSC is $\frac{1}{\zeta} \in (0.15, 0.35)$, and the entanglement error range of DE-MPFSC is $\zeta \in (0.015, 0.1)$. When the number of individuals is 200, the entanglement error rate ranges of JDE, OBDE, DEGL and SADE are $\varepsilon \in (\pm 0.4\%, \pm 0.45\%)$, $\varepsilon = \pm 1\%$, $\varepsilon = \pm 1\%$, $\varepsilon \in (\pm 0.3\%, \pm 0.4\%)$, compared with DE-MPFSC, the former has a higher entanglement error rate, which shows that the entanglement degree of DE-MPFSC (since we have done a detailed analysis of the entanglement effect of the DE-MPFSC algorithm in the case of the same number of individuals, we will not repeat it here. For all comparative data, see Section 5.3) is significantly higher than that of JDE, OBDE, DEGL and SADE, and the algorithm has good stability and effectiveness. We also found a special case, when the number of individuals is 200, the entanglement degree and entanglement degree error of JDE, OBDE, DEGL are infinite or non-existent. The reason for the case is that the entangled branches are too discrete or the edges of the entangled branches are too severely differentiated. Obviously, their entanglement efficiency is significantly lower than DE-MPFSC. The entanglement range and entanglement error of SADE are $\frac{1}{\zeta} \in (2.43 \times 10^{-2}, 1.06 \times 10^{-1})$ and $\zeta \in (2.43 \times 10^{-2}, 1.06 \times 10^{-1})$, which is also lower than the entanglement effect of DE-MPFSC, which shows that the degree of entanglement of DE-MPFSC is significantly higher than that of JDE, OBDE, DEGL and SADE. The algorithm has small fluctuations and good convergence. When the number of individuals is 400, the entanglement error rate ranges of JDE, OBDE, DEGL and SADE are $\varepsilon \in (\pm 0.4\%, \pm 0.45\%)$, $\varepsilon = \pm 1\%$, $\varepsilon = \pm 0.25\%$, $\varepsilon = \pm 0.15\%$, compared with DE-MPFSC, the entanglement error rate is higher, which shows that the degree of entanglement of DE-MPFSC is significantly higher than that of JDE, OBDE, DEGL and SADE, and the algorithm is better stability and effectiveness. The entanglement ranges of JDE, OBDE, DEGL and SADE are $\frac{1}{\zeta} \in (5.44 \times 10^{-2}, 2.33 \times 10^{-1})$, $\frac{1}{\zeta} \in (0.53 \times 10^{0}, 4.22 \times 10^{-2})$, $\frac{1}{\zeta} \in (1.73 \times 10^{-1}, 2.04 \times 10^{-1})$, $\frac{1}{\zeta} \in (6.23 \times 10^{-2}, 2.43 \times 10^{-1})$, which shows that the degree of entanglement of the DE-MPFSC is significantly higher than that of JDE, OBDE, DEGL, SADE, and the DE-MPFSC algorithm has less volatility. The entanglement degree error ranges of JDE, OBDE, DEGL and SADE are $\zeta \in (1.13 \times 10^{-3}, 1.49 \times 10^{-1})$, $\zeta \in (0.59 \times 10^{0}, 0.69 \times 10^{-1})$, $\zeta \in (1.34 \times 10^{-1}, 3.70 \times 10^{-2})$, $\eta \in (2.95 \times 10^{-2}, 1.08 \times 10^{-1})$, which shows that the degree of entanglement of DE-MPFSC is significantly higher than that of JDE, OBDE, DEGL and SADE, and the algorithm has strong convergence. When the number of individuals is 600, 800, 1000, the comparison between DE-MPFSC and JDE, OBDE, DEGL and SADE is similar, and the data has been marked in the Table 6. On the whole, the effectiveness, stability, and convergence of DE-MPFSC are higher than JDE, OBDE, DEGL and SADE on average.

Table 6. Numerical Analysis of Entanglement Degree $\frac{1}{\xi}$ and Entanglement Degree Error ζ of the DE algorithm and the DE-MPFSC algorithm Several Times.

NP	JDE					OBDE					DEGL					SADE				
	F	CR	ξ	ζ	$\varepsilon\%$	F	CR	ξ	ζ	$\varepsilon\%$	F	CR	ξ	ζ	$\varepsilon\%$	F	CR	ξ	ζ	$\varepsilon\%$
200	0.2	0.3	∞	∞	±0.40	0.2	0.3	–	–	±1	0.2	0.3	∞	∞	±1	0.2	0.3	1.04×10^{-1}	1.04×10^{-1}	±0.30
200	/	/	∞	∞	±0.45	/	/	–	–	±1	/	/	∞	∞	±1	/	/	1.06×10^{-1}	1.06×10^{-1}	±0.40
200	/	/	∞	∞	±0.40	/	/	∞	∞	±1	/	/	–	–	±1	/	/	0.81×10^{-1}	0.81×10^{-1}	±0.35
200	/	/	–	–	±0.40	/	/	–	–	±1	/	/	∞	∞	±1	/	/	2.43×10^{-2}	2.43×10^{-2}	±0.35
200	/	/	–	–	±0.45	/	/	∞	∞	±1	/	/	–	∞	±1	/	/	2.77×10^{-2}	2.77×10^{-2}	±0.40
NP	JDE					OBDE					DEGL					SADE				
	F	CR	ξ	ζ	$\varepsilon\%$	F	CR	ξ	ζ	$\varepsilon\%$	F	CR	ξ	ζ	$\varepsilon\%$	F	CR	ξ	ζ	$\varepsilon\%$
400	0.2	0.3	0.88×10^{0}	0.49×10^{0}	±0.45	0.2	0.3	0.53×10^{0}	0.60×10^{0}	±1	0.2	0.3	2.04×10^{-1}	1.34×10^{-1}	±0.25	0.2	0.3	2.42×10^{-1}	1.06×10^{-1}	±0.15
400	/	/	5.44×10^{-2}	1.49×10^{-1}	±0.40	/	/	0.58×10^{0}	0.59×10^{0}	±1	/	/	1.74×10^{-1}	1.41×10^{-1}	±0.25	/	/	2.41×10^{-1}	1.05×10^{-1}	±0.15
400	/	/	6.05×10^{-2}	0.52×10^{0}	±0.45	/	/	1.61×10^{-1}	4.72×10^{-2}	±1	/	/	1.72×10^{-1}	3.72×10^{-2}	±0.25	/	/	6.23×10^{-2}	2.95×10^{-2}	±0.15
400	/	/	2.15×10^{-1}	0.47×10^{0}	±0.40	/	/	4.22×10^{-2}	4.51×10^{-2}	±1	/	/	1.73×10^{-1}	1.38×10^{-1}	±0.25	/	/	6.49×10^{-2}	3.02×10^{-1}	±0.15
400	/	/	2.33×10^{-1}	1.13×10^{-3}	±0.40	/	/	1.67×10^{-1}	1.69×10^{-1}	±1	/	/	5.16×10^{-2}	3.70×10^{-2}	±0.25	/	/	2.43×10^{-1}	1.08×10^{-1}	±0.15
NP	JDE					OBDE					DEGL					SADE				
	F	CR	ξ	ζ	$\varepsilon\%$	F	CR	ξ	ζ	$\varepsilon\%$	F	CR	ξ	ζ	$\varepsilon\%$	F	CR	ξ	ζ	$\varepsilon\%$
600	0.2	0.3	–	–	±1	0.2	0.3	4.44×10^{-2}	0.49×10^{0}	±0.45	0.2	0.3	1.62×10^{-1}	3.71×10^{-2}	±0.15	0.2	0.3	2.20×10^{-1}	1.11×10^{-1}	±0.1
600	/	/	–	–	±1	/	/	1.62×10^{-1}	0.51×10^{0}	±0.46	/	/	4.30×10^{-2}	1.38×10^{-1}	±0.1	/	/	2.18×10^{-1}	1.10×10^{-1}	±0.1
600	/	/	∞	∞	±1	/	/	1.62×10^{-1}	0.47×10^{0}	±0.40	/	/	1.56×10^{-1}	3.70×10^{-2}	±0.1	/	/	2.17×10^{-1}	2.67×10^{-2}	±0.15
600	/	/	–	–	±1	/	/	4.41×10^{-2}	1.38×10^{-1}	±0.40	/	/	4.28×10^{-2}	1.36×10^{-1}	±0.15	/	/	5.91×10^{-2}	1.02×10^{-1}	±0.15
600	/	/	∞	∞	±1	/	/	4.49×10^{-2}	1.16×10^{-1}	±0.38	/	/	1.59×10^{-1}	3.69×10^{-2}	±0.1	/	/	2.17×10^{-1}	1.00×10^{-1}	±0.1
NP	JDE					OBDE					DEGL					SADE				
	F	CR	ξ	ζ	$\varepsilon\%$	F	CR	ξ	ζ	$\varepsilon\%$	F	CR	ξ	ζ	$\varepsilon\%$	F	CR	ξ	ζ	$\varepsilon\%$
800	0.2	0.3	5.49×10^{-2}	1.41×10^{-1}	±0.6	0.2	0.3	2.28×10^{-1}	0.30×10^{0}	±0.40	0.2	0.3	2.01×10^{-2}	1.31×10^{-1}	±0.30	0.2	0.3	2.14×10^{-1}	1.01×10^{-1}	±0.15
800	/	/	5.12×10^{-2}	1.44×10^{-1}	±0.6	/	/	2.25×10^{-1}	0.35×10^{0}	±0.30	/	/	2.03×10^{-1}	1.38×10^{-1}	±0.40	/	/	2.15×10^{-1}	1.06×10^{-1}	±0.15
800	/	/	15.70×10^{-3}	3.84×10^{-2}	±0.6	/	/	2.36×10^{-1}	2.54×10^{-2}	±0.40	/	/	2.03×10^{-1}	3.72×10^{-2}	±0.50	/	/	2.08×10^{-1}	2.82×10^{-2}	±0.2
800	/	/	2.06×10^{-1}	3.99×10^{-2}	±0.6	/	/	2.17×10^{-1}	0.37×10^{0}	±0.30	/	/	2.05×10^{-1}	1.34×10^{-1}	±0.60	/	/	5.57×10^{-2}	0.95×10^{-1}	±0.2
800	/	/	15.86×10^{-3}	1.54×10^{-1}	±0.6	/	/	2.22×10^{-1}	1.01×10^{-1}	±0.40	/	/	5.55×10^{-2}	3.62×10^{-2}	±0.75	/	/	1.52×10^{-1}	0.96×10^{-1}	±0.2
NP	JDE					OBDE					DEGL					SADE				
	F	CR	ξ	ζ	$\varepsilon\%$	F	CR	ξ	ζ	$\varepsilon\%$	F	CR	ξ	ζ	$\varepsilon\%$	F	CR	ξ	ζ	$\varepsilon\%$
1000	0.2	0.3	–	–	±1	0.2	0.3	∞	∞	±1	0.2	0.3	∞	∞	±1	0.2	0.3	1.93×10^{-1}	1.03×10^{-1}	±0.1
1000	/	/	–	–	±0.1	/	/	∞	∞	±1	/	/	∞	∞	±1	/	/	1.88×10^{-1}	0.96×10^{-1}	±0.1
1000	/	/	∞	∞	±1	/	/	–	–	±1	/	/	∞	∞	±1	/	/	1.90×10^{-1}	2.60×10^{-2}	±0.1
1000	/	/	∞	∞	±1	/	/	–	–	±1	/	/	–	–	±1	/	/	5.09×10^{-2}	0.87×10^{-1}	±0.1
1000	/	/	–	–	±1	/	/	∞	∞	±1	/	/	∞	∞	±1	/	/	1.91×10^{-1}	0.84×10^{-1}	±0.1

Then, we analyze from the state of entangled branches. When the number of individuals is 200, 400 or 1000, the entanglement degree and entanglement degree error of JDE do not exist, and the entanglement branches are in a discrete state. As the number of individuals increases, the entanglement effect appears fluctuating. When the number of individuals is 200 or 1000, the entanglement degree and entanglement degree error of OBDE do not exist, and the entangled branches are in a discrete state. When the number of individuals is 200, 400, 600 or 800, the integration of entangled branches gradually increases. However, the increase of OBDE is weaker than that of DE-MPFSC. The entanglement of the former shows a coexistence state of gradualism and fluctuation. When the number of individuals is 200 or 1000, the entanglement degree and entanglement degree error of DEGL do not exist, and the entangled branches are in a discrete state. When the number of individuals is 400 and 800, the integration of entangled branches begins to appear and it reaches the best state when the number of individuals is 600. The entanglement of DEGL shows coexistence of symmetry and fluctuation. The case is not conducive to the stable development of population evolution. When the number of individuals gradually increases, the entanglement degree and entanglement degree error of SADE are always integrated and increase with the number of individuals. However, due to the dispersion of the edges of the entangled branches, the entanglement increases slowly. Overall, DE-MPFSC has better stability and effectiveness than JDE, OBDE, DEGL and SADE.

6. Empirical Analysis

6.1. Verification Data Sets

We consider the imbalanced datasets, the UCI machine learning datasets [51] (the UCI Machine Learning Database is available at the following URL. https://archive.ics.uci.edu/ml/index.php), as integrated datasets to test the practical application level of the DE-MPFSC algorithm. First, we select three types of testing data from the UCI machine learning datasets, which are numerical value datasets (N.V. datasets), classify value datasets (C.V. datasets) and mixed value datasets(M.V. datasets). At the same time, according to the data integration level, the above three types of datasets are divided into level I, level II and level III, their proportions are 1, 2 and 3, respectively. The specific information on the three types of datasets is shown in Table 7 .

Table 7. The situation description of various types of data about the UCI machine learning datasets.

Datasets	Sample Capacity	N.V. Datasets	C.V. Datasets	M.V. Datasets
Dermatology	400	5	30	8
Credit Approval	572	9	13	4
Automobile	382	15	27	17
Sponge	108	6	32	28
Contraceptive Method Choice (CMC)	1458	4	21	16
Soybean	453	2	39	14
Glass	314	5	4	12

In this experiment, we divided the three types of data integration into single-link data integration (SLCE) and complete-link data integration (CLCE) according to the DE-MPFSC algorithm. The results obtained by the two integrated methods are compared with the results obtained by the KNN-SK (KNN data filling) [52–54] and the SKNN-SK (SKNN data filling) [52–54]. To ensure that there is single-random for the data in Table 1, we performed random deletion in a five percent ratio on the computer. Then, we conducted the algorithm test and obtained the results by performing the test 200 times on the computer in the same integrated method for the same datasets.

6.2. Verification Indicator

To interpret the effect of the DE-MPFSC algorithm on the imbalanced data integrated and classified, we need to verify the consistency of the imbalanced data. First, we should establish the verification indicator of classification accuracy (CA) [55,56], adjusted rand index (ARI) [55,56], normalized mutual information (NMI) [55,56] to test the effect of the DE-MPFSC algorithm, which are described as follows.

(1) CA: CA is a statistical mathematics variable measuring the imbalanced data samples proportion of the DE-MPFSC algorithm in making up for data integrated defects, which is expressed as follows.

$$CA = \frac{\sum_{i=1}^{k} \alpha_i}{n}, \tag{34}$$

where α_i is a subsample of the DE-MPFSC algorithm in making up for data integrated defects, k is the number of integrated classes, n is the sample capacity.

(2) ARI: ARI is a statistical mathematics variable measuring the imbalanced data samples proportion of the DE-MPFSC algorithm in making up for data integrated defects after considering the same data class and different data classes, which is expressed as follows.

$$ARI = \frac{\sum_{i=1}^{I} \sum_{j=1}^{J} C_{n_{ij}}^{2} - \eta}{\frac{1}{2}(\rho + \phi) - \eta} \tag{35}$$

$$\rho = \sum_{i=1}^{I} C_{n_i}^{2}, \phi = \sum_{j=1}^{J} C_{n_j}^{2}, \eta = \frac{2\rho\phi}{n(n-1)}. \tag{36}$$

(3) NMI: NMI is a similarity variable measuring sample data and integrated data after comparing and correctly integrating and classing imbalanced data within being integrated defects, which is expressed as follows.

$$NMI = \frac{\sum_{i=1}^{I} \sum_{j=1}^{J} \frac{n_{ij}^{2} n}{n_i n_j}}{\sqrt{\sum_{i=1}^{I} \frac{n_i^2}{n} \sum_{j=1}^{J} \frac{n_j^2}{n}}}, \tag{37}$$

where n_{ij} is the sample size in the data integrated result where the ith datasets contains the original datasets j. n_i is the sample size of the ith datasets in the data integrated result. n_j is the sample size of the original datasets j. n is the samples. I, J is the number of different types of datasets and the number of original datasets obtained in the data integrated result. The limit value of the three types of verification indicators is set to 1. If the data structure is close to the original data structure after imbalanced integrated data, which means that the value of the dataset is larger compared to the original datasets. That is, the DE-MPFSC algorithm has a better effect on imbalanced data integration.

6.3. The Test Analysis

The number of neighbors selected by the KNN-SK and SKNN-SK is $K = 10$. The testing results of the (Differential Evolution Algorithm Based on Mixed Penalty Function Screening Criteria)DE-MPFSC algorithm and other algorithms under the verification indicators are shown in Table 8 and Figures 16–18 about Classification Accuracy (CA), Adjusted Rand Index (ARI), and Normalized Mutual Information (NMI) datasets.

Table 8. Data integration Comparison about CA, NMI and ARI Values of the DE-MPFSC Algorithm.

Data Sets	*CA*				*ARI*				*NMI*			
	SLCE	*CLCE*	*KNN*	*SKNN*	*SLCE*	*CLCE*	*KNN*	*SKNN*	*SLCE*	*CLCE*	*KNN*	*SKNN*
Der.	0.430	0.772	0.674	0.710	0.152	0.667	0.666	0.458	0.351	0.728	0.588	0.579
C.A.	0.576	0.753	0.726	0.756	0.261	0.275	0.215	0.274	0.206	0.217	0.167	0.213
Aut.	0.523	0.537	0.529	0.516	0.168	0.148	0.157	0.153	0.265	0.257	0.237	0.269
Spo.	0.780	0.790	0.738	0.726	0.424	0.438	0.523	0.447	0.705	0.701	0.707	0.747
■(CMC)	0.428	0.429	0.427	0.410	0.016	0.501	0.516	0.424	0.032	0.039	0.020	0.015
Soy.	0.545	0.668	0.631	0.634	0.431	0.436	0.356	0.346	0.622	0.637	0.638	0.696
■Glass	0.473	0.546	0.524	0.525	0.199	0.154	0.158	0.168	0.323	0.302	0.315	0.226
Average	*CA*				*ARI*				*NMI*			
	0.322	0.262	0.267	0.183	0.783	0.203	0.358	0.283	0.437	0.392	0.438	0.395
Variance	*CA*				*ARI*				*NMI*			
	0.002	0.0012	0.120	0.050	0.030	0.070	0.090	0.030	0.0013	0.0042	0.0912	0.095

Where according to the data analysis in Table 8, we know that since the single data integrated method KNN-SK and SKNN-SK do not have fitting characteristics inside the data integration and consistency are better than single integrated methods.

The CA value, ARI value and NMI value of UCI data about DE-MPFSC algorithm are as follows: the optimal value and sub-optimal value tend to distribution in probability 1, while the other data points tend to distribution in probability 0, in which the data points tended to distribution in probability 1 are in the feasible region and the data points tended to distribution in probability 0 are in the non-feasible region. The CA value, ARI value and NMI value of Credit Approval data about DE-MPFSC algorithm are as follows: the optimal value and sub-optimal value tend to distribution in probability 1, while the other data points tend to distribution in probability 0, in which the data points tended to distribution in probability 1 are in the feasible region and the data points tended to distribution in probability 0 are in the non-feasible region. For the data (CMC) and Glass marked with ■, the values of CA, ARI and NMI are as follows: the optimal value and the sub-optimal value are stable around the probability of 0.5, and the error does not exceed ± 0.2, which accord with the stability requirements of data.

The CA value, ARI value and NMI value of all data about DE-MPFSC algorithm are as follows: the average optimal value and the average sub-optimal quality are stable around the probability of 0.5 and the error does not exceed ± 0.2. The variance optimal value and the variance sub-optimal value are basically stable around 0.01, which accord with the data stability standard and the error does not exceed 0.0001.From the datasets in Table 8, the optimal value of the DE-MPFSC algorithm in both mean and variance shows the superiority of the algorithm.

It can be seen from the Figures 16–18 that through the two-dimensional reduced order analysis of UCI machine learning data, it is found that the three data integration effects of *CA*, *NMI* and *ARI* show a trend of highly integrated convergence within an acceptable error range.

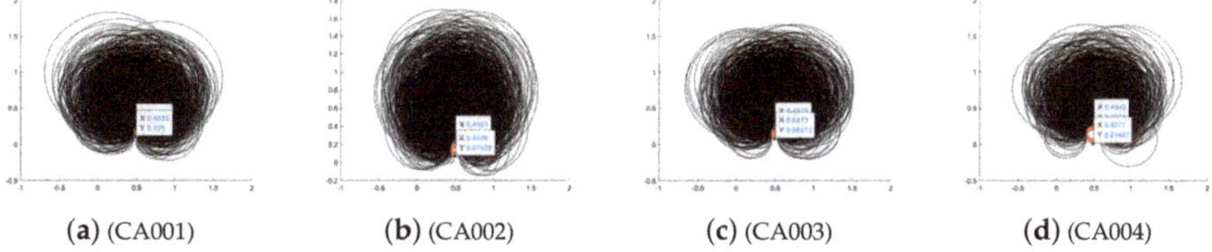

(a) (CA001) **(b)** (CA002) **(c)** (CA003) **(d)** (CA004)

Figure 16. Convergence integration of DE-MPFFC algorithm for CA.

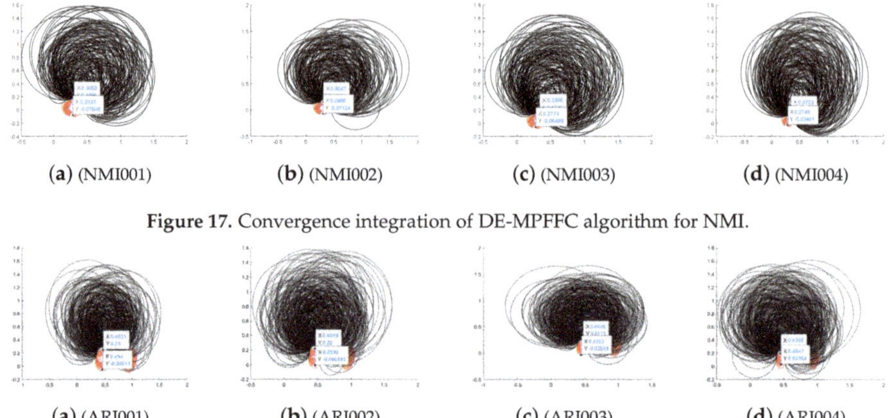

(a) (NMI001) **(b)** (NMI002) **(c)** (NMI003) **(d)** (NMI004)

Figure 17. Convergence integration of DE-MPFFC algorithm for NMI.

(a) (ARI001) **(b)** (ARI002) **(c)** (ARI003) **(d)** (ARI004)

Figure 18. Convergence integration of DE-MPFFC algorithm for ARI.

First, we analyze from the precision. The integration effect on CA data is as follows: the minimum upper bound of the reduced-order 2D data in the X direction is $\Omega_{min} = 0.4$ and the error does not exceed ± 0.01. The maximum lower bound of the reduced-order 2D data in the Y direction is $\Omega_{max} = 0.1$ and the error does not exceed ± 0.02. The entanglement error of the imbalanced data set in both X and Y directions is $\frac{\Omega_{max}}{\Omega_{min}} = 0.25$. It can be seen that the precision in both directions of X and Y is in a fully controllable and highly integrated range, which fully shows that the DE-MPFSC algorithm has a good integration advantage for improving the integration accuracy of CA data. The integration effect on NMI data is as follows—the minimum upper bound of the reduced-order 2D data in the X direction is $\Omega_{min} = 0.3$ and the error does not exceed ± 0.03. The maximum lower bound of the reduced-order 2D data in the Y direction is $\Omega_{max} = 0.02$ and the error does not exceed ± 0.01. The entanglement error of the imbalanced data set in both X and Y directions is $\frac{\Omega_{max}}{\Omega_{min}} = 0.06$. It can be seen that the precision in both directions of X and Y is in a fully controllable and highly integrated range, which fully demonstrates that the DE-MPFSC algorithm has a good integration advantage for improving the integration accuracy of NMI data. The integration effect on ARI data is as follows: the minimum upper bound of the reduced-order 2D data in the X direction is $\Omega_{min} = 0.45$ and the error does not exceed ± 0.01. The maximum lower bound of the reduced-order 2D data in the Y direction is $\Omega_{max} = 0.006$ and the error does not exceed ± 0.01. The entanglement error of the imbalanced data set in both X and Y directions is $\frac{\Omega_{max}}{\Omega_{min}} = 0.013$. It can be seen that the precision in both directions of X and Y is in a fully controllable and highly integrated range, which fully shows that the DE-MPFSC algorithm has a good integration advantage for improving the integration accuracy of ARI data.

Then we analyze the density of data integration. The integration effect on CA data is as follows: The integration curve L_{CA} integrated with $(Average_X = 0.4911, Average_Y = 0.0851)$ as the center. The entanglement error is $\varpi = \frac{Average_Y}{Average_X} = 0.17$ and the integration curve radius is $R_{CA} \doteq 1.5$. The data integration density is higher than before, and the imbalanced data that presents the red part and the balanced data that presents the black part are completely separated, which fully shows that the DE-MPFSC algorithm has a better distribution advantage for improving the accuracy of CA data integration. The integration effect on NMI data is as follows: The integration curve L_{NMI} integrated with $(Average_X = 0.2864, Average_Y = -0.0671)$ as the center. The entanglement error is $\varpi = \frac{Average_Y}{Average_X} = -0.23$ and the integration curve radius is $R_{NMI} \doteq 1.6$. The data integration density is higher than before, and the imbalanced data that presents the red part and the balanced data that presents the black part are completely separated, which fully shows that the DE-MPFSC algorithm has a better distribution advantage for improving the accuracy of NMI data integration. The integration effect on ARI data is as follows: The integration curve L_{ARI} integrated with $(Average_X = 0.4510,$

$Average_Y = -0.0283$) as the center. The entanglement error is $\varpi = \frac{Average_Y}{Average_X} = -0.06$ and the integration curve radius is $R_{ARI} \doteq 1.5$. The data integration density is higher than before and the imbalanced data that presents the red part and the balanced data that presents the black part are completely separated, which fully shows that the DE-MPFSC algorithm has a better distribution advantage for improving the accuracy of ARI data integration.

Finally, we analyze the degree of data separation. From the distribution of Figures 16–18, CA data, NMI data and ARI data have been completely separated by the DE-MPFSC algorithm. The data integration effect presents the following trend: the data set that is balanced is in the black curve part and tends to converge toward the center of the integration curve, and the imbalanced data set is in the red curve part and tends to converge toward the center of the integration curve.

The above analysis shows that the DE-MPFSC algorithm has a good natural advantage in integrating imbalanced data.

7. Conclusions

In this paper, some processing problems of imbalanced data. We design the algorithm based on the DE algorithm, constructs a differential evolution algorithm based on a mixed penalty function screening criterion for imbalanced data integration. The method can improve the classification problem of imbalanced data and construct an empirical analysis through UCI machine learning datasets. In addition, the verification result is in accordance with the imbalanced data classification expectation and the integration effect. The main work of this article is as follows:

1. Based on the normalization of internal and external penalty functions, we established the mixed penalty function screening criteria on DE algorithm.

2. We established a differential evolution integration algorithm (DE-MPFSC algorithm) based on the mixed penalty function screening criterion, which broadens the algorithm foundation for the efficient integration of imbalanced data.

3. We constructed the Markov process of the DE-MPFSC algorithm and proved the theoretical validity and evolution mechanism of DE-MPFSC algorithm.

4. We creatively introduce the entanglement and entanglement error and compare the performance of the DE-MPFSC algorithm through data analysis.

5. Based on the empirical analysis of the UCI machine learning data, we further illustrated the effectiveness of the DE-MPFSC algorithm in the efficient integration of the imbalanced data, leading to multimode imbalanced data integration.

Author Contributions: Y.G. provided the research ideas of the paper; K.W. solved the overall framework, algorithm framework, paper structure, and result analysis of the research ideas of the paper; C.G. provides data analysis and corresponding analysis results; Y.S. provides the overall research ideas and main methods of the paper; T.L. has helped the writing of the paper.

Funding: This work was supported by the Major Scientific Research Special Projects of North Minzu University (No.ZDZX201901), the NSFC (No.61561001;11961001), Graduate Innovation Project of North Minzu University (No:YCX19120), First-Class Disciplines Foundation of Ningxia (No:NXYLXK2017B09) and the by National Natural Science Foundation of Shanxi Province (2019JM-425), China Postdoctoral Science Foundation Funded Project (2019M653567).

Acknowledgments: Acknowledgment for the financial support of the National Natural Science Foundation Project, the University-level Project of Northern University for Nationalities and the District-level Project of Ningxia, the Major Scientific Research Special Projects of North Minzu University and all authors' efforts. And the reviewers and instructors for the paper.

Conflicts of Interest: The authors declare no conflict of interest.

References

1. Everitt, B. Cluster Analysis. *Qual. Quant.* **1980**, *14*, 75–100. [CrossRef]
2. Chunyue, S.; Zhihuan, S.; Ping, L.; Wenyuan, S. The study of Naive Bayes algorithm online in data mining. In Proceedings of the World Congress on Intelligent Control and Automation, Hangzhou, China, 15–19 June 2004.
3. Samanta, B.; Al-Balushi, K.R.; Al-Araimi, S.A. Artificial neural networks and genetic algorithm for bearing fault detection. *Soft Comput. Fusion Found. Methodol. Appl.* **2006**, *10*, 264–271. [CrossRef]
4. Díez-Pastor, J.F.; Rodríguez, J.J.; García-Osorio, C.; Kuncheva, L.I. Random Balance: Ensembles of variable priors classifiers for imbalanced data. *Knowl.-Based Syst.* **2015**, *85*, 96–111. [CrossRef]
5. Maldonado, S.; Weber, R.; Famili, F. Feature selection for high-dimensional class-imbalanced data sets using Support Vector Machines. *Inf. Sci.* **2014**, *286*, 228–246. [CrossRef]
6. Chawla, N.V.; Bowyer, K.W.; Hall, L.O.; Kegelmeyer, W.P. SMOTE: Synthetic minority over-sampling technique. *J. Artif. Intell. Res.* **2002**, *16*, 321–357. [CrossRef]
7. Vorraboot, P.; Rasmequan, S.; Chinnasarn, K.; Lursinsap, C. Improving classification rate constrained to imbalanced data between overlapped and non-overlapped regions by hybrid algorithms. *Neurocomputing* **2015**, *152*, 429–443. [CrossRef]
8. Krawczyk, B.; Woźniak, M.; Schaefer, G. Cost-sensitive decision tree ensembles for effective imbalanced classification. *Appl. Soft Comput.* **2014**, *14*, 554–562. [CrossRef]
9. López, V.; Río, S.D.; Benítez, J.M.; Herrera, F. Cost-sensitive linguistic fuzzy rule-based classification systems under the MapReduce framework for imbalanced big data. *Fuzzy Sets Syst.* **2015**, *258*, 5–38. [CrossRef]
10. Santucci, V.; Milani, A.; Caraffini, F. An Optimisation-Driven Prediction Method for Automated Diagnosis and Prognosis. *Mathematics* **2019**, *7*, 1051. [CrossRef]
11. Mikalef, P.; Boura, M.; Lekakos, G.; Krogstie, J. Big data analytics and firm performance: Findings from a mixed-method approach. *J. Bus. Res.* **2019**, *98*, 261–276. [CrossRef]
12. Thai-Nghe, N.; Gantner, Z.; Schmidt-Thieme, L. Cost-sensitive learning methods for imbalanced data. In Proceedings of the International Joint Conference on Neural Networks, Barcelona, Spain, 18–23 July 2010; pp. 1–8.
13. Freund, Y. Boosting a weak learning algorithm by majority. *Inf. Comput.* **1995**, *121*, 256–285. [CrossRef]
14. Breiman, L. Bagging predictors, machine learning research: Four current directors. *ResearchGate* **1996**, *24*, 123–140.
15. Sun, Z.; Song, Q.; Zhu, X.; Sun, H.; Xu, B.; Zhou, Y. A novel ensemble method for classifying imbalanced data. *Pattern Recognit.* **2015**, *48*, 1623–1637. [CrossRef]
16. Chawla, N.V.; Lazarevic, A.; Hall, L.O.; Bowyer, K.W. SMOTEBoost: Improving Prediction of the Minority Class in Boosting. Knowledge Discovery in Databases: Pkdd 2003. In Proceedings of the European Conference on Principles and Practice of Knowledge Discovery in Databases, Cavtat-Dubrovnik, Croatia, 22–26 September 2003; pp. 107–119.
17. Flach, P.A.; Lachiche, N. Naive Bayesian Classification of Structured Data. *Mach. Learn.* **2004**, *57*, 233–269. [CrossRef]
18. Fernández, A.; López, V.; Galar, M.; Del Jesus, M; Herrera, F. Analysing the classification of imbalanced data-sets with multiple classes: Binarization techniques and ad-hoc approaches. *Knowl.-Based Syst.* **2013**, *42*, 97–110. [CrossRef]
19. Storn, R.; Price, K. Differential Evolution—A Simple and Efficient Heuristic for global Optimization over Continuous Spaces. *J. Glob. Optim.* **1997**, *11*, 341–359. [CrossRef]
20. Wang, K.; Gao, Y. Topology Structure Implied in $\beta - Hilbert$ Space, Heisenberg Uncertainty Quantum Characteristics and Numerical Simulation of the DE Algorithm. *Mathematics* **2019**, *7*, 330. [CrossRef]
21. Das, S.; Mullick, S.S.; Suganthan, P.N. Recent advances in differential evolution—An updated survey. *Swarm Evol. Comput.* **2016**. [CrossRef]
22. Neri, F.; Tirronen, V. Recent advances in differential evolution: A survey and experimental analysis. *Artif. Intell. Rev.* **2010**, *33*, 61–106. [CrossRef]
23. Das, S.; Suganthan, P.N. Differential Evolution: A Survey of the State-of-the-Art. *IEEE Trans. Evol. Comput.* **2011**, *15*, 4–31. [CrossRef]

24. Brest, J.; Zumer, V.; Maucec, M.S. Self-Adaptive Differential Evolution Algorithm in Constrained Real-Parameter Optimization. In Proceedings of the IEEE Congress on Evolutionary Computation (CEC 2006), Vancouver, BC, Canada, 16–21 July 2006.
25. Brest, J.; Zamuda, A.; Bo šković, B.; Greiner, S.; Zumer, V. An Analysis of the Control Parameters' Adaptation in DE. *Stud. Comput. Intell.* **2008**._3. [CrossRef]
26. Rahnamayan, S.; Tizhoosh, H.R.; Salama, M.M.A. Opposition-based differential evolution. *IEEE Trans. Evol. Comput.* **2008**, *12*, 64–79. [CrossRef]
27. Das, S.; Abraham, A.; Chakraborty, U.K.; Konar, A. Differential Evolution Using a Neighborhood-Based Mutation Operator. *IEEE Trans. Evol. Comput.* **2009**, *13*, 526–553. [CrossRef]
28. Mallipeddi, R.; Suganthan, P.N. Ensemble of Constraint Handling Techniques. *IEEE Trans. Evol. Comput.* **2010**, *14*, 561–579. [CrossRef]
29. Qu, B.Y.; Suganthan, P.N. Constrained multi-objective optimization algorithm with an ensemble of constraint handling methods. *Eng. Optim.* **2011**, *43*, 403–416. [CrossRef]
30. Qin, A.K.; Suganthan, P.N. Self-adaptive differential evolution algorithm for numerical optimization. In Proceedings of the IEEE Congress on Evolutionary Computation (CEC 2005), Edinburgh, UK, 2–4 September 2005.
31. Zou, D.; Liu, H.; Gao, L.; Li, S. A modified differential evolution algorithm for unconstrained optimization problems. *Neurocomputing* **2011**, *120*, 1608–1623. [CrossRef]
32. Ghosh, A.; Das, S.; Chowdhury, A.; Giri, R. An improved differential evolution algorithm with fitness-based adaptation of the control parameters. *Inf. Sci.* **2011**, *181*, 3749–3765. [CrossRef]
33. Caraffini, F.; Kononova, A.V.; Corne, D. Infeasibility and structural bias in Differential Evolution. *Inf. Sci.* **2019**, *496*, 161–179. [CrossRef]
34. Storn, R. System design by constraint adaptation and differential evolution. *IEEE Trans. Evol. Comput.* **1999**, *3*, 22–34. [CrossRef]
35. Thomsen, R. Flexible ligand docking using differential evolution. In Proceedings of the 2003 Congress on Evolutionary Computation (CEC '03), Canberra, ACT, Australia, 8–12 December 2003; pp. 2354–2361; Volume 4.
36. Ali, M.; Pant, M.; Singh, V.P. An improved differential evolution algorithm for real parameter optimization problems. *Int. J. Recent Trends Eng.* **2009**, *1*, 63–65.
37. Yang, M.; Li, C.; Cai, Z.; Guan, J. Differential evolution with auto-enhanced population diversity. *IEEE Trans. Cybern.* **2015**, *45*, 302–315. [CrossRef] [PubMed]
38. Iri, M.; Imai, H. Theory of the multiplicative penalty function method for linear programming. *Discret. Algorithms Complex.* **1987**, *30*, 417–435.
39. Rao, S.S.; Bard, J. *Engineering Optimization: Theory and Practice*, 4th ed.; A I I E Transactions; John Wiley & Sons: New York, NY, USA, 1997; Volume 29; pp. 802–803.
40. Wright, M. The interior-point revolution in optimization: History, recent developments, and lasting consequences. *Bull. Am. Math. Soc.* **2005**, *42*, 39–57. [CrossRef]
41. Deb, K. An efficient constraint handling method for genetic algorithms. *Comput. Methods Appl. Mech. Eng.* **2000**, *186*, 311–338. [CrossRef]
42. Venkatraman, S.; Yen, G.G. *A Generic Framework for Constrained Optimization Using Genetic Algorithms*; IEEE Press: Piscataway, NJ, USA, 2005.
43. Xie, X.F.; Zhang, W.J.; Bi, D.C. Handling equality constraints by adaptive relaxing rule for swarm algorithms. In Proceedings of the 2004 Congress on Evolutionary Computation (IEEE Cat. No.04TH8753), Portland, OR, USA, 19–23 June 2004; Volume 2.
44. Schwefel, H.P. *Evolution and Optimum Seeking*; Wiley: New York, NY, USA, 1995.
45. Kramer, O. Premature Convergence in Constrained Continuous Search Spaces. In Proceedings of the Parallel Problem Solving from Nature—PPSN X, Dortmund, Germany, 13–17 September 2008; Springer: Berlin/Heidelberg, Germany, 2008; pp. 62–71.
46. Gasparini, M. *Markov Chain Monte Carlo in Practice*; Markov chain monte carlo in practice; Chapman and Hall: London, Uk, 1996; pp. 9236–9240.
47. Eiben, A.E.; Aarts, E.H.L.; Hee, K.M.V. *Global Convergence of Genetic Algorithms: A Markov Chain Analysis*; Parallel Problem Solving from Nature; Springer: Berlin/Heidelberg, Germany, 1991; pp. 3–12.

48. Rudolph, G. Convergence of evolutionary algorithms in general search spaces. In Proceedings of the IEEE International Conference on Evolutionary Computation, Nagoya, Japan, 20–22 May 1996; pp. 50–54.
49. Cerf, R. Asymptotic Convergence of Genetic Algorithms. *Adv. Appl. Probab.* **1998**, *30*, 521–550. [CrossRef]
50. Xu, Z.B.; Nie, Z.K.; Zhang, W.X. Almost sure strong convergence of a class of genetic algorithms with parent-offspring competition. *Acta Math. Appl. Sin.* **2002**, *25*, 167–175.
51. Unified Communications Irvine Machine Learning Repository(UCI). Available online: https://archive.ics.uci.edu/ml/index.php (accessed on 1 August 2019).
52. Silva, L.O.; Zárate, L.E. A brief review of the main approaches for treatment of missing data. *Intell. Data Anal.* **2014**, *18*, 1177–1198. [CrossRef]
53. Batista, G.A.P.A. MariaCarolinaMonard. An analysis of four missing data treatment methods for supervised learning. *Appl. Artif. Intell.* **2003**, *17*, 519–533. [CrossRef]
54. Liang, J.; Bai, L.; Dang, C.; Cao, F. The *K*-Means-Type Algorithms Versus Imbalanced Data Distributions. *IEEE Trans. Fuzzy Syst.* **2012**, *20*, 728–745. [CrossRef]
55. Strehl, A.; Ghosh, J. Cluster ensembles—A knowledge reuse framework for combining multiple partitions. *J. Mach. Learn. Res.* **2002**, *3*, 583–617.
56. Zhao, X.; Liang, J.; Dang, C. Clustering ensemble selection for categorical data based on internal validity indices. *Pattern Recognit.* **2017**, *69*, 150–168. [CrossRef]

Article

Developing a New Robust Swarm-Based Algorithm for Robot Analysis

Abubakar Umar [1], Zhanqun Shi [1,*], Alhadi Khlil [1] and Zulfiqar I. B. Farouk [2]

1 School of Mechanical Engineering, Hebei University of Technology, Tianjin 300401, China; 201541201002@stu.hebut.edu.cn (A.U.); 201740000010@stu.hebut.edu.cn (A.K.)

2 School of Mechanical engineering, Tianjin University, Tianjin 300350, China; zulfiqarbibifarouk@yahoo.co.uk

* Correspondence: z_shi@hebut.edu.cn; Tel.: +86-2260438217

Received: 1 November 2019; Accepted: 2 January 2020; Published: 22 January 2020

Abstract: Metaheuristics are incapable of analyzing robot problems without being enhanced, modified, or hybridized. Enhanced metaheuristics reported in other works of literature are problem-specific and often not suitable for analyzing other robot configurations. The parameters of standard particle swarm optimization (SPSO) were shown to be incapable of resolving robot optimization problems. A novel algorithm for robot kinematic analysis with enhanced parameters is hereby presented. The algorithm is capable of analyzing all the known robot configurations. This was achieved by studying the convergence behavior of PSO under various robot configurations, with a view of determining new PSO parameters for robot analysis and a suitable adaptive technique for parameter identification. Most of the parameters tested stagnated in the vicinity of strong local minimizers. A few parameters escaped stagnation but were incapable of finding the global minimum solution, this is undesirable because accuracy is an important criterion for robot analysis and control. The algorithm was trained to identify stagnating solutions. The algorithm proposed herein was found to compete favorably with other algorithms reported in the literature. There is a great potential of further expanding the findings herein for dynamic parameter identification.

Keywords: PSO; robot; manipulator; analysis; kinematic parameters; identification

1. Introduction

The quest for developing improved techniques for parameter identification of industrial robots has resulted in the novel concept of a mutating particle swarm optimization (MuPSO) based algorithm for analyzing multi-degree of freedom robot manipulators which was briefly introduced in [1], the research sought to employ artificial intelligence, particularly population-based Evolutionary Algorithms (EA), and computational methods for solving kinematic and dynamic problems of industrial manipulators. A robot manipulator is an electro-mechanical device that depicts the upper human limb. It was originally used in industrial workspaces to carry out tasks that were deemed boring, repetitive, highly monotonous, or dangerous, and therefore not suitable for human labor. Recent applications of robot manipulators include aeronautics and medicine, where the tolerance is very tight and human errors could be fatal. A robot manipulator comprises of solid non-moveable links connected by joints which allow either rotational or translational motion between successive links.

The increasing demand for robot manipulators has required that the manipulators become more autonomous and therefore increased accuracy and stability. The kinematic problems of robot manipulators were traditionally computed using analytical techniques which sometimes required finding the derivative of computationally expensive functions. These problems were found to sometimes possess multiple solutions or even no solution at all. Recently, swarm-based techniques have been studied which promises improved computational efficiency.

1.1. Swarm Intelligence

Evolutionary computation algorithms (EA) are stochastic optimization methods which have proven suitable for solving complex structured optimization and combinatory problems typical of robot analysis. They are biologically inspired population-based techniques that have relatively simple structures which are robust and computationally efficient. A variety of these algorithms have been developed over the years but based on simple implementation and the ability to readily combine with other algorithms, the PSO algorithm stands out. PSO was initially introduced by [2], despite being amongst the earliest EA algorithms, PSO remains relevant as it is still being improved, enhanced, and modified for solving real-world optimization problems.

In additive manufacturing (3D printing), constructing over-hanging features can only be achieved by introducing some support structures beneath the overhang which can be removed afterward to get the desired shape, [3] used a hybrid variation of PSO with greedy algorithm to reduce the volume of the support structure thereby save printing time, material and minimize budget. The recent hype in groundbreaking fifth-generation (5G) wireless communication technology presents the need to improve the quality of service with massive multiple-input multiple-output (MIMO) antenna arrays, [4] used a contraction adaptive PSO to optimize the design and positions of antenna array elements. Inspired by the success achieved by the proportional integral derivative (PID) controller in automation and its vast industrial applications [5–7] attempted using PID techniques to improve the performance of PSO. Reference [8] proposed the novel PID based strategy PSO (PBSPSO) algorithm based on the PID controller which was found to improve convergence while reducing stagnation of the PSO algorithm. A new variant of PSO with cross-over operation (PSOCO) was introduced by [9] which improved divergent search abilities of the PSO while avoiding stagnation by implementing a new learning model for the particles' velocity formula and two cross-over operations, while [10] used PSO and multi-objective PSO to develop a two-stage auto-tuning technique for PID controllers. Automation of logistics, maintenance/support, storage, and others have led to rapid improvements in the vehicle routing problem (VRP) algorithm. The pick-up and delivery problem (PDP) is an extension of the VRP which collects goods from suppliers or pick-up points and conveys them to the delivery points, [11] introduced a novel pick-up and delivery problem with transfers (PDPT) algorithm using a hybrid PSO and local search algorithm to minimize distance and maximize profit. A PSO variant based on random perturbation (RP-PSO) was used to identify the parameters of a model pressurized water reactor nuclear power plant in [12]. The estimation of distribution algorithm (EDA) framework has been demonstrated in [13] to have high performance despite little memory requirements, it was combined with PSO in [14] and was used to estimate and preserve the distribution information of particles' historical memories (personal best positions) to help the algorithm break out of local minimum solutions. The particle swarm estimation of distribution algorithms (PSDA) was also implemented in [15] for optimal-driven-projection of automated medical diagnosis and prognosis. Medical diagnosis is a key process in clinical medicine for identifying diseases, reducing cost, and enhancing accuracy. Enhanced algorithms were also exploited in [16–18] for diagnosis. Still, in medical sciences, minimally invasive surgery is a cost-effective alternative to open surgery where specialized instruments are used to operate by inserting them into several tiny punctures instead of one large incision. Reference [19] combined PSO with a back-propagation neural network (BPNN) algorithm to optimize the target position of the medical puncture robot.

1.2. Particle Swarm Optimization

The PSO consists of population members known as particles. Each particle refers to a bird in a flock, fish in a swarm, or in this case a possible solution to an optimization problem. The algorithm is initiated by populating it with n random particles, each particle has m dimensions, for the sake of robot analysis, the dimensions would be regarded as the degree of freedom (DOF). The position and velocity vectors of the ith particle can be defined as X_i and V_i in Equations (1) and (2) below. The position and velocity of every particle in the swarm is updated according to Equations (3) and (4) through every iteration, the first part of the Equation (3) is the previous velocity which describes the particles previous experience, the second part of the equation is the cognitive component which describes the particles personal experience while the third part of the equation is the social component which describes the entire swarms best experience. The inertia weight w is a learning coefficient associated with the previous velocity, while c_1 and c_2 are learning coefficients associated with the cognitive and social components respectively. The fitness function is a mathematical representation of the real-world problem to be analyzed, it evaluates how well the particles adapt to the actual solution of the problem. The leader of the swarm (fittest particle) is the particle that best adapts to the solution as it is updated through every iteration. The algorithm keeps a record of the solution of the fittest particle and also that of the best position achieved by each particle. The personal best position ever achieved by ith particle and global best position of the swarm is defined as P_{iBest} and G_{iBest} in Equations (5) and (6). The particles in the swarm would be seen to consistently move towards the solution of the problem through every iteration by updating the particle's position, Equation (4), towards these two best memories (P_{iBest} and G_{iBest}).

$$X_i = (x_{i1}, x_{i2}, \ldots, x_{im}), \tag{1}$$

$$V_i = (v_{i1}, v_{i2}, \ldots, v_{im}), \tag{2}$$

$$V_i(t+1) = w \cdot r \cdot V_i(t) + c_1 \cdot r_1(P_{iBest}(t)) - X_i(t) + c_2 \cdot r_2(G_{iBest}(t) - X_i(t)), \tag{3}$$

$$X_i(t+1) = X_i(t) + V_i(t+1), \tag{4}$$

$$P_{iBest} = (p_{iBest}, p_{iBest}, \ldots, p_{iBest}), \tag{5}$$

$$G_{iBest} = (g_{iBest}, g_{iBest}, \ldots, g_{iBest}), \tag{6}$$

where r_j is a randomly generated number between [0, 1] and $j \in [0, 1, 2]$. PSO has a high convergence speed which is very desirable for robot applications, but this convergence speed sometimes results in stagnation which is a major limitation of the PSO algorithm. The characteristic of the PSO algorithm that allows it to define promising regions in search space is referred to as exploration while exploitation allows it to refine solutions within the defined promising region. These are the major characteristics of any PSO algorithm, a good algorithm balances these properties to find the best solution to a problem while avoiding stagnation. The contributions of [20] showed that these properties can be tuned by carefully selecting the value of w of the PSO algorithm. The concept of constriction coefficient was introduced, setting the inertia weight at *0.712* while the cognitive and social coefficient were both *1.494*. This version of the PSO has come to be known today as the standard PSO (SPSO). The biological background of SPSO is believed to have evolved from the bird-like objects or BOIDS introduced by [21] to simulate flocking birds or animals in virtual reality studios. The BOIDS was governed by three basic rules; separation, alignment, and cohesion. The SPSO ignored the alignment and cohesion rules to reduce computational cost and increase convergence speed. Reference [22] proposed re-incorporating these rules reduces the convergence speed which is advantageous in pushing the algorithm out of stagnation. The effects of topologies on PSO were studied in [23], the SPSO has a star topology, where every individual is connected to other individuals such that information or direction of search is communicated and implemented throughout the swarm. Observing that this topology allows too much communication between the swarm particles and may be responsible for the quick convergence and stagnation of the SPSO, Reference [24] investigated the circle, wheel, and random topologies which isolate the individual particles of the swarm at different degrees so that information is communicated to the swarm by the focal individual or short-cuts between the isolated groups thereby causing a buffering effect which reduces the convergence speed of the swarm and improves search results. In [25], it was shown that the success of individual particles is not as a result of only the particle with the best fitness but by the influence of the entire swarm to a certain degree. An algorithm was presented that allowed every particle to have a weighted influence on other particles, based on their fitness such that particles with higher fitness exerted more influence. References [26,27] proposed the combination of SPSO with other computational search methods like conjugate gradient and steepest descent respectively. These variations of PSO also aimed at slowing down the convergence speed by allowing the algorithm to stop and search for promising regions in the local space, while [28] successfully merged PSO with ABC for nonlinear statistical analysis.

Swarms are best suited for analyzing static search spaces with one global solution (unimodal). In reality, most robot analysis problems contain more than one local solution (multimodal) and the search space is sometimes dynamic, therefore maintaining diversity is crucial for the performance of PSO. Various strains of PSO involving sub-swarms have been developed for this purpose, PSO was combined with expanding neighborhood topology in [29] to solve the permutation flow-shop scheduling problem. The algorithm is initiated with sub-swarms of small size neighborhoods, slowly expanding through every iteration, absorbing other particles, taking advantage of both the global and local neighborhood structures to increase the performance of the PSO algorithm. Competitive strategy was used in [30] to manage convergence, while entropy measurement was employed to maintain the diversity of the swarm. In [31], an adaptive multi-swarm competition PSO was proposed where the swarm is adaptively divided into sub-swarms and a competition mechanism is used to maintain diversity in the swarms. The sub-swarms slowly converge, adaptively reducing the number of swarms while balancing between exploration and exploitation tendencies. Other algorithms that employed sub-swarms include [32–35]. Although multi-swarm based algorithms were found to be efficient for solving multimodal problems, these algorithms have very high computational cost [33]. The new trends of adaptive SPSO where the inertia weight and acceleration components are altered during the search process is capable of improving exploration and exploitation tendencies with less computational cost [36]. In [37–40] the parameters of the adaptive PSO were dependent on the quality of the solution and tailored to the specific problem, the parameters were updated by comparing the values of the best particles (P_{iBest} *and* G_{iBest}). This technique was found favorable in analyzing both static and dynamic search spaces without incurring too much computational cost.

1.3. PSO and Robot Parameter Identification

Over the years, the use of PSO for robot parameter identification has been researched, a comparison between the linear least squares (LLS) method and the PSO was presented in [41] for the dynamic parameter identifications of a 3DOF Staubli RX-60 robot manipulator, where the PSO was found to produce better results. Reference [42] combined the linear simplification of the LLS and the non-linear optimization of the PSO for online and offline parameter identification of space robots. Space robots encounter changes in their kinematic parameters while running in the orbit. The non-linear model is first used for parameter identification in the offline mode while the LLS is used for online identification in the follow-up mission knowing that the parameters would not change much. These works and other similar research works exhibit the superiority of intelligent swarm-based techniques over traditional methods. A hybridized genetic algorithm and PSO (GAPSO) was implemented by [43] for parameter identification of a SCARA robot, [44] also implemented a hybridized BPNN and PSO for determining the kinematic parameters of a 6DOF robot manipulator, while [45] investigated the performance of seven PSO variants in solving the inverse kinematics of 2DOF robots. In [46,47], a combination of PSO and simulated annealing (SA) was used to optimize the geometric structure of non-redundant 6DOF manipulators. In [48] the Elitist Learning Strategy PSO (ELS-PSO) for dynamic parameter analysis of a 3DOF Staubli RX-60 robot manipulator. The Quantum-Behaved PSO (QPSO) was implemented for parameter identification of a puma 560 robot by [49] in two steps by first optimizing the individual joint parameters so that the identified values are close to the theoretical values, then all the joint parameters are further optimized simultaneously around the previously converged values. During the course of this research, it was observed that the solution for robot kinematic parameter identification problem did not converge under the basic parameters of the PSO especially when there were more than three degrees of freedom, and most published works implementing the PSO for robot parameter identification either used lower DOF robots or the algorithms were enhanced, modified and hybridized usually for specific robot manipulators, these algorithms are often not applicable for other robot configurations. Therefore, the concept of a novel Mutating PSO (MuPSO) algorithm was conceived for analyzing robot kinematics. To the best of our knowledge, there has not been any research tailored at determining the best range of PSO parameters to develop an algorithm for robot analysis, therefore this work aims to develop a new PSO variant capable of analyzing all robot configurations and least likely to fall into stagnation. This was achieved by first studying the behavior of PSO under various robot configurations and determining a new range of parameters for robot kinematic analysis, then a suitable adaptive technique was investigated and finally, the mutation function was implemented. A total of 54 different PSO parameters were tested on 6 robot manipulators. The rest of this paper is organized as follows; Section 2 presents the kinematic model of the robot configurations to be studied and the fitness function for the algorithm was formulated. Experiments studying the behavior of these robot configurations under various parameters were conducted in Section 3, the new adaptive strategy is presented in Section 4, comparing it with other variations of PSO. The results are presented in Section 5, the Mutation function was introduced in Section 6 and in Section 7 conclusions were drawn, while Section 8 presents the future thrust.

2. Kinematic Model of Robots

To determine a new set of parameters for robot analysis, the effect of four popular robot configurations was studied under various parameters. The robot configurations include the Articulate, Stanford, SCARA, and Dual-Arm robot configurations. These robot configurations were implemented on six different robot manipulators; the articulate configuration was implemented on a 3DOF robot manipulator because it is regarded as the most complex spatial robot configuration. The articulate robot configuration was also implemented on two different 6DOF robot manipulators of different sizes to study the effect of size on the PSO parameters.

2.1. Robot Configurations

Industrial manipulators usually have 6DOF as this gives the manipulator optimum dexterity, allowing it to complete most tasks in an industrial workspace. A robot with less than 6DOF is deficient, as it is easier to analyze and control but it cannot reach all the possible positions and orientations in its workspace. A redundant robot possesses more than 6DOF, they are more flexible, capable of maneuvering behind obstacles but more expensive to analyze and control. The SCARA manipulator is an example of a deficient robot manipulator, articulate manipulators are usually 6DOF while the dual-arm robot is a redundant manipulator (usually greater than 6DOF). It would be keen to note that the presence of a prismatic joint in any robot configuration simplifies the analysis while complicating the solution, it requires less computation but, like redundant configurations, there is a possibility of having numerous or even infinite solutions to every problem. The joints and end-effector of robots are always oriented along the z-axis of the coordinate frame.

- Scara Robot Configuration: The Selective Compliance Assembly Robot (SCARA) is one of the earliest industrial manipulators patented in 1981, it is usually a 4- or 5-DOF manipulator with all revolute joint, except one. In this analysis, a 4DOF SCARA manipulator shall be analyzed. The first joint of the manipulator being the prismatic joint, the other joints are revolute, parallel to each other and pointing along the direction of gravity. The SCARA manipulator is shown in Figure 1a and the DH-parameters are tabulated in Table 1.
- Articulate Robot Configuration: The articulate manipulator is the most popular robot configurations used in industrial workspaces, its analysis and solutions are trivial, therefore a very high accuracy can be achieved. All the joints of the articulated robot are revolute with the first joint pointing along the direction of gravity.

  The second third and fifth joints are parallel to each other and perpendicular to the first joint axis. The fourth and sixth joints are coincident and perpendicular to all the other joints. Three articulated manipulators were used in this analysis, one 3DOF articulated manipulator and two 6DOF articulated manipulators with significantly different sizes. The 3DOF articulate manipulator is configured exactly like the first three joints of the 6DOF articulate manipulator previously described. Figure 1b shows the 3DOF while Figure 2a shows the 6DOF robot configurations, their D-H parameters are tabulated in Tables 3–5.

- Stanford Robot Configuration: The Stanford manipulator is also a 6DOF manipulator with five revolute joints and one prismatic joint. It is configured very much like the articulate arm except that the third arm is prismatic. The Stanford manipulator is shown in Figure 2b with its detailed D-H parameters defined in Table 2.
- Dual-Arm Robot Configuration: The dual-arm robot is the most dexterous of the manipulators with a total of 17DOF which are all revolute. It is configured such that the first joint at the 'base' of the structure is pointing along the direction of gravity. The third joint is parallel to the first joint and coincident with the fourth joint. The fifth, seventh, and ninth joints are also coincident with the third joint.

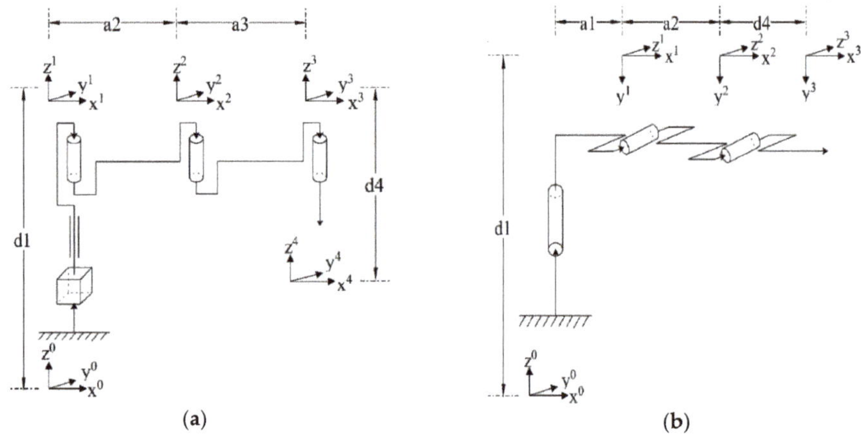

Figure 1. (**a**) 4DOF SCARA arm. (**b**) 3DOF Articulate arm.

Table 1. D-H Parameters for 4DOF SCARA arm.

Joint	Link	Off-Set	Joint Displacement [1]	Off-Set Displacement [2]
1	0	Variable (0–300)	0	0
2	350	0	Variable (−pi/2–pi/2)	0
3	350	0	Variable (−pi/2–pi/2)	0
4	0	200	Variable (−pi/2–pi/2)	0

For all DH parameters; [1] joint displacement is Theta, [2] off-set displacement of joint is Alfa, see Equations (7) and (8).

Table 2. D-H parameters for 6DOF Stanford arm.

Joint	Link	Off-Set	Joint Displacement	Off-Set Displacement
1	154	412	Variable (−pi/2–pi/2)	−pi/2
2	0	0	Variable (−pi/2–pi/2)	0
3	0	Variable (50–154)	−pi/2	−pi/2
4	0	0	Variable (−pi/2–pi/2)	pi/2
5	0	0	Variable (−pi/2–pi/2)	−pi/2
6	0	263	Variable (−pi/2–pi/2)	0

Table 3. D-H Parameters for 3DOF articulate arm.

Joint	Link	Off-Set	Joint Displacement	Off-Set Displacement
1	70	200	Variable (−pi/2–pi/2)	−pi/2
2	150	0	Variable (−pi/2–pi/2)	0
3	150	0	Variable (−pi/2–pi/2)	0

Table 4. D-H Parameters for 6DOF articulate arm.

Joint	Link	Off-Set	Joint Displacement	Off-Set Displacement
1	64.5	170	Variable (−pi/2–pi/2)	−pi/2
2	305	0	Variable (−pi/2–pi/2)	0
3	0	0	Variable (−pi/2–pi/2)	pi/2
4	0	−220	Variable (−pi/2–pi/2)	−pi/2
5	0	0	Variable (−pi/2–pi/2)	pi/2
6	0	−36	Variable (−pi/2–pi/2)	0

Table 5. D-H parameters for large 6DOF articulate arm

Joint	Link	Off-Set	Joint Displacement	Off-Set Displacement
1	300	320	Variable (−pi/2–pi/2)	−pi/2
2	700	0	Variable (−pi/2–pi/2)	0
3	0	0	Variable (−pi/2–pi/2)	−pi/2
4	0	697.5	Variable (−pi/2–pi/2)	pi/2
5	0	0	Variable (−pi/2–pi/2)	−pi/2
6	0	127.5	Variable (−pi/2–pi/2)	0

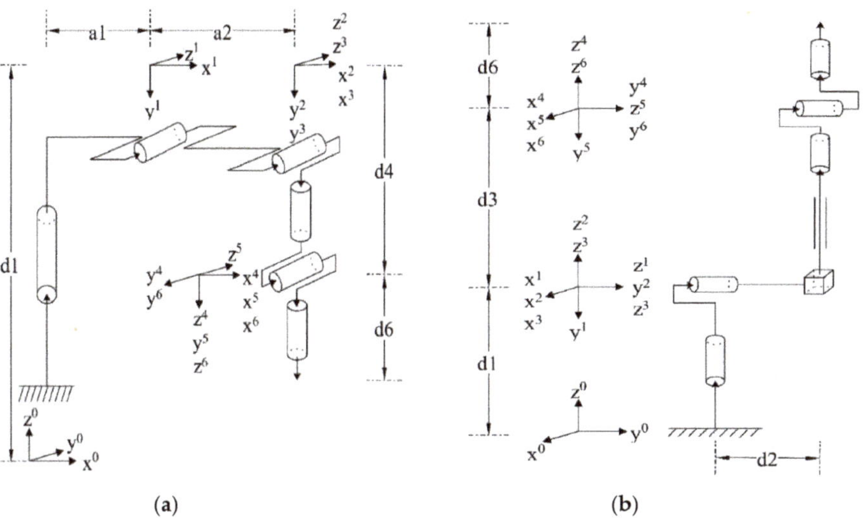

Figure 2. (**a**) 6DOF articulate arm. (**b**) 6DOF Stanford arm.

The fourth, sixth, and eight joints are parallel to each other and perpendicular to the first joint while the second joint is perpendicular to all the other joints. Figure 3 illustrates the dual-arm robot manipulator while Table 6 shows the D-H parameters of the robot.

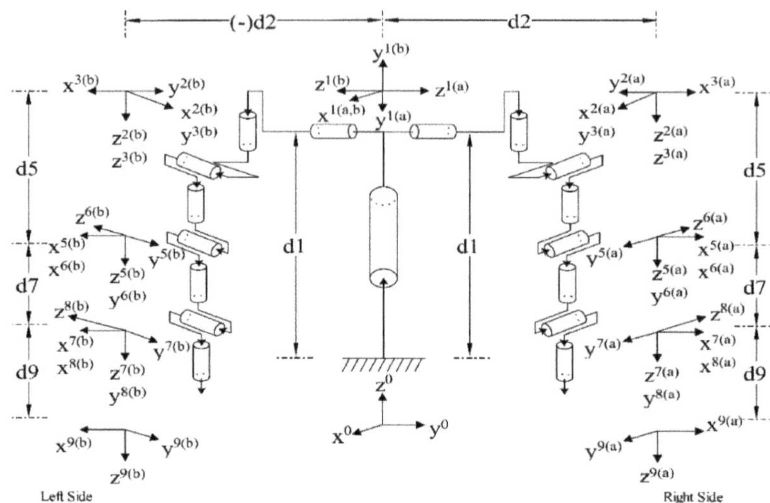

Figure 3. 17DOF dual-arm.

Table 6. D-H parameters for 17DOF dual-arm.

Joint	Link	Off-Set	Joint Displacement	Off-Set Displacement
1	0	400	th1	pi/2
2	0	200 −200	Variable (−pi/2–pi/2) (a) Variable (−pi/2–pi/2) (b)	pi/2 −pi/2
3	0	0 0	pi/2) (a) −pi/2) (b)	0 0
4	0	0 0	Variable (−pi/2–pi/2) (a) Variable (−pi/2–pi/2) (b)	−pi/2 pi/2
5	0	150 −150	Variable (−pi/2–pi/2) (a) Variable (−pi/2–pi/2) (b)	pi/2 −pi/2
6	0	0 0	Variable (−pi/2–pi/2) (a) Variable (−pi/2–pi/2) (b)	−pi/2 pi/2
7	0	150 −150	Variable (−pi/2–pi/2) (a) Variable (−pi/2–pi/2) (b)	pi/2 −pi/2
8	0	0 0	Variable (−pi/2–pi/2) (a) Variable (−pi/2–pi/2) (b)	−pi/2 pi/2
9	0	150 −150	Variable (−pi/2–pi/2) (a) Variable (−pi/2–pi/2) (b)	0 0

2.2. Fitness Function

The homogeneous matrix of each successive pair of frames can be obtained using the formula in (7) below from the D-H parameters. the transformation matrix T for the robot manipulator's end effector is a product of post multiplication as shown in the formula in (8).

$$T_{k-1}^{k} = \begin{bmatrix} \cos\theta_k & -\sin\theta_k & 0 & a_k \\ \sin\theta_k\cos\alpha_{k-1} & \cos\theta_k\cos\alpha_{k-1} & -\sin\alpha_{k-1} & -d_k\sin\alpha_{k-1} \\ \sin\theta_k\sin\alpha_{k-1} & \cos\theta_k\sin\alpha_{k-1} & \cos\alpha_{k-1} & d_k\cos\alpha_{k-1} \\ 0 & 0 & 0 & 1 \end{bmatrix}, \tag{7}$$

$$T_0^{dof} = \begin{bmatrix} t_{11} & t_{12} & t_{13} & t_{14} \\ t_{21} & t_{22} & t_{23} & t_{24} \\ t_{31} & t_{32} & t_{33} & t_{34} \\ 0 & 0 & 0 & 1 \end{bmatrix} = T_0^1 T_1^2 \cdots T_{dof-1}^{dof}, \tag{8}$$

$$A = \begin{bmatrix} -0.5161 & 0.2261 & 0.8262 & 349.5064 \\ 0.7185 & 0.6394 & 0.2739 & 1419.7 \\ -0.4663 & 0.7349 & -0.4924 & -516.5116 \\ 0 & 0 & 0 & 1 \end{bmatrix}, \quad (9)$$

$$f_{ij} = t_{ij} - a_{ij}, \quad (10)$$

$$Fitness = \sum_{i}^{n=3} \sum_{j}^{m=4} (f_{ij} - E), \quad (11)$$

If the actual values of T read from sensors attached to the robot's end-effector is given in A, then the fitness function (*Fitness*) would be as described in Equations (10) and (11). Where $k \in [1, 2, \ldots, \text{dof}]$, subscripts $i, j \in [1, 2, \ldots, 4]$ and $E = 1 \times 10^{-8}$. Most robots are usually fitted with encoders, gyroscopes, and current controllers which can measure joint position, end-effector orientation, and actuator currents (torque) respectively. There are a variety of sensors that can also be mounted manually on robots.

3. Determining New PSO Parameters

An experiment aimed at studying the behavior of all popular robot configurations on different PSO parameters was performed to identify the best values of w and c that balances out exploration and exploitation tendencies while ensuring convergence of results and also to determine the solutions with the best computational efficiency. The initial value of w (w_i) was set to 0.7, increasing with an interval of 0.4 to a final value (w_f) at 1.5, such that w = 0.7:0.4:1.5. The initial value of c (c_i) was set at 1.5, increasing with an interval of 0.3 to a final value (c_f) at 3.9, such that c = 1.5:0.3:3.9 as elaborated in Figure 4. Then the experiment was repeated for c = 1.4:0.6:2.6 and w = 0.6:0.3:3.0. The 6 robot configurations were tested with 54 sets of PSO parameters in 30 generations and 2000 iterations. The mutation function was not implemented in this experiment, the results were tabulated and the performance of the PSO plotted. In Table 7, the average and standard deviation of the best solutions after thirty runs and the solution that best minimizes the problem were presented, the average number of iterations required to find the best solution for each of the six robot manipulators was also presented at w = *0.7*. Tables 8 and 9 present a similar set of results for w at 1.1 and 1.5, respectively. To ease comparison and visualization of the results, a summary is presented in Table 10 showing the averages of normalized values obtained in Tables 7–9 while Figure 5a–f shows a pictorial plot of the performance of PSO for each of the robot manipulators. Similarly, Tables 11–14 and Figure 6a–f present the results and plots for the seco–nd experiment. The minimum solution for each robot manipulator problem was reported in this analysis because the average solutions (after 30 runs) usually reported in other works of literature does not completely capture the results obtained from the experiment especially in the SCARA, Stanford and dual-arm configurations where there is a possibility of multiple solutions to every problem. It can be shown that the average best solution alone is not enough to make a good comparison between the different scenarios. If the minimum solution presented in the tables represent the probability for the given parameters to find the minimum solution whereas the average best solution represents the probability for the solution to run into stagnation, then it can be shown that some parameters with very competitive average solutions are not capable of finding the minimum (global best) solution.

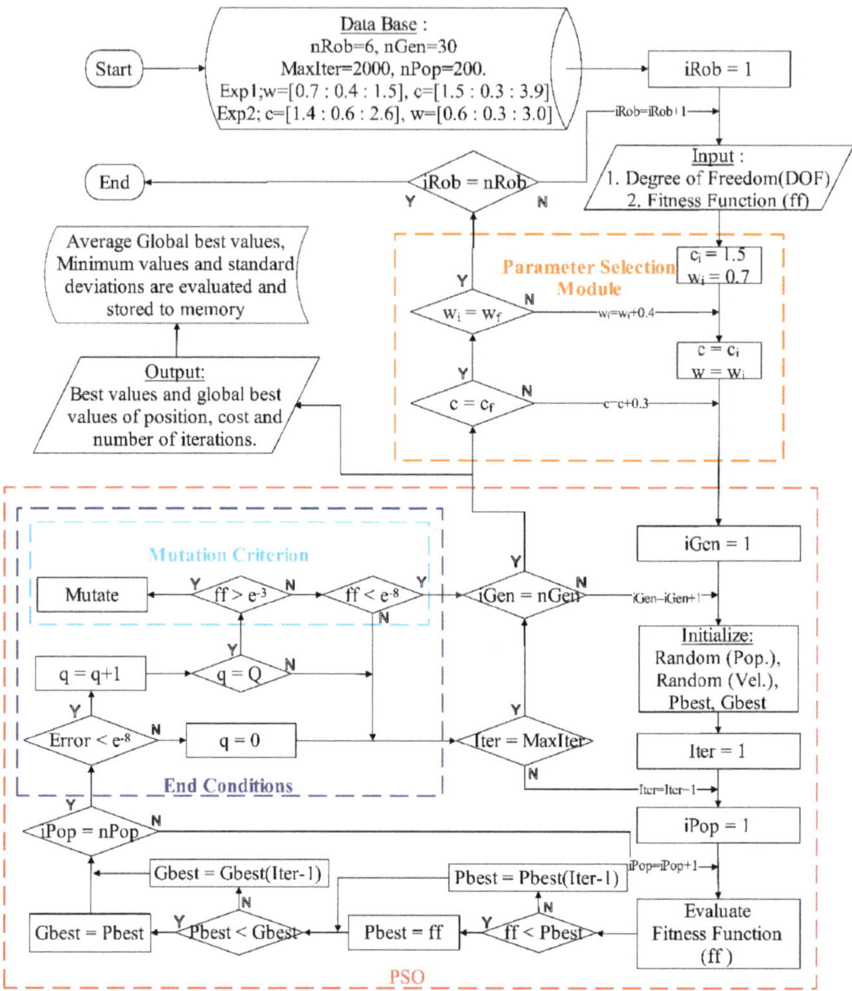

Figure 4. Flowchart of parameter selection experiment for the proposed PSO variant elaborating the end conditions and mutation criterion.

Table 7. Performance of PSO when $w = 0.7$.

Robot Configuration		PSO Parameters @ $w = 0.7$								
		$c = 1.5$	$c = 1.8$	$c = 2.1$	$c = 2.4$	$c = 2.7$	$c = 3.0$	$c = 3.3$	$c = 3.6$	$c = 3.9$
3DOF Articulate Robot Arm	*Average (Std)*	1.1999 (0.996)	0.66662 (0.958)	0.99993 (1.017)	0.99993 (1.017)	0.8666 (1.008)	1.0666 (1.015)	0.99993 (1.017)	0.79994 (0.996)	0.8666 (1.008)
	Minimum	6.10×10^{-9}	6.10×10^{-9}	6.10×10^{-9}	6.10×10^{-9}	6.10×10^{-9}	6.10×10^{-9}	6.10×10^{-9}	6.10×10^{-9}	6.10×10^{-9}
	Iteration	74.8	74.1	74.333	74.4	75.233	75.867	77.133	78.4	81.333
4DOF SCARA Robot Arm	*Average (Std)*	2.1333 (2.029)	1.7333 (2.016)	1.4667 (1.960)	2.1333 (2.029)	2.1333 (2.029)	2.2667 (2.016)	2.2667 (2.016)	1.7333 (2.02)	1.3333 (1.918)
	Minimum	9.73×10^{-10}	9.73×10^{-10}	9.73×10^{-10}	9.73×10^{-10}	9.73×10^{-10}	9.73×10^{-10}	9.73×10^{-10}	9.73×10^{-10}	9.73×10^{-10}
	Iteration	78.967	78.067	78.467	79.233	79.5	81.233	82.733	85.433	87.267
6DOF Stanford Robot Arm	*Average (Std)*	2.4859 (2.145)	1.2563 (0.992)	1.2951 (1.049)	0.85651 (1.123)	0.28667 (0.751)	0.18892 (0.590)	0.10886 (0.509)	0.08696 (0.388)	0.015589 (0.0401)
	Minimum	4.67×10^{-4}	3.21×10^{-2}	1.05×10^{-4}	4.31×10^{-7}	1.68×10^{-6}	1.32×10^{-6}	5.80×10^{-9}	5.80×10^{-9}	5.80×10^{-9}
	Iteration	533.4	508.3	554.1	511.8	417.4	601.4	567.6	651.57	643.47
6DOF Articulate Robot Arm (small)	*Average (Std)*	3.7837 (1.759)	1.5927 (1.237)	1.1403 (0.633)	0.80981 (0.506)	0.59953 (0.780)	0.24822 (0.412)	0.27092 (0.457)	0.17026 (0.385)	0.31949 (0.596)
	Minimum	1.11×10^{0}	3.48×10^{-1}	9.21×10^{-2}	8.27×10^{-2}	2.83×10^{-9}	2.83×10^{-9}	2.83×10^{-9}	2.83×10^{-9}	2.83×10^{-9}
	Iteration	261.87	235.43	265.87	260.93	296.7	214.53	166.63	170.67	157.63
6DOF Articulate Robot Arm (big)	*Average (Std)*	3.4266 (1.732)	2.9774 (1.649)	1.5511 (0.902)	1.3035 (1.180)	0.86441 (1.095)	0.43759 (0.784)	0.30934 (0.458)	0.35057 (0.680)	0.35187 (0.456)
	Minimum	6.53×10^{-1}	7.02×10^{-1}	5.70×10^{-2}	7.09×10^{-2}	6.98×10^{-4}	5.53×10^{-9}	5.53×10^{-9}	5.53×10^{-9}	5.53×10^{-9}
	Iteration	244.33	230.5	285.63	343.43	416.5	375.47	327.27	323.77	377.8
17DOF Dual-Arm Robot Arm	*Average (Std)*	2.873 (2.307)	1.7774 (1.672)	0.71947 (0.803)	0.51867 (0.569)	0.25626 (0.425)	0.13317 (0.388)	0.14343 (0.469)	0.14844 (0.678)	0.13608 (0.419)
	Minimum	2.35×10^{-2}	1.81×10^{-2}	3.90×10^{-3}	1.99×10^{-4}	4.43×10^{-5}	3.82×10^{-9}	3.82×10^{-9}	3.82×10^{-9}	3.82×10^{-9}
	Iteration	326.07	378.17	364.87	390.77	451.3	437.33	447.07	457.43	468.83

Table 8. Performance of PSO when $w = 1.1$

Robot Configuration		PSO Parameters @ $w = 1.1$								
		$c = 1.5$	$c = 1.8$	$c = 2.1$	$c = 2.4$	$c = 2.7$	$c = 3.0$	$c = 3.3$	$c = 3.6$	$c = 3.9$
3DOF Articulate Robot Arm	*Average (Std)*	1.0666 (1.015)	1.1999 (0.996)	0.79994 (0.996)	0.99993 (1.017)	0.66662 (0.959)	1.3332 (0.959)	1.2666 (0.980)	1.1333 (1.008)	1.0666 (1.015)
	Minimum	6.10×10^{-9}	6.10×10^{-9}	6.10×10^{-9}	6.10×10^{-9}	6.10×10^{-9}	6.10×10^{-9}	6.10×10^{-9}	6.10×10^{-9}	6.10×10^{-9}
	Iteration	81.6	81.7	81.267	82.267	83.9	85.733	87.667	91.1	96.533
4DOF SCARA Robot Arm	*Average (Std)*	2 (2.034)	2.2667 (2.016)	1.7333 (2.016)	2.1333 (2.03)	2.2667 (2.016)	1.8667 (2.03)	2.2667 (2.016)	2 (2.034)	2.2667 (2.016)
	Minimum	9.73×10^{-10}	9.73×10^{-10}	9.73×10^{-10}	9.73×10^{-10}	9.73×10^{-10}	9.73×10^{-10}	9.73×10^{-10}	9.73×10^{-10}	9.73×10^{-10}
	Iteration	88.667	87.7	88.633	90.133	91.6	93.367	97.767	101.97	106
6DOF Stanford Robot Arm	*Average (Std)*	0.75869 (1.099)	0.71229 (1.079)	0.37784 (1.009)	0.27387 (0.758)	0.28313 (0.852)	0.02518 (0.059)	0.53394 (1.008)	0.19628 (0.561)	0.17413 (0.519)
	Minimum	7.02×10^{-5}	1.06×10^{-5}	4.14×10^{-6}	5.80×10^{-9}	5.80×10^{-9}	5.80×10^{-9}	5.80×10^{-9}	5.80×10^{-9}	5.80×10^{-9}
	Iteration	661.17	637.47	581.6	590.93	548.53	648.57	585.13	648.67	810.17
6DOF Articulate Robot Arm (small)	*Average (Std)*	1.3059 (1.047)	0.71658 (0.695)	0.27506 (0.580)	0.46958 (0.945)	0.22999 (0.387)	0.32675 (0.611)	0.30032 (0.452)	0.46418 (0.779)	0.22069 (0.412)
	Minimum	9.23×10^{-3}	3.23×10^{-3}	2.83×10^{-9}	2.83×10^{-9}	2.83×10^{-9}	2.83×10^{-9}	2.83×10^{-9}	2.83×10^{-9}	2.83×10^{-9}
	Iteration	375.67	385.77	367.23	193.1	186.47	162.57	178.2	190.97	246.33
6DOF Articulate Robot Arm (big)	*Average (Std)*	1.5539 (1.113)	0.82201 (0.625)	0.46599 (0.473)	0.30729 (0.433)	0.52439 (1.167)	0.49672 (0.635)	0.33982 (0.588)	0.48452 (0.488)	0.70028 (1.07)
	Minimum	6.30×10^{-2}	2.71×10^{-2}	3.28×10^{-7}	5.53×10^{-9}	5.53×10^{-9}	5.53×10^{-9}	5.53×10^{-9}	5.53×10^{-9}	5.53×10^{-9}
	Iteration	341.5	532.1	553.77	405.8	311.9	293.23	352.83	315.17	483.03
17DOF Dual-Arm Robot Arm	*Average (Std)*	0.4248 (0.48)	0.09798 (0.131)	0.14656 (0.316)	0.11974 (0.248)	875.8 (4796.4)	0.068428 (0.151)	723.89 (3964.6)	0.18718 (0.324)	0.12041 (0.269)
	Minimum	5.31×10^{-3}	3.60×10^{-4}	5.10×10^{-5}	3.82×10^{-9}	3.82×10^{-9}	3.82×10^{-9}	3.82×10^{-9}	3.82×10^{-9}	3.82×10^{-9}
	Iteration	501.6	562.43	483.93	444.2	432	422.9	422.1	440	611.93

Table 9. Performance of PSO when $w = 1.5$.

Robot Configuration		PSO Parameters @ $w = 1.5$								
		$c = 1.5$	$c = 1.8$	$c = 2.1$	$c = 2.4$	$c = 2.7$	$c = 3.0$	$c = 3.3$	$c = 3.6$	$c = 3.9$
3DOF Articulate Robot Arm	*Average (Std)*	0.8666 (1.008)	0.9999 (1.017)	0.9333 (1.015)	1.0666 (1.015)	0.7999 (0.996)	1.3999 (0.932)	1.0666 (1.015)	0.7333 (0.980)	1.1999 (0.996)
	Minimum	6.10×10^{-9}	6.10×10^{-9}	6.10×10^{-9}	6.10×10^{-9}	6.10×10^{-9}	6.10×10^{-9}	6.10×10^{-9}	6.10×10^{-9}	6.10×10^{-9}
	Iteration	93.633	94.333	95.1	97.6	101.53	104.93	111.3	119.9	130.03
4DOF SCARA Robot Arm	*Average (Std)*	2.4 (1.993)	1.8667 (2.03)	2.1333 (2.03)	2 (2.034)	2.2667 (2.016)	2.5333 (1.960)	1.6 (1.993)	1.7333 (2.016)	2 (2.034)
	Minimum	9.73×10^{-10}	9.73×10^{-10}	9.73×10^{-10}	9.73×10^{-10}	9.73×10^{-10}	9.73×10^{-10}	9.73×10^{-10}	9.73×10^{-10}	9.73×10^{-10}
	Iteration	103.9	103.9	105.03	110.07	113.53	117	126.2	139.67	156.7
6DOF Stanford Robot Arm	*Average (Std)*	0.0169 (0.048)	0.1762 (0.58)	0.0927 (0.387)	0.0793 (0.382)	0.2655 (0.69)	0.1456 (0.462)	0.1750 (0.603)	0.2623 (0.748)	0.0996 (0.241)
	Minimum	5.80×10^{-9}	5.80×10^{-9}	5.80×10^{-9}	5.80×10^{-9}	5.80×10^{-9}	5.80×10^{-9}	5.86×10^{-9}	6.77×10^{-9}	6.72×10^{-7}
	Iteration	784.03	764.1	728.07	689.4	664.9	839.4	1143.2	1273.1	1667.7
6DOF Articulate Robot Arm (small)	*Average (Std)*	0.4613 (0.735)	0.2363 (0.411)	0.3917 (0.629)	0.27185 (0.456)	0.7003 (0.841)	0.29195 (0.6)	0.38275 (0.474)	0.24172 (0.434)	0.42294 (0.499)
	Minimum	2.83×10^{-9}	2.83×10^{-9}	2.83×10^{-9}	2.83×10^{-9}	2.83×10^{-9}	2.83×10^{-9}	2.83×10^{-9}	2.83×10^{-9}	2.90×10^{-9}
	Iteration	235.23	236.67	208.27	213.37	208.53	251.27	331.77	377.17	558.5
6DOF Articulate Robot Arm (big)	*Average (Std)*	0.3755 (0.616)	0.3122 (0.429)	0.2982 (0.602)	0.54127 (0.736)	0.24539 (0.422)	0.33632 (0.472)	0.33908 (0.666)	0.54117 (0.492)	0.60112 (0.79)
	Minimum	5.53×10^{-9}	5.53×10^{-9}	5.53×10^{-9}	5.53×10^{-9}	5.53×10^{-9}	5.53×10^{-9}	5.53×10^{-9}	5.56×10^{-9}	6.22×10^{-9}
	Iteration	490.57	434.43	405.5	375.57	438.6	509.8	695.57	935.57	1374.1
17DOF Dual-Arm Robot Arm	*Average (Std)*	0.03969 (0.091)	0.1555 (0.349)	0.58298 (1.275)	0.12749 (0.280)	0.24011 (0.726)	0.25313 (0.850)	0.089832 (0.219)	0.31591 (0.629)	0.27926 (0.494)
	Minimum	3.82×10^{-9}	3.82×10^{-9}	3.82×10^{-9}	3.82×10^{-9}	3.82×10^{-9}	3.82×10^{-9}	3.82×10^{-9}	3.84×10^{-9}	4.09×10^{-9}
	Iteration	566.6	419.9	412.63	538.6	571.17	587.53	911.87	945.53	1338

Table 10. Aggregate of performance for PSO for different values of inertia weight (w).

Robot Configuration		PSO Parameters								
		$c = 1.5$	$c = 1.8$	$c = 2.1$	$c = 2.4$	$c = 2.7$	$c = 3.0$	$c = 3.3$	$c = 3.6$	$c = 3.9$
3DOF Articulate Robot Arm	$w = 0.7$	0.02513	0.14763	0.06318	0.06297	0.09043	0.04511	0.05457	0.09739	0.07168
	$w = 1.1$	0.08921	0.06846	0.14458	0.09944	0.17200	0.04226	0.04449	0.05379	0.05053
	$w = 1.5$	0.16746	0.14006	0.15103	0.12242	0.16699	0.06913	0.09608	0.14758	0.04076
4DOF SCARA Robot Arm	$w = 0.7$	0.03849	0.08687	0.12197	0.03773	0.03696	0.01897	0.01468	0.06577	0.11672
	$w = 1.1$	0.07029	0.04539	0.10203	0.05269	0.03619	0.07446	0.02166	0.0389	0.00224
	$w = 1.5$	0.10249	0.15062	0.12251	0.12707	0.09746	0.07244	0.14586	0.10839	0.05263
6DOF Stanford Robot Arm	$w = 0.7$	0.29170	0.31301	0.53404	0.58656	0.72349	0.68145	0.71189	0.69607	0.74687
	$w = 1.1$	0.04599	0.28527	0.45162	0.55492	0.54350	0.77809	0.41407	0.60752	0.57449
	$w = 1.5$	0.84829	0.52364	0.67211	0.69225	0.41774	0.58066	0.46007	0.30967	0.32558
6DOF Articulate Robot Arm (small)	$w = 0.7$	0.02935	0.44219	0.58987	0.63601	0.59946	0.74424	0.77674	0.79026	0.76129
	$w = 1.1$	0.00654	0.35941	0.57088	0.55936	0.74264	0.68628	0.71899	0.60129	0.69970
	$w = 1.5$	0.26829	0.44414	0.33662	0.42852	0.16347	0.36172	0.33072	0.37207	0.20066
6DOF Articulate Robot Arm (big)	$w = 0.7$	0.12076	0.15641	0.56486	0.50314	0.52856	0.62945	0.71490	0.68191	0.68178
	$w = 1.1$	0.10744	0.38614	0.57372	0.67457	0.52483	0.65161	0.65997	0.67519	0.44003
	$w = 1.5$	0.33744	0.43315	0.38945	0.25154	0.46242	0.39587	0.29945	0.22558	0
17DOF Dual-Arm Robot Arm	$w = 0.7$	0.07612	0.26993	0.61418	0.68276	0.69043	0.71311	0.69829	0.66969	0.69280
	$w = 1.1$	0.54493	0.75326	0.79984	0.81848	0.32351	0.82719	0.41427	0.82017	0.74995
	$w = 1.5$	0.62593	0.55313	0.18953	0.55625	0.41449	0.38161	0.51477	0.33042	0.28333

Table 11. Performance of PSO when $c = 1.4$.

Robot Configuration		PSO Parameters @ $c = 1.4$								
		$w = 0.6$	$w = 0.9$	$w = 1.2$	$w = 1.5$	$w = 1.8$	$w = 2.1$	$w = 2.4$	$w = 2.7$	$w = 3.0$
3DOF Articulate Robot Arm	*Average (Std)*	0.73328 (0.980)	0.79994 (0.996)	0.66662 (0.959)	1.2666 (0.980)	0.73328 (0.980)	0.93327 (1.015)	0.99994 (1.017)	0.73819 (0.978)	1.0239 (1.018)
	Minimum	6.10×10^{-9}	6.10×10^{-9}	6.10×10^{-9}	6.10×10^{-9}	6.10×10^{-9}	6.10×10^{-9}	1.31×10^{-8}	3.26×10^{-5}	7.19×10^{-4}
	Iteration	74.3	78.033	83.933	93.667	111.1	163.37	232.83	194.1	148.43
4DOF SCARA Robot Arm	*Average (Std)*	1.8667 (2.03)	1.7333 (2.016)	2.2667 (2.016)	2.2667 (2.016)	2 (2.034)	1.7333 (2.016)	2.1337 (2.03)	2.0962 (2.004)	3.1436 (2.001)
	Minimum	9.73×10^{-10}	9.73×10^{-10}	9.73×10^{-10}	9.73×10^{-10}	9.73×10^{-10}	9.75×10^{-10}	1.55×10^{-6}	1.79×10^{-3}	1.40×10^{-2}
	Iteration	79.5	83.367	91	105.17	130.5	206.57	308.47	187.27	157.93
6DOF Stanford Robot Arm	*Average (Std)*	2.5438 (1.583)	1.2003 (0.786)	0.52798 (0.625)	0.097716 (0.141)	0.35949 (0.813)	0.32893 (0.483)	0.77691 (0.806)	1.7795 (1.523)	2.3766 (1.732)
	Minimum	1.51×10^{-1}	9.36×10^{-2}	3.03×10^{-4}	5.80×10^{-9}	5.91×10^{-9}	3.63×10^{-7}	4.73×10^{-2}	3.86×10^{-3}	5.43×10^{-2}
	Iteration	262.63	306.27	642.43	435.93	499.73	1470.9	1207.9	310.93	248.13
6DOF Articulate Robot Arm (small)	*Average (Std)*	4.6704 (1.794)	2.1154 (1.661)	0.82515 (0.557)	0.22069 (0.412)	0.4114 (1.166)	0.51296 (0.829)	1.2454 (1.115)	4.4305 (2.404)	4.6366 (1.966)
	Minimum	1.21×10^{-1}	3.65×10^{-1}	1.47×10^{-3}	2.83×10^{-9}	2.83×10^{-9}	2.99×10^{-9}	2.33×10^{-2}	4.50×10^{-1}	1.44×10^{0}
	Iteration	235.63	275.37	484.83	287.63	285.87	966.1	958.5	195.5	149.53
6DOF Articulate Robot Arm (big)	*Average (Std)*	4.1872 (1.933)	3.1769 (1.566)	1.2053 (0.941)	0.33989 (0.610)	0.44367 (0.68)	0.68313 (1.127)	4.4449 (2.091)	4.6878 (2.046)	5.856 (2.395)
	Minimum	1.64×10^{-1}	2.00×10^{-1}	2.27×10^{-2}	5.53×10^{-9}	5.53×10^{-9}	8.06×10^{-8}	2.35×10^{-1}	1.87×10^{0}	1.89×10^{0}
	Iteration	215.27	302.93	744.57	518.87	555.8	1648.6	316.07	177.73	153.57
17DOF Dual-Arm Robot Arm	*Average (Std)*	3.498 (2.36)	1.446 (1.482)	0.57004 (1.077)	0.32323 (1.048)	0.1926 (0.663)	0.31277 (0.748)	1.253 (1.894)	3.564 (2.658)	3.7033 (2.702)
	Minimum	3.11×10^{-1}	9.79×10^{-2}	1.88×10^{-4}	3.82×10^{-9}	3.82×10^{-9}	3.76×10^{-8}	4.11×10^{-3}	1.56×10^{-1}	6.29×10^{-2}
	Iteration	372.1	390.1	749.33	641.27	717.17	1487.3	840.23	222.47	167.8

Table 12. Performance of PSO when $c = 2.0$.

Robot Configuration		PSO Parameters @ $c = 2.0$								
		$w = 0.6$	$w = 0.9$	$w = 1.2$	$w = 1.5$	$w = 1.8$	$w = 2.1$	$w = 2.4$	$w = 2.7$	$w = 3.0$
3DOF Articulate Robot Arm	*Average (Std)*	1.2666 (0.980)	1.0666 (1.015)	0.79994 (0.996)	1.1333 (1.008)	1.2666 (0.980)	1.1333 (1.008)	1.0668 (1.015)	0.87109 (1.009)	1.2906 (0.996)
	Minimum	6.10×10^{-9}	6.10×10^{-9}	6.10×10^{-9}	6.10×10^{-9}	6.10×10^{-9}	6.14×10^{-9}	4.64×10^{-6}	3.20×10^{-4}	2.23×10^{-4}
	Iteration	72.2	77.133	83.2	94.767	116.83	188.97	234.67	177.8	160.8
4DOF SCARA Robot Arm	*Average (Std)*	2.5333 (1.960)	1.6 (1.993)	1.4667 (1.960)	1.4667 (1.960)	1.3333 (1.918)	2.1333 (2.03)	1.7349 (2.017)	2.3599 (2.030)	2.766 (2.038)
	Minimum	9.73×10^{-10}	9.73×10^{-10}	9.73×10^{-10}	9.73×10^{-10}	9.73×10^{-10}	1.09×10^{-9}	3.84×10^{-7}	3.16×10^{-3}	2.99×10^{-2}
	Iteration	76.533	82.733	91.7	105.2	135.63	241.73	241.67	168.1	172
6DOF Stanford Robot Arm	*Average (Std)*	1.3284 (1.109)	0.81005 (0.726)	0.21514 (0.525)	0.14129 (0.178)	0.18694 (0.239)	0.35786 (0.465)	1.0369 (0.939)	3.3027 (8.697)	2.3353 (1.637)
	Minimum	1.11×10^{-4}	2.88×10^{-4}	6.62×10^{-7}	5.80×10^{-9}	2.84×10^{-8}	4.72×10^{-7}	5.75×10^{-2}	3.22×10^{-1}	2.23×10^{-1}
	Iteration	262.17	343.6	482.7	407.93	686.8	1528.3	922.13	433.7	230.27
6DOF Articulate Robot Arm (small)	*Average (Std)*	1.8589 (1.261)	0.77962 (0.727)	0.47691 (0.630)	0.29102 (0.703)	0.42741 (0.633)	0.37187 (0.476)	2.0606 (1.968)	4.4253 (1.984)	4.2691 (1.984)
	Minimum	3.62×10^{-1}	4.90×10^{-2}	2.83×10^{-9}	2.83×10^{-9}	2.83×10^{-9}	4.88×10^{-9}	3.23×10^{-2}	2.99×10^{-1}	6.28×10^{-1}
	Iteration	274.63	288.7	248.77	224.4	302.47	1291.5	698.8	222.47	170.67
6DOF Articulate Robot Arm (big)	*Average (Std)*	2.791 (1.543)	1.018 (0.532)	0.41474 (0.622)	0.34059 (0.61)	0.36141 (0.621)	1.1712 (1.458)	3.9099 (2.242)	4.9021 (2.015)	5.8541 (2.339)
	Minimum	7.32×10^{-1}	1.39×10^{-2}	5.53×10^{-9}	5.53×10^{-9}	5.53×10^{-9}	3.10×10^{-2}	5.18×10^{-1}	1.41×10^{0}	1.28×10^{0}
	Iteration	217.1	339.4	518.1	430.33	600.13	1428.2	293.03	185	157.67
17DOF Dual-Arm Robot Arm	*Average (Std)*	725.92 (3965.7)	0.56706 (0.482)	0.12174 (0.205)	0.10393 (0.219)	0.1511 (0.329)	0.28911 (0.646)	0.84244 (1.004)	3.791 (2.441)	3.6224 (2.501)
	Minimum	1.03×10^{-1}	2.38×10^{-3}	3.82×10^{-9}	3.82×10^{-9}	3.82×10^{-9}	2.09×10^{-4}	7.55×10^{-3}	5.07×10^{-1}	9.86×10^{-2}
	Iteration	343.93	426.93	486.23	500.5	787.8	1574.7	676.1	205.9	172.47

Table 13. Performance of PSO when $c = 2.6$.

Robot Configuration		PSO Parameters @ $c = 2.6$								
		$w = 0.6$	$w = 0.9$	$w = 1.2$	$w = 1.5$	$w = 1.8$	$w = 2.1$	$w = 2.4$	$w = 2.7$	$w = 3.0$
3DOF Articulate Robot Arm	*Average (Std)*	1.1333 (1.008)	0.66662 (0.959)	0.93327 (1.015)	0.93327 (1.015)	0.8666 (1.008)	0.9333 (1.015)	1.0005 (1.017)	0.9414 (1.012)	0.74984 (0.980)
	Minimum	6.10×10^{-9}	6.10×10^{-9}	6.10×10^{-9}	6.10×10^{-9}	6.10×10^{-9}	6.48×10^{-9}	8.80×10^{-7}	1.47×10^{-3}	7.17×10^{-7}
	Iteration	72.833	79.533	86.8	99.633	132.5	240.2	178.6	139.67	126.83
4DOF SCARA Robot Arm	*Average (Std)*	2.6667 (1.918)	1.4667 (1.960)	2.2667 (2.016)	2.2667 (2.016)	2.4 (1.993)	1.8667 (2.03)	1.8716 (2.03)	2.3128 (2.103)	2.5558 (1.99)
	Minimum	9.73×10^{-10}	9.73×10^{-10}	9.73×10^{-10}	9.73×10^{-10}	9.75×10^{-10}	9.97×10^{-10}	4.60×10^{-6}	1.37×10^{-4}	9.94×10^{-3}
	Iteration	77.867	84.367	93.467	110.87	149.87	276.57	225.33	153.43	158.63
6DOF Stanford Robot Arm	*Average (Std)*	0.4103 (0.361)	0.1916 (0.253)	0.1814 (0.25)	0.1011 (0.164)	0.3338 (0.887)	0.5239 (0.755)	1.2597 (0.836)	2.2256 (2.137)	2.3583 (1.831)
	Minimum	5.62×10^{-5}	2.36×10^{-8}	5.80×10^{-9}	5.80×10^{-9}	4.34×10^{-6}	1.70×10^{-2}	7.80×10^{-2}	1.55×10^{-1}	7.98×10^{-2}
	Iteration	263.4	394.53	381.87	493.23	1202.9	1434.6	672.2	388.73	202.37
6DOF Articulate Robot Arm (small)	*Average (Std)*	0.5845 (0.664)	0.39707 (0.609)	0.39503 (1.16)	0.23352 (0.413)	0.56577 (0.935)	0.66547 (0.888)	2.8483 (2.109)	4.8635 (2.095)	4.2852 (2.107)
	Minimum	7.65×10^{-3}	2.83×10^{-9}	2.83×10^{-9}	2.83×10^{-9}	2.83×10^{-9}	1.02×10^{-3}	2.51×10^{-1}	3.28×10^{-1}	1.27×10^{0}
	Iteration	321.03	240.93	174.77	233.2	407.57	1274.5	379.97	186.7	162.97
6DOF Articulate Robot Arm (big)	*Average (Std)*	1.4178 (1.265)	0.42641 (0.437)	0.38365 (0.686)	0.41372 (0.678)	0.41043 (1.08)	1.603 (1.249)	4.4837 (2.419)	5.165 (1.76)	6.3959 (2.660)
	Minimum	2.18×10^{-2}	5.53×10^{-9}	5.53×10^{-9}	5.53×10^{-9}	5.55×10^{-9}	3.18×10^{-1}	4.05×10^{-1}	1.42×10^{0}	1.22×10^{0}
	Iteration	233.9	473.1	344.93	443.6	1276.8	1122	225.47	181.97	139.43
17DOF Dual-Arm Robot Arm	*Average (Std)*	0.2944 (0.450)	0.0372 (0.081)	0.0088 (0.035)	875.85 (4796.4)	0.2173 (0.630)	0.6797 (1.129)	1449.2 (5508.9)	3.3124 (2.753)	3.3555 (2.487)
	Minimum	6.63×10^{-4}	1.06×10^{-6}	3.82×10^{-9}	3.82×10^{-9}	4.53×10^{-9}	1.33×10^{-3}	1.55×10^{-2}	1.05×10^{-1}	1.86×10^{-1}
	Iteration	335.73	496.2	504.8	536.53	1252.9	1246.7	434.8	209.63	162.83

Table 14. Aggregate of Performance for PSO for Different Values of Learning coefficient (c).

Robot Configuration		PSO Parameters								
		$w = 0.6$	$w = 0.9$	$w = 1.2$	$w = 1.5$	$w = 1.8$	$w = 2.1$	$w = 2.4$	$w = 2.7$	$w = 3.0$
3DOF Articulate Robot Arm	$c = 1.4$	0.5347464	0.513582	0.542799	0.4086851	0.4952326	0.3911338	0.3028495	0.3943838	0.138528
	$c = 2.0$	0.436254	0.461214	0.5109207	0.4312077	0.3887085	0.3308507	0.28977	0.1432765	0.158528
	$c = 2.6$	0.4264314	0.53446	0.454324	0.4409674	0.4231626	0.2946653	0.3432588	0.1481972	0.461408
4DOF SCARA Robot Arm	$c = 1.4$	0.5376697	0.546828	0.4982226	0.4867385	0.4851827	0.4469782	0.3308146	0.4031722	0.126061
	$c = 2.0$	0.4513523	0.525295	0.5320692	0.5181073	0.5039198	0.308167	0.3458049	0.3372666	0.072116
	$c = 2.6$	0.4515854	0.553144	0.46332	0.4475889	0.4025616	0.3336784	0.3793495	0.3910511	0.130394
6DOF Stanford Robot Arm	$c = 1.4$	0.226814	0.561708	0.7481638	0.8958934	0.7622983	0.6479014	0.5237504	0.5460761	0.384375
	$c = 2.0$	0.8245952	0.861384	0.889674	0.9174587	0.866636	0.7095383	0.6990268	0.1790552	0.565292
	$c = 2.6$	0.8682516	0.881334	0.8852368	0.8841414	0.6511562	0.5787048	0.525432	0.1963255	0.371534
6DOF Articulate Robot Arm (small)	$c = 1.4$	0.4815013	0.579652	0.77219	0.8708856	0.7827311	0.6363195	0.565344	0.3842927	0.258683
	$c = 2.0$	0.5388443	0.788992	0.8455001	0.8515363	0.8375048	0.6689601	0.4875258	0.3378529	0.225812
	$c = 2.6$	0.8267822	0.860148	0.8079613	0.8933018	0.7801186	0.6103401	0.4797442	0.400593	0.248079
6DOF Articulate Robot Arm (big)	$c = 1.4$	0.5651641	0.628611	0.7344366	0.8431123	0.8258427	0.6031733	0.5129578	0.3123379	0.226712
	$c = 2.0$	0.548103	0.837771	0.8250971	0.8449657	0.8131663	0.5386513	0.4502542	0.2929035	0.245574
	$c = 2.6$	0.7761238	0.849648	0.8529644	0.8332735	0.6324973	0.5443506	0.4819496	0.3473131	0.257056
17DOF Dual-Arm Robot Arm	$c = 1.4$	0.233272	0.621082	0.7357603	0.7733828	0.8050895	0.6597047	0.5956085	0.3511421	0.421283
	$c = 2.0$	0.394751	0.93082	0.9227511	0.9204908	0.8748558	0.7497568	0.8885874	0.7158518	0.922621
	$c = 2.6$	0.9320448	0.900978	0.8992706	0.5241844	0.7499339	0.7492772	0.3923669	0.8158875	0.716818

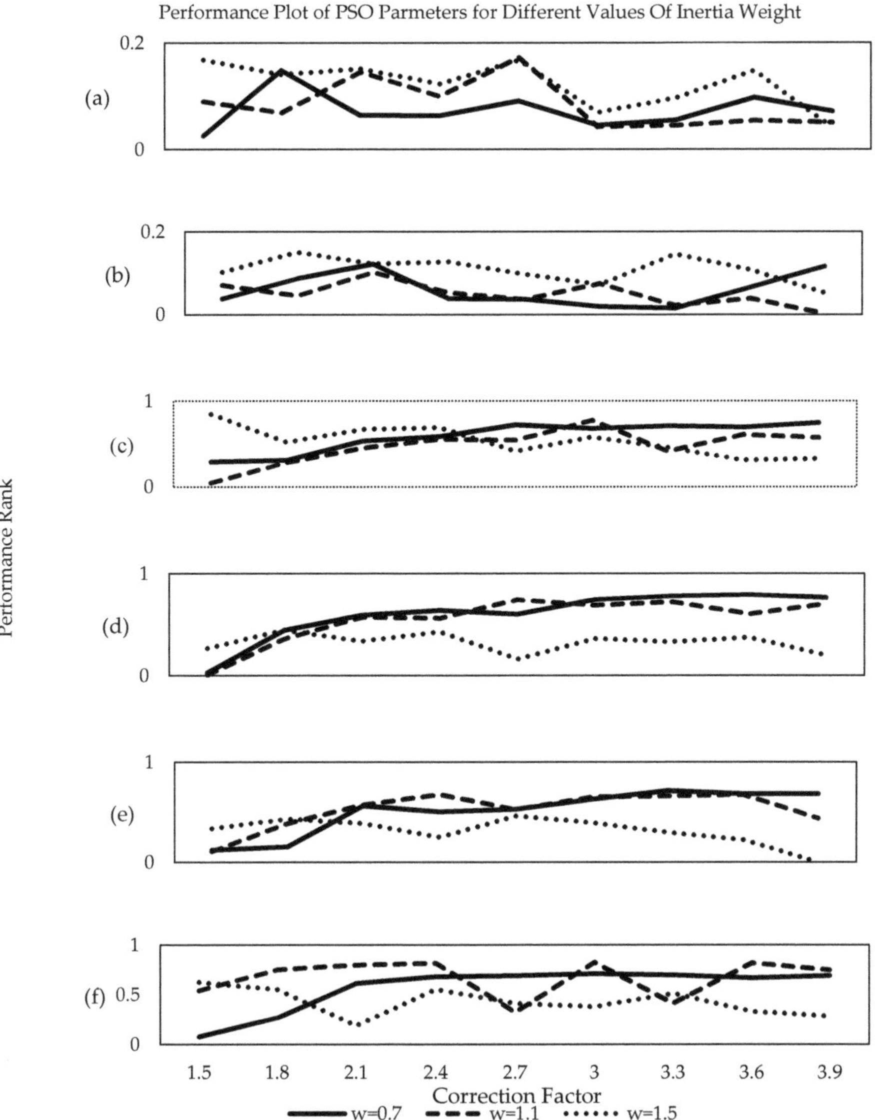

Figure 5. Performance of PSO for different values of w (**a**) 3DOF articulate, (**b**) 4DOF SCARA, (**c**) 6DOF Stanford, (**d**) small 6DOF articulate, (**e**) big 6DOF articulate, (**f**) 17DOF dual-arm.

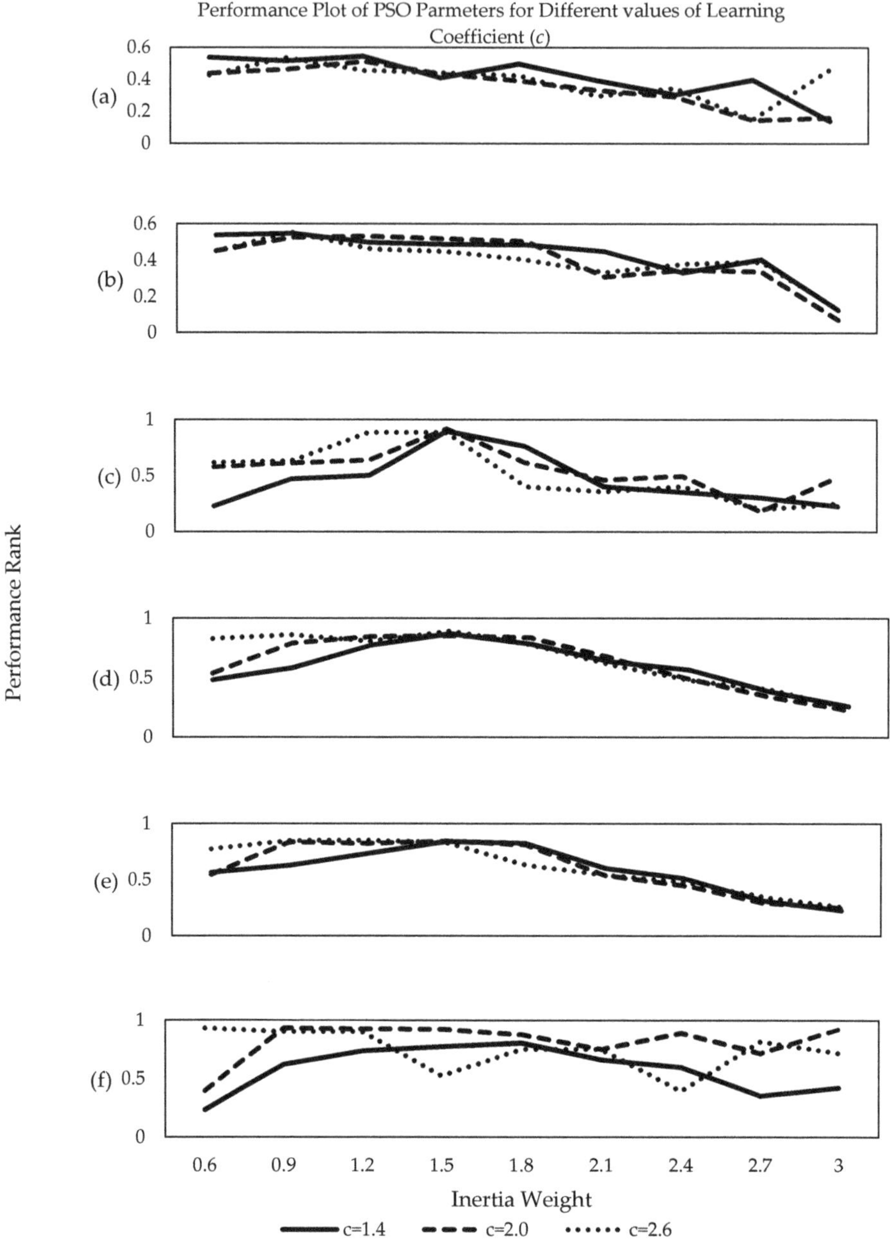

Figure 6. Performance of PSO for Different Values of c for (**a**) 3DOF Articulate (**b**) 4DOF SCARA (**c**) 6DOF Stanford (**d**) 6DOF Articulate (**e**) Larger 6DOF Articulate (**f**) 17DOF Dual-Arm.

While some other parameters with very poor average best solutions are capable of finding the minimum solution. For example, the results for the SCARA robot configuration in Table 11 shows that when c is 1.4 and w is between 2.4–2.7 the average best solution dominates the results obtained when w was between 1.2–1.5, yet the solution of the former cannot find the global minimum solution, and many more instances can be sited. Accuracy is very important in robot analysis, therefore the best solution should be able to find the global minimum solution always, followed by the solution that can find the global minimum at least once. Algorithms that may produce competitive averages yet incapable of locating the global minimum are regarded as poor solutions. Therefore, the minimum solution achieved by every pair of parameters was reported in the tables as the minimum solution. The aggregate performance of the PSO presented in Table 10, and Table 14 is an average of the normalized values obtained in the experiments such that the variables with higher normalized values have better performance, also a penalty was introduced when solving the average of normalized values, where binary probability distribution was used to replace the normalized minimum solution such that the solutions capable of finding the global minimum solution were assigned the value 1 while other undesirable results were assigned the value 0.

3.1. Observations

For the 3DOF Articulate robot configuration in Table 10, a random-like fluctuation in the performance of the PSO can be observed. When w was at 0.7 the best result was achieved at $c = 1.8$, it can also be observed that when w was increased to 1.1 the best results occurred at around $c = 2.7$ then when w was further increased to 1.5 the best result occurred at $c = 1.5$. However, Figure 5a shows that the performance of the PSO increases with increasing w and decreasing learning coefficient (c).

The performance of the PSO is somewhat stable when c was between 1.8–2.4 and w at 1.5 was found to be dominant. On the other hand, it can be seen from the Table 14 that the best result obtained for the PSO was when w was 1.2 and c was 1.4, then when c was increased to 2.6 the best result was observed at $w = 0.9$ which suggests that an improved result was achieved with an increasing c and a decreasing w. Figure 6a shows that the performance of the PSO algorithm deteriorates with increasing c. From Table 14 and Figure 6a, it can also be observed that the algorithm was stable when w was between 0.6 and 1.2 with $c = 1.4$ being dominant.

From Table 7, that when $w = 0.7$ and $c < 2.7$, the PSO algorithm was incapable of finding the global minimum solution for the higher DOF robots, likewise in Table 12 when $c = 2.0$ and $w < 1.2$. These observations confirm that the standard PSO (SPSO) is only capable of analyzing robot manipulator configurations with lower degrees of freedom. Because the dominant solutions are within the range of the SPSO, but as the DOF increases, the dominant solutions would be seen to deviate from the range of the SPSO.

In the 4DOF SCARA robot configuration, it can be observed from Table 10 that when w was equal to 0.7, the best result was obtained at $c = 2.1$ when w was increased and c decreased, the performance of the PSO was seen to increases as made evident in Figure 5b. The algorithm can be seen to be stable when c is between 1.8–2.4 and w at 1.5 dominating other solutions. Likewise, from Table 14 and Figure 6b, although it can be observed that the algorithm was stable when w was between 0.6–1.8 the best solution was recorded when w was between 0.9–1.2. In the initial stages when w was small $c = 1.4$ was dominant, but as w increased, $c = 2.0$ became the dominant solution.

From Table 10 it would be observed that the best result obtained for a 6DOF Stanford robot occurred when $w = 0.7$ and $c = 3.9$. When w was increased to 1.5 the best result was obtained with a decreasing c at 1.5. From Figure 5c the performance of the algorithm can also be observed to deteriorate with increasing w. the algorithm was stable when c was between 1.8–3.6 with $w = 1.5$ being the dominant solution when c was small, then $w = 0.7$ becomes dominant when c increases beyond 2.4. In Table 14 when $c = 1.4$ the best result for the 6DOF Stanford manipulator was obtained at $w = 1.5$, this value decreases slightly with an increase in c such that at $c = 2.6$ the best result occurred at $w = 1.2$. The algorithm was found stable between $w = 0.6$–1.5 and the solutions of $c = 2.0$ and $c = 2.6$ can be seen to compete for dominance. The dominant solution would lay be between $c = 2.0$–2.6.

The trend of decreasing inertia weight (w) and increasing learning coefficient (c) is observed to continue for the 6DOF Articulate manipulator configurations (both small and big) it can be seen from Table 10 that the best results were obtained at $c = 3.6$ when w was 0.7 for both configurations, this value reduced to $c = 1.8$ and $c = 2.7$ for the small and big manipulators respectively while w increased to 1.5. From Figure 5d,e, it can be observed that the algorithm remains stable when c was between 2.7–3.9 for both configurations with $w = 0.7$ being the dominant solution. It can also be observed from both plots that the performance of the algorithm also deteriorated with increasing w. An even stronger correlation is observed from Table 14 where the best results occurred at $w = 1.5$ for all values of c with a slightly decreasing w observed in the larger Articulated robot configuration. From Figure 6d,e, it would be seen that the algorithm is stable when w was between 0.9–1.8 and the dominant solution of c lies between 2.0–2.6. It would, therefore, be safe to deduce that the size of the robot manipulator has little effect on the PSO solution, therefore any PSO algorithm that can analyze a given configuration is most likely able to analyze the different varieties of sizes within the same configuration.

From Table 8 when $w = 1.1$ at $c = 2.3$ and $c = 3.3$, fluctuating results were recorded in the dual-arm configuration which is believed to be a result of stagnation in the algorithm.

Similar results were recorded in Table 13 when $c = 2.6$ at $w = 1.5$ and $w = 2.4$, and also in Table 12 when $c = 2.0$ at $w = 0.6$. Otherwise, in Table 10 the best results were obtained at $c = 3.0$ for $w = 0.7$ and $c = 1.5$ when w increased to 1.5. the best result occurred with a decreased value of c at 1.5. Figure 5f shows the performance of the algorithm deteriorates with an increasing w. although the algorithm remained consistent for almost all values of c, a sharp deterioration can be observed when c was 1.5–2.1. $w = 0.7$ is the dominant solution.

In Table 14 the best results were obtained when $c = 1.4$ at $w = 1.8$, the best results were further observed to occur at a decreased w and increased c such that when $c = 2.0$ the best result was obtained at $w = 0.9$ and when $c = 2.6$, the best result was obtained at $w = 0.6$ which also supports the theory of decreasing w with an increasing c. From Figure 6f, the performance was stable when w was between 0.9–2.1 with $c = 2.0$ being the dominant solution.

3.2. Deductions

The performance of the PSO algorithm was seen to improve with a decreasing inertia weight (w) and an increasing learning coefficient (c) in all the robot manipulator configuration except in the 3DOF Articulate manipulator. Fortunately, our analysis is more concerned with higher DOF robot manipulators, therefore techniques for decreasing w and increasing c shall be investigated. From the observations above for all robot manipulator configurations, the optimal w was 0.6–2.1 while c was 1.8–3.9. These best values were plotted and a fitted curve generated which shall be used to determine a suitable adaptive technique in the next section.

4. Adaptive Computation Technique

In robot analysis, the multi-swarm variations of the PSO are best suited for trajectory analysis especially in mobile robots where the robot would be required to track or follow a moving target. Kinematic/dynamic analysis of industrial robots generally have static search spaces, so this solution is not suitable considering the increased computational cost. Although the variation of the adaptive PSO which is dependent on the best solution (P_{Best} or G_{Best}) seems most promising as demonstrated in [45] but this requires knowledge of the established range of values for w and c that ensures exploitation and exploration. It was previously observed that the robot optimization problem does not converge under the conditions of basic parameters for most known EA. Therefore, this research was aimed at establishing a new set of parameters that ensure convergence. As such, the time-dependent variation of the Adaptive PSO shall be implemented for this analysis. Several time-varying algorithms have been reported in theory. A few of which shall be implemented in the foregoing experiment, a total of 13 distinct PSO algorithms shall be used for the experiment.

- PSO_1: The linear decreasing inertia weight was reported in [40] where inertia weight (w) decreases linearly from 0.9 to 0.4, the governing equation for updating the w is

$$w_{iter} = w_{\max} - \frac{w_{\min} - w_{\max}}{t_{\max}} \times t_{iter}, \tag{12}$$

$$c_{iter} = 2.05, \tag{13}$$

where w_{max} and w_{min} are values of the initial and final inertia weight, t_{max} is the maximum number of iterations while t_{iter} is the current iteration.

- $PSO_{2\text{–}3}$: A non-linear decreasing inertia weight was also reported in [50] with w decreasing linearly from 0.9 to 0.4. The governing equation for updating the inertia weight is

$$w_{iter} = \frac{(t_{\max} - t_{iter})^n}{t_{\max}^n} \times (w_{initial} - w_{final}) + w_{final}, \tag{14}$$

Observe that when n = 1, the inertia weight would be linearly decreasing as shown in Figure 7a, where n is a constant ranging from 0.9 to 1.3, the value n = 1.2 was reported as the recommended value for n in [50] but n = 3 was found to be more suitable for robot analysis. Therefore, the results for the two values n = 1.2 and n = 3.0 shall be presented in this experiment as PSO_2 and PSO_3, respectively.

- PSO_4: A novel non-linear decreasing w and non-linear increasing c is hereby proposed. The parameters recorded for the best performance of PSO from the previous experiment were plotted and a fitted curve generated as shown in Figure 8. The non-linear technique presented in (15)–(17) exploits the experimental range of values for w and c, where n and m are problem dependent variables.

$$w_{iter} = w_{initial} \times n^{iter}, \tag{15}$$

$$c_1 = 2.24, \tag{16}$$

$$c_2 = \frac{c_{initial}}{m^{iter}}, \tag{17}$$

If the maximum number of iterations is 3000, then the values of the coefficients n and m can be easily determined. The parameter w in PSO_4 shall be updated according on Equation (15) while c_1 and c_2 are updated according to Equation (16), where the learning coefficients are not adaptive therefore a reduced computation cost can be achieved, while the parameters in PSO_{11} shall be updated according to Equations (15)–(17) as originally proposed.

- $PSO_{5\text{–}6}$: The concept of multi-stage decreasing inertia weight was introduced in [51], where w was decreased linearly from 0.9 to 0.4 in three distinct stages. The inertia weight first decreases from the initial value to a predetermined value w_m where it remains constant for a while before decreasing further to the final value. As shown in Figure 7b, five different scenarios were presented, and the governing equation for updating the value of the inertia weight in each of the scenarios is given in Equations (18)–(22)

$$t_1 = \left[\ \tfrac{1}{5}t_{\max}, \ \tfrac{2}{5}t_{\max}, \ \tfrac{1}{5}t_{\max}, \ \tfrac{2}{5}t_{\max}, \ \tfrac{1}{5}t_{\max}, \ \tfrac{2}{5}t_{\max} \ \right], \tag{18}$$

$$t_2 = \left[\ \tfrac{4}{5}t_{\max}, \ \tfrac{3}{5}t_{\max}, \ \tfrac{4}{5}t_{\max}, \ \tfrac{3}{5}t_{\max}, \ \tfrac{4}{5}t_{\max}, \ \tfrac{3}{5}t_{\max} \ \right], \tag{19}$$

$$w_n = \left[\ \tfrac{4(w_{\max}-w_{\min})}{5} + w_{\min}, \ \tfrac{2.5(w_{\max}-w_{\min})}{5} + w_{\min}, \ \tfrac{(w_{\max}-w_{\min})}{5} + w_{\min} \ \right], \tag{20}$$

$$w_m = \left[\ w_n(1) \ \ w_n(1) \ \ w_n(2) \ \ w_n(2) \ \ w_n(3) \ \ w_n(3) \ \right], \tag{21}$$

$$w(i) = \begin{cases} (w_s - w_m)(t_1(i) - t)/t_1(i) + w_m(i) & 0 \le t \le t_1 \\ w_m(i) & t_1 \lhd t \le t_2 \\ (w_m(i) - w_e)(t_{\max} - t)/(t_{\max} - t_2(i)) + w_e & t_2 \lhd t \le t_{\max} \end{cases}, \quad (22)$$

The parameter for $MLDIW_5$ was recommended in [51] for inertia weight but the parameters of $MLDIW_3$ were found to be more suitable for the robot analysis. The results for the two values $MLDIW_5$ and $MLDIW_3$ shall also be presented as PSO_4 and PSO_5 respectively.

- PSO_7: All the aforementioned algorithms exploited only the inertia weight, leaving the learning factor constantly at 2.05. In [52] and [36], a linear decreasing and linear increasing inertia weights were proposed respectively, both with decreasing cognitive component and increasing social component, thereby these techniques exploited both the inertia weight and learning coefficients. The technique reported in [52] shall be utilized in this experiment as it incorporates a linear decreasing inertia weight which is in line with our objectives and its parameters are updated according to Equation (12) above while the cognitive and social components shall be updated as

$$c_{cognitive} = \left(c_{initial} - c_{final}\right)\frac{t_{iter}}{t_{\max}} + c_{final}, \quad (23)$$

$$c_{social} = \left(c_{final} - c_{initial}\right)\frac{t_{iter}}{t_{\max}} + c_{initial}, \quad (24)$$

The inertia weight of $PSO_{8\text{–}13}$ shall be updated according to the equations for $PSO_{1\text{–}6}$ respectively, while the learning coefficients shall be updated according to Equations (16) and (17). For the sake of fair comparison, the adaptive values of w for all the aforementioned techniques shall decrease from the initial value of 2.1 to a final value of 0.6, the cognitive component remains at 2.24 while the social component is nonlinearly increasing from 1.8 to 3.9.

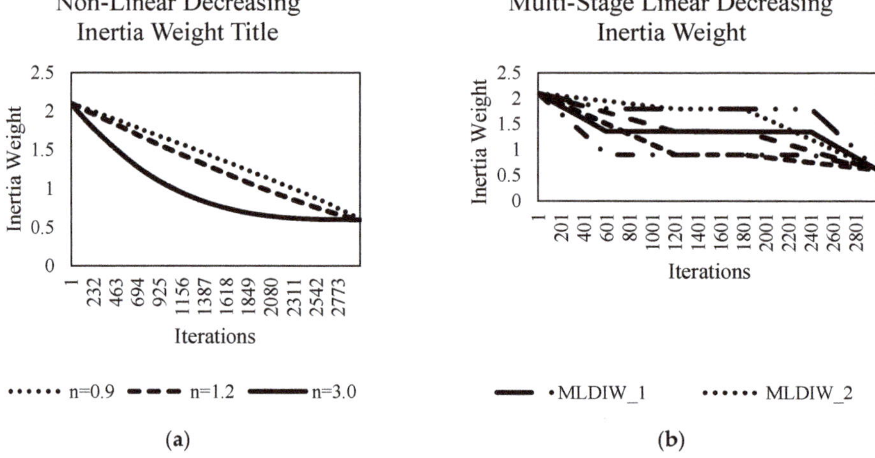

Figure 7. (**a**) Plot showing the rate of change of inertia weight at different values of n (**b**) plot showing various techniques for multi-staged decreasing inertia weights.

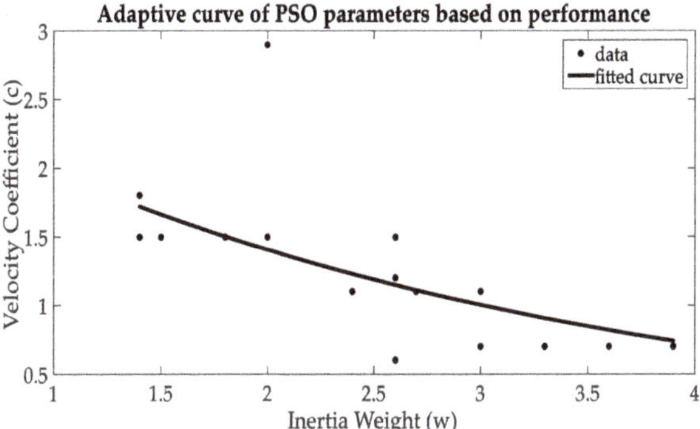

Figure 8. Fitted curve of PSO parameters.

5. Results

For this experiment, the swarm size was again maintained at 200, a total of 13 adaptive PSO techniques were tested on all six robot manipulators with the maximum number of iterations for each run set at 3000 and a total of 30 runs each. As earlier presented in previous tables, the solution that best minimizes the problem (global minimum) was presented in Table 15 for every pair of robot configuration and PSO technique, the average and standard deviation of the best solution after 30 runs and the average number of iterations required to find the best solution are also presented. These values were normalized, summed, averaged, then ranked (sorted) such that the solution with the least rank possesses the best performance. The ranking of all tested PSO algorithms is presented in Table 16. The first column of Table 16 presents the PSO techniques according to their ranks. The second to the seventh column of Table 16 shows the individual ranks of all the PSO techniques against the six robot manipulators. In the second column, under 3DOF Articulate robot configuration, PSO_5 has the best solution for the particular robot manipulator followed by PSO_6, then PSO_{10} has the third-best solution, etc. The last column of Table 16 is the sum of all ranks presented in columns 2–7, PSO_{13} having the lowest total rank is considered the best result overall. During the course of the experiment, it was observed that:

- Most algorithms stagnate above 1×10^{-3}, also most of the algorithms tested were found to either find the global minimum solution or run into stagnation. Accuracy is an important criterion for robot analysis and control, some algorithms were found to sometimes avoid stagnation but still incapable of finding the global minimum result even after 3000 iterations. This is an anomaly that was unfavorable in this analysis because when taking averages, these solutions gave competitive results, which is capable of confusing the algorithm. Therefore, another penalty was introduced such that if a solution is found to escape stagnation yet incapable of finding the global minimum solution, then the best solution for that run is recorded as 5.5. This signifies that the run gave bad results, and this problem is easily captured when taking averages.
- It can be seen from Table 15 that using the parameters derived from the previous experiment in place of the parameters of the PSO enhances the results as all the adaptive PSO techniques implemented so far found the global minimum solution except in PSO_5 and PSO_{12}. When the robot configuration has more than 4DOF PSO_5 was not capable of finding the global minimum solution while PSO_{12} could not find the global minimum for the 6DOF Stanford robot configuration. Since almost all the algorithms were capable of finding the minimum solution, the quest was therefore reduced to finding the algorithm with the least computations and less likely to fall into stagnation.

Table 15. Performance of various adaptive PSO techniques for different robot configurations.

Robot Configuration		PSO Algorithms												
		PSO_1	PSO_2	PSO_3	PSO_4	PSO_5	PSO_6	PSO_7	PSO_8	PSO_9	PSO_{10}	PSO_{11}	PSO_{12}	PSO_{13}
3DOF Articulate Robot Arm	*Minimum*	6.10×10^{-9}	6.10×10^{-9}	6.10×10^{-9}	6.10×10^{-9}	6.10×10^{-9}	6.10×10^{-9}	6.10×10^{-9}	6.10×10^{-9}	6.10×10^{-9}	6.10×10^{-9}	6.10×10^{-9}	6.10×10^{-9}	6.10×10^{-9}
	Average	1.0221	0.91104	0.9555	1.0666	1.0222	1.0444	1.0833	0.88883	0.89991	0.82216	1.0444	1.0888	0.84438
	Std	1.0087	0.99104	1.0079	1.0132	1.0102	1.0017	1.104	1.001	1.0012	0.99385	1.0125	0.99689	0.99459
	Iteration	342.95	337.83	314.22	329.42	222.5	222.63	427.43	330.79	323.28	305.74	320.52	300.66	312.12
4DOF SCARA Robot Arm	*Minimum*	9.73×10^{-10}	9.73×10^{-10}	9.73×10^{-10}	9.73×10^{-10}	9.73×10^{-10}	9.73×10^{-10}	9.73×10^{-10}	9.73×10^{-10}	9.73×10^{-10}	9.73×10^{-10}	9.73×10^{-10}	9.73×10^{-10}	9.73×10^{-10}
	Average	1.8222	1.5555	2.1333	1.8222	2	1.9111	1.9556	2.1333	1.9111	1.7778	1.4667	1.5556	1.7333
	Std	2.0066	1.9757	2.0175	2.019	1.9975	2.0129	2.0053	2.0205	2.0312	2.0206	1.9605	1.9822	2.0068
	Iteration	383.48	377.17	343.6	363.21	226.72	227.24	526.11	371.67	357.73	337.02	352.49	326.02	346.79
6DOF Stanford Robot Arm	*Minimum*	5.80×10^{-9}	5.80×10^{-9}	5.80×10^{-9}	5.80×10^{-9}	0.00503	5.80×10^{-9}	5.80×10^{-9}	5.80×10^{-9}	5.80×10^{-9}	5.80×10^{-9}	5.80×10^{-9}	6.85×10^{-5}	5.80×10^{-9}
	Average	0.73517	0.56363	1.4628	1.0201	1.4391	0.80209	0.57368	0.66049	0.7891	0.89401	0.48719	1.3465	0.36397
	Std	1.501	1.4168	2.2052	1.7473	3.1713	1.228	1.3985	2.881	3.3545	1.7495	1.2585	2.2116	0.87091
	Iteration	1296.4	1336.9	992.82	1216.1	612.91	746.29	1593.2	1355.6	1247.6	999.88	1187.5	1201.9	960.5
6DOF Articulate Robot Arm	*Minimum*	2.83×10^{-9}	2.83×10^{-9}	2.83×10^{-9}	2.83×10^{-9}	0.05314	2.83×10^{-9}	2.83×10^{-9}	2.83×10^{-9}	2.83×10^{-9}	2.83×10^{-9}	2.83×10^{-9}	2.83×10^{-9}	2.83×10^{-9}
	Average	0.40148	0.46231	0.40928	0.41159	1.1148	0.84461	0.32191	0.41798	0.37357	0.34845	0.50466	0.34546	0.47359
	Std	0.54642	0.7966	0.66896	0.53455	0.85375	1.0268	0.55764	0.57991	0.77192	0.55917	0.97798	0.72753	0.65956
	Iteration	730.8	683.83	530.39	643.9	541.98	590.92	1030.6	710.09	689.53	534.81	618.8	510.45	547.89
6DOF Articulate Robot Arm	*Minimum*	5.53×10^{-9}	5.53×10^{-9}	5.53×10^{-9}	5.53×10^{-9}	0.07665	5.53×10^{-9}	5.53×10^{-9}	5.53×10^{-9}	5.53×10^{-9}	5.53×10^{-9}	5.53×10^{-9}	5.53×10^{-9}	5.53×10^{-9}
	Average	0.57103	0.58342	0.4679	0.63143	1.5126	1.2445	0.59572	0.46228	0.52115	0.52637	0.47206	0.8547	0.47416
	Std	0.77659	0.92347	0.64328	1.0937	1.1692	0.9791	0.72651	0.79572	0.9415	0.8635	0.87465	1.3579	0.7217
	Iteration	1160.2	1083.9	938.25	966.36	604.29	548.72	1508.2	1175.9	1092.1	975.23	987.9	980.3	853.53
17DOF Dual-Arm Robot Arm	*Minimum*	3.82×10^{-9}	3.82×10^{-9}	3.82×10^{-9}	3.82×10^{-9}	0.00432	3.82×10^{-9}	3.82×10^{-9}	3.82×10^{-9}	3.82×10^{-9}	3.82×10^{-9}	3.82×10^{-9}	3.82×10^{-9}	3.82×10^{-9}
	Average	0.32223	241.76	0.4961	0.36216	0.54445	1016	0.33813	0.238	0.24413	0.75383	0.36512	0.67878	0.27086
	Std	0.73876	1322.3	1.1388	0.88827	0.62487	3807.8	0.8389	0.48422	0.65331	1.4823	1.0622	1.5277	0.66004
	Iteration	1156.6	1101.9	909.2	1021.2	721.7	947.77	1526.4	1162.5	1118.4	1038	1020.4	1042.9	926.02

- The proposed algorithm in PSO_{11} was found to produce better results than the PSO techniques reported in other works of literature and was only surpassed by PSO_{13} which is a modification of PSO_6 and contested keenly with PSO_{10} which is also a modification of PSO_3.
- From the results presented in Table 16, it can also be deduced that varying both the inertia weight and learning coefficient produces better results in adaptive PSO algorithms compared to only varying the inertia weight.
- It would also be observed that the proposed algorithm (PSO_{11}) does not perform well at lower DOFs but performed better than PSO_{10} at higher DOF configurations. PSO_{13} perform well across all robot configurations. producing the best result for all techniques tested in this experiment.
- In the NLDIW adaptive technique, although PSO_2 was recommended in [50], it can be seen from Table 16 that PSO_3 performed better. Likewise, in the MLDIW adaptive technique, PSO_5 was recommended in [51] but PSO_6 performed better for robot analysis.
- Modifying the adaptive PSO techniques presented in literature (PSO_{1-6}) with a non-linear increasing learning coefficient as defined in Equations (16) and (17) greatly improved the performance of the PSO algorithm as it can be seen that the modified algorithms in PSO_{8-13} performed better than their counterparts with constant learning factors (PSO_{1-6}).

Table 16. Ranking of the various adaptive PSO techniques.

Rank PSO Algorithms	3DOF Articulate Robot	4DOF SCARA Robot	6DOF Stanford Robot	6DOF Articulate Robot	6DOF Articulate Robot	17DOF Dual-Arm Robot	Sum of Ranks
PSO_{13}	4	7	1	4	1	1	18
PSO_{10}	3	6	4	2	4	9	28
PSO_{11}	10	4	3	7	3	2	29
PSO_9	5	9	8	6	6	3	37
PSO_6	2	1	2	12	10	12	39
PSO_3	7	11	11	3	2	7	41
PSO_{12}	9	3	12	1	11	6	42
PSO_4	12	8	10	5	5	4	44
PSO_2	8	5	5	10	8	11	47
PSO_8	6	12	7	8	7	8	48
PSO_1	11	10	6	9	9	5	50
PSO_5	1	2	13	13	13	13	55
PSO_7	13	13	9	11	12	10	68

6. Mutation Function

Structural bias in population-based algorithms is a characteristic that confines the search in a constrained search space. Replacing randomly selected samples with newly generated random samples enhances unbiased coverage of the search space [53]. In the mutating PSO algorithm, when the algorithm stagnates at a local optimum solution, a mutation operation is used to replace the swarm with new samples. The mutation function is an artificial perturbation of the system used to push the algorithm out of stagnation. The mutation probability was set at 100%, because if any solution from the previous iteration should remain, then that solution shall be the global best solution for the next iteration, therefore, the entire swarm would converge on that solution causing a recurring stagnating cycle. Four variables and two end conditions were introduced to train the algorithm to identify a stagnating solution. When the two conditions are satisfied, the algorithm terminates the iteration signifying that the actual solution has been identified, while if only one condition is satisfied, it signifies that the algorithm has run into stagnation and the mutation function is initiated. The abandonment threshold (E) is the global minimum solution. The *Fitness error* (e) is the difference between the current *Fitness* and the previous *Fitness* as elaborated in Equation (25), and the abandonment counter (q) monitors the second differential of *Fitness error*. When the second differential of the *Fitness error* becomes small than E then the algorithm is assumed to have slowed down, therefore, the condition in

Equation (26) states that when the difference in e is less than E then q begins to count consecutively through every iteration, and if the condition in (26) is broken then counter in q is reset to zero.

$$e = Fitness_{t-1} - Fitness_t, \tag{25}$$

$$q = \begin{cases} q+1, & if, (e_{i-1} - e_i) < E \\ 0, & Else \end{cases}, \tag{26}$$

$$f(MuPSO) = \begin{cases} end & if \quad q \geq Q \quad and \quad Fitness \leq E \\ mutate & if \quad q \geq Q \quad and \quad Fitness > 1e^{-3} \end{cases}, \tag{27}$$

The two end conditions in Equation (27) states that when q is equal to or greater than the abandonment limit (Q) and E is less than 1×10^{-8} then the algorithm has found the global minimum solution and should be terminated, but when only the first condition is met then this signifies that the algorithm has run into stagnation. Q should be large enough so as not to prematurely terminate a solution allowing the algorithm to break out of stagnation but must also not be too large to allow a failed solution to continue. Table 17 shows the performance of the a few variants of PSO modified with the mutating operator. The proposed mutating PSO algorithm is presented in the first column, followed by the basic PSO (w = 1.0 and c = 2.05). The MLDIW-PSO with the enhanced parameters is in the third column as Mu-MLDIW, while the basic MLDIW (w = 0.9–0.4) is in the fourth. Likewise, the NLDIW-PSO with enhanced parameters is presented in the fifth column as Mu-NLDIW, and the basic NLDIW is in the last column. All these algorithms were further enhanced with the mutating operation.

Observe that all the basic PSO algorithms were able to analyze the 3DOF articulate and 4DOF SCARA manipulators, but unable to find the minimum solution for the higher DOF manipulators. At lower DOFs, although the basic PSO algorithms gave better results, the results from the proposed MuPSO are sufficient for robot analysis as seen in Table 18 where the joint parameters of the robots are identified with an accuracy of three decimal places.

Converging results were not obtained for the Stanford and the Dual-arm configurations. In a real experiment, the robot would be required to move from a known initial position and orientation T_i to a final destination T_f, therefore introducing more constraints like minimizing the distance traveled by joints or the energy consumption of joints may improve the convergence of prismatic and redundant configurations.

Table 17. Performance of the mutating particle swarm optimization.

Robot Configuration		PSO Algorithms					
		MuPSO	**PSO**	**Mu-MLDIW**	**MLDIW**	**Mu-NLDIW**	**NLDIW**
3 DOF Articulate Robot Arm	*Min*	6.10×10^{-9}	6.10×10^{-9}	6.10×10^{-9}	6.10×10^{-9}	6.10×10^{-9}	6.10×10^{-9}
	Average	6.78×10^{-9}	6.10×10^{-9}	6.93×10^{-9}	6.10×10^{-9}	6.23×10^{-9}	6.10×10^{-9}
	std	1.21×10^{-9}	7.90×10^{-21}	1.37×10^{-9}	7.48×10^{-21}	3.15×10^{-10}	6.45×10^{-21}
	Iteratn	354.83	169.03	271.93	122.8	333.83	160.07
4 DOF SCARA Robot Arm	*Min*	9.73×10^{-10}	9.73×10^{-10}	9.73×10^{-10}	9.73×10^{-10}	9.73×10^{-10}	9.73×10^{-10}
	Average	1.78×10^{-9}	9.73×10^{-10}	1.48×10^{-9}	9.73×10^{-10}	1.07×10^{-9}	9.73×10^{-10}
	std	2.07×10^{-9}	2.90×10^{-21}	1.55×10^{-9}	1.57×10^{-21}	2.94×10^{-10}	2.19×10^{-21}
	Iteratn	375.1	144.2	375.9	193.13	344.27	108.67
6 DOF Stanford Robot Arm	*Min*	5.80×10^{-9}	0.002297	5.80×10^{-9}	0.042038	5.80×10^{-9}	0.22311
	Average	0.18333	3.6594	2.74×10^{-5}	36.556	3.6698	2.8953
	std	1.0042	2.4823	0.00015	191.83	2.6327	5.7772
	Iteratn	1893.9	3000	1615.7	3000	2577.1	3000
6 DOF Articulate Robot Arm (small)	*Min*	2.83×10^{-9}	2.83×10^{-9}	2.83×10^{-9}	0.3014	2.83×10^{-9}	0.78448
	Average	2.84×10^{-9}	22.959	2.83×10^{-9}	3.1822	2.83×10^{-9}	2.9146
	std	2.13×10^{-11}	119.19	1.36×10^{-12}	2.6104	2.76×10^{-13}	1.9617
	Iteratn	641.47	2907.5	613.8	3000	540.97	3000
6 DOF Articulate Robot Arm (big)	*Min*	5.53×10^{-9}	0.002899	5.53×10^{-9}	0.9546	5.53×10^{-9}	0.66512
	Average	5.53×10^{-9}	1.021	5.61×10^{-9}	621.59	0.44417	5538.1
	std	4.36×10^{-12}	0.93764	4.43×10^{-10}	2520.5	1.4376	29995
	Iteratn	1136.6	3000	1020.1	3000	1574.9	3000
17 DOF Dual-Arm Robot	*Min*	3.82×10^{-9}	0.006624	3.82×10^{-9}	0.10807	3.82×10^{-9}	0.00634
	Average	0.000172	0.50215	0.18333	1212.1	0.5695	Inf
	std	0.000942	1.1307	1.0042	5014.6	1.6748	NaN
	Iteratn	1222.8	3000	1315	3000	1621.9	3000

Table 18. Identified joint parameters for MuPSO and SPSO.

Robot Configurations		Identified Joint Parameters					
		Joint 1	**Joint 2**	**Joint 3**	**Joint 4**	**Joint 5**	**Joint 6**
Ideal Parameters		80°/(200 mm)	20°	−60°/(60 mm)	270°	50°	10°
3DOF Articulate	MuPSO (PSO)	80° (80)	20° (20°)	−60° (−60°)	NA	NA	NA
4DOF SCARA	MuPSO (PSO)	200 mm (200 mm)	−17.124° (−57.156°)	−34.876° (17.146°)	−30° (−66.000°)	NA	NA
6DOF Stanford	MuPSO (PSO)	73.633° (52.043°)	3.260° (−0.661°)	60.396 mm (34.905 mm)	−72.909° (−43.846°)	56.575° (40.649°)	−8.424° (−19.521°)
6DOF Articulate (small)	MuPSO (PSO)	80.000° (80.903°)	20.000° (18.656°)	−60.001° (−57.726°)	−89.998° (−54.286°)	50.001° (27.956°)	9.9983° (−16.874°)
6DOF Articulate (big)	MuPSO (PSO)	80.000° (77.829°)	20.002° (17.475°)	−60.004° (−56.446°)	−89.997° (−46.180°)	50.001° (20.163°)	9.9959° (−15.677°)

In the ideal parameters' row, the values in parenthesis represent prismatic joints where applicable.

7. Conclusions

Research into developing an intelligent swarm-based algorithm for robot analysis and parameter identification was proposed, experiments were performed to study the behavior of four popular robot configurations under various PSO parameters, and two anomalies were identified from the experimental results and successfully resolved. The anomalies were capable of masking poor solutions as good solutions. In this experiment, all the strong local minimizers have been identified; the 3DOF articulate configuration has only one strong local minimizer with Fitness = 2.0 at a position vector of $[80, -40, -60]^T$. The 4DOF SCARA configuration also has only one local minimizer at Fitness = 4.0 with an infinite possibility of position vectors, while the other higher DOF robot configurations have at least five minimizers each. The minimizers are very sensitive, they are shifted by the slightest change in parameters therefore it is almost impossible to identify the stagnation points in real-time. An average solution is capable of breaking out of weak local minimizer but even the best solutions are helpless in the vicinity of a strong local minimizer. This is the basis of introducing the mutation function, to help break algorithms out of stagnation.

Since the algorithm can be taught to identify stagnating solutions, then the best solutions either find the global minimizer or stagnate at local minimum, but not linger without a solution. The two anomalies observed were capable of disguising poor solutions and confusing the algorithm therefore two penalties were introduced to help unmask poor solutions while distinctively identifying the best solutions. Some correlations were observed between the robot configurations and the various PSO parameters, a non-linear decreasing inertia weight and a non-linear increasing correction factor were adopted based on the experimental results. A new range of adaptive parameters were identified and implemented on the PSO algorithm. The algorithm was found to be capable of solving the robot kinematic problem for all four robot configurations. Algorithms from other works of literature were also modified with the newly identified adaptive parameters and compared with the proposed algorithm for solving robot kinematic problems. The proposed algorithm was found to dominate the other algorithms reported in the literature, succumbing to only the modified MLDIW-PSO that had the best overall performance, surpassing the runner up with large margins while the modified NLDIW and the proposed MuPSO algorithm closely contested the second position. More emphasis is on higher DOF configurations, therefore, if the lower DOF manipulators are ignored, the MuPSO would completely dominate the NLDIW-PSO in PSO_{10}.

8. Future Thrust

The future aspirations of this work are to implement the algorithm in dynamic parameter identification of these robot manipulators, and also to compare the performance of the proposed algorithm with other metaheuristics on standard benchmark function. Testing the algorithm on benchmark functions would hopefully shed more light on the complex phenomena of modeling and control of non-linear dynamic systems. The algorithm described herein utilizes a time-dependent adaptive technique, a solution dependent (P_{Best} or G_{Best} based) adaptive technique seems more promising with better maneuverability, therefore since the range of parameters which ensure convergence of the robot dynamic problem has been established, it would be worthwhile to investigate a solution dependent algorithm for robot analysis. It has been established that even the best solution runs into stagnation, studying the initial conditions of the randomly populated swarm may give more insights on early identification of stagnating solutions so that the algorithm can be trained to completely avoid them.

Author Contributions: Conceptualization, A.U. and Z.I.B.F.; Methodology, A.U. and Z.I.B.F.; Software, A.U.; Validation, A.U., Z.I.B.F., and A.K.; Formal analysis, A.U.; Investigation, A.U.; Resources, A.K.; Data curation, A.U. and A.K.; Writing—original draft preparation, A.U.; Writing—review and editing, A.U. and Z.S.; Visualization, A.U., Z.S., and A.K.; Supervision, Z.S.; Project administration, Z.S.; Funding acquisition, Z.S. All authors have read and agreed to the published version of the manuscript.

Funding: This research was funded by the Natural Science Foundation of Hebei Province of China, grant number F2017202243.

Conflicts of Interest: The authors declare no conflict of interest.

References

1. Umar, A.; Shi, Z.; Wang, W.; Farouk, Z.I.B. A Novel Mutating PSO Based Solution for Inverse Kinematic Analysis of Multi Degree-Of-Freedom Robot Manipulators. In Proceedings of the 2019 IEEE International Conference on Artificial Intelligence and Computer Applications, Dalian, China, 29–31 March 2019; pp. 554–559. [CrossRef]
2. Kennedy, J.; Eberhart, R.C. Particle Swarm Optimization. In *Proceeding of International Conference on Neural Networks*; IEEE Press: Perth, Australia, 1995; pp. 1942–1948. [CrossRef]
3. Zhu, L.; Feng, R.; Li, X.; Xi, J.; Wei, X. A Tree-Shaped Support Structure for Additive Manufacturing Generated by Using a Hybrid of Particle Swarm Optimization and Greedy Algorithm. *J. Comput. Inf. Sci. Eng.* **2019**, *19*, 1–12. [CrossRef]
4. Zhang, X.; Lu, D.; Zhang, X.; Wang, Y. Antenna Array Design by a Contraction Adaptive Particle Swarm Optimization Algorithm. *EURASIP J. Wirel. Commun. Netw.* **2019**, *2019*, 1–7. [CrossRef]
5. Cai, X.; Cui, Z.; Zeng, J.; Tan, Y. Self-adaptive PID-controlled Particle Swarm Optimization. In Proceedings of the Chinese Control Conference, Hunan, China, 26–31 July 2007; pp. 799–803. [CrossRef]
6. Cui, Z.; Cai, X.; Zeng, J.; Yin, Y. PID-Controlled Particle Swarm Optimization. *Mult. Valued Log. Soft Comput.* **2010**, *16*, 585–609.
7. Lu, Y.; Yan, D.; Zhang, J.; Levy, D. A Variant with a Time Varying PID Controller of Particle Swarm Optimizers. *Inf. Sci.* **2015**, *297*, 21–49. [CrossRef]
8. Xiang, Z.; Ji, D.; Zhang, H.; Wu, H.; Li, Y. A Simple PID-based Strategy for Particle Swarm Optimization Algorithm. *Inf. Sci.* **2019**, *502*, 558–574. [CrossRef]
9. Chang, C.; Wu, X. An Improved Particle Swarm Optimization Algorithm. *Adv. Intell. Sys. Comp.* **2020**, *928*, 1406–1410. [CrossRef]
10. Zidan, A.; Tappe, S.; Ortmaier, T. Auto-tuning of PID Controllers for Robotic Manipulators using PSO and MOPSO. *Lect. Notes Electr. Eng.* **2020**, *495*, 339–354. [CrossRef]
11. Peng, Z.; Al Chami, Z.; Manier, H.; Manier, M.-A. A Hybrid Particle Swarm Optimization for the Selective Pickup and Delivery Problem with Transfers. *Eng. Appl. Artif. Intell.* **2019**, *85*, 99–111. [CrossRef]
12. Jiang, G.; Luo, M.; Bai, K.; Chen, S. A Precise Positioning Method for a Puncture Robot Based on a PSO-Optimized BP Neural Network Algorithm. *Appl. Sci.* **2017**, *7*, 969. [CrossRef]
13. Iacca, G.; Caraffini, F.; Neri, F. Compact Differential Evolution Light: High Performance Despite Limited Memory Requirement and Modest Computational Overhead. *J. Comput. Sci. Technol.* **2012**, *27*, 1056–1076. [CrossRef]
14. Li, J.; Zhang, J.; Jiang, C.; Zhou, M. Composite Particle Swarm Optimizer with Historical Memory for Function Optimization. *IEEE Trans. Cybern.* **2015**, *45*, 2350–2363. [CrossRef] [PubMed]
15. Santucci, V.; Milani, A.; Caraffini, F. An Optimisation-Driven Prediction Method for Automated Diagnosis and Prognosis. *Mathematics* **2019**, *7*, 1051. [CrossRef]
16. Hu, J.; Chen, D.; Liang, P. A Novel Interval Three-Way Concept Lattice Model with Its Application in Medical Diagnosis. *Mathematics* **2019**, *7*, 103. [CrossRef]
17. Hu, J.; Yang, Y.; Chen, X. A Novel TODIM Method-based Three-Way Decision Model for Medical Treatment Selection. *Int. J. Fuzzy Syst.* **2017**, *33*, 3405–3417. [CrossRef]
18. Yao, J.T.; Azam, N. Web-based Medical Decision Support Systems for Three-Way Medical Decision Making with Game-theoretic Rough Sets. *IEEE Trans. Fuzzy Syst.* **2015**, *23*, 3–15. [CrossRef]
19. Wang, L.; Zhao, J.; Liu, D.; Lin, Y.; Zhao, Y.; Lin, Z.; Zhao, T.; Lei, Y. Parameter Identification with the Random Perturbation Particle Swarm Optimization Method and Sensitivity Analysis of an Advanced Pressurized Water Reactor Nuclear Power Plant Model for Power Systems. *Energies* **2017**, *10*, 173. [CrossRef]
20. Clerc, M.; Kennedy, J. The Particle Swarm—Explosion, Stability, and Convergence in a Multi-Dimensional Complex Space. *IEEE Trans. Evol. Comput.* **2002**, *6*, 58–73. [CrossRef]
21. Reynolds, W.C. *Flocks, Herds, and Schools: A Distributed Behavioral Model*; ACM: New York, NY, USA, 1987; Volume 21, pp. 25–34. [CrossRef]

22. Cui, Z.; Shi, Z. Boids Particle Swarm Optimization. *Int. J. Innov. Comput. Appl.* **2009**, *2*, 78–85. [CrossRef]
23. Kennedy, J. Small Worlds and Mega-Minds: Effects of Neighborhood Topology on Particle Swarm Performance. In Proceedings of the 1999 Congress on Evolutionary Computation-CEC99 (Cat. No. 99TH8406), Washington, DC, USA, 6–9 July 1999; Volume 3, pp. 1931–1938. [CrossRef]
24. Kennedy, J. The Particle Swarm: Social Adaptation of Knowledge. In Proceedings of the 1997 International Conference on Evolutionary Computation, Indianapolis, IN, USA, 13–16 April 1997; pp. 303–308. [CrossRef]
25. Mendes, R.; Kennedy, J.; Neves, J. Watch Thy Neighbor or How the Swarm Can Learn from Its Environment. In Proceedings of the 2003 IEEE Swarm Intelligence Symposium, Indianapolis, IN, USA, 24–26 April 2003; pp. 88–94. [CrossRef]
26. Qteish, A.; Hamdan, M. Hybrid Particle Swarm and Conjugate Gradient Optimization Algorithm. In *Advances in Swarm Intelligence*; ICSI 2010; Lecture Notes in Computer Science; Springer: Berlin/Heidelberg, Germany, 2010; Volume 6145, pp. 582–588. [CrossRef]
27. Qin, J.; Yin, Y.; Ban, X. A Hybrid of Particle Swarm Optimization and Local Search for Multimodal Functions. In *Advances in Swarm Intelligence*; ICSI 2010; Lecture Notes in Computer Science; Springer: Berlin/Heidelberg, Germany, 2010; Volume 6145, pp. 589–596. [CrossRef]
28. De-los-Cobos-Silva, S.G.; Andrade, M.Á.G.; Lara-Velázquez, P.; García, E.A.R.; Mora-Gutiérrez, R.A.; Ponsich, A. ABC-PSO: An Efficient Bioinspired Metaheuristic for Parameter Estimation in Nonlinear Regression. In *Advances in Soft Computing, Mexican International Conference on Artificial Intelligence MICAI 2016*; Pichardo-Lagunas, O., Miranda-Jiménez, S., Eds.; Lecture Notes in Computer Science; Springer: Cham, Switzerland, 2016; Volume 10062, pp. 388–400. [CrossRef]
29. Marinakis, Y.; Marinaki, M. Particle Swarm Optimization with Expanding Neighborhood Topology for the Permutation Flowshop Scheduling Problem. *Soft Comput.* **2013**, *17*, 1159–1173. [CrossRef]
30. Guo, W.; Zhu, L.; Wang, L.; Wu, Q.; Kong, F. An Entropy-Assisted Particle Swarm Optimizer for Large-Scale Optimization Problem. *Mathematics* **2019**, *7*, 414. [CrossRef]
31. Kong, F.; Jiang, J.; Huang, Y. An Adaptive Multi-Swarm Competition Particle Swarm Optimizer for Large-Scale Optimization. *Mathematics* **2019**, *7*, 521. [CrossRef]
32. Ahmad, N.; Mohammed, M.E.; Reza, S. DNPSO: A Dynamic Niching Particle Swarm Optimization for Multi-Modal Optimization. In Proceedings of the 2008 IEEE Congress on Evolutionary Computation, Hong Kong, China, 1–6 June 2008; pp. 26–32. [CrossRef]
33. Seo, J.H.; Im, C.H.; Heo, C.G.; Kim, J.K.; Jung, H.K.; Lee, C.G. Multimodal Function Optimization Based on Particle Swarm Optimization. *IEEE Trans. Magn.* **2006**, *42*, 1095–1098. [CrossRef]
34. Blackwell, T.M. Particle swarms and population diversity. *Soft Comput.* **2005**, *9*, 793–802. [CrossRef]
35. Blackwell, T.; Branke, J. Multi-swarm Optimization in Dynamic Environments. In *Applications of Evolutionary Computing*; Springer: Berlin/Heidelberg, Germany, 2004; Volume 3005, pp. 489–500. [CrossRef]
36. Ratnaweera, A.; Halgamuge, S.K.; Watson, H.C. Self-Organizing Hierarchical Particle Swarm Optimizer with Time-Varying Acceleration Coefficients. *IEEE Trans. Evol. Comput.* **2004**, *8*, 240–255. [CrossRef]
37. Qin, Z.; Yu, F.; Shi, Z.; Wang, Y. Adaptive Inertia Weight Particle Swarm Optimization. In *Artificial Intelligence and Soft Computing—ICAISC 2006*; Rutkowski, L., Tadeusiewicz, R., Zadeh, L.A., Żurada, J.M., Eds.; Lecture Notes in Computer Science; Springer: Berlin/Heidelberg, Germany, 2006; Volume 4029, pp. 450–459. [CrossRef]
38. Yang, X.; Yuan, J.; Mao, H. A Modified Particle Swarm Optimizer with Dynamic Adaptation. *Appl. Math. Comput.* **2007**, *189*, 1205–1213. [CrossRef]
39. Arumugam, M.S.; Rao, M.V.C. On the Improved Performances of the Particle Swarm Optimization Algorithms with Adaptive Parameters, Cross-over Operators and Root Mean Square (RMS) Variants for Computing Optimal Control of a Class of Hybrid Systems. *Appl. Soft Comput.* **2008**, *8*, 324–336. [CrossRef]
40. Pandey, B.B.; Debbarma, S.; Bhardwaji, P. Particle Swarm Optimization with varying Inertia Weight for solving nonlinear optimization problem. In Proceedings of the International Conference on Electrical, Electronics, Signals, Communication and Optimization, EESCO, Visakhapatnam, India, 24–25 January 2015. [CrossRef]
41. Bingul, Z.; Karahan, O. Dynamic identification of Staubli RX-60 Robot using PSO and LS Methods. *Expert Syst. Appl.* **2011**, *38*, 4136–4149. [CrossRef]
42. Xue-qian, W.; Hou-de, L.; Ye, S.; Bin, L.; Ying-chun, Z. Research on Identification Method of Kinematics for Space Robot. *Procedia Eng.* **2012**, *29*, 3381–3386. [CrossRef]

43. Feng, F.; Hu, H.; Guo, Z. Application of Genetic Algorithm PSO in Parameter Identification of SCARA Robot. In Proceedings of the Chinese Automation Congress, CAC, Jinan, China, 20–22 October 2017; pp. 923–927. [CrossRef]
44. Gao, G.; Lin, F.; San, H.; Wu, X.; Wang, W. Hybrid Optimal Kinematic Parameter Identification for an Industrial Robot Based on BPNN-PSO. *Complexity* **2018**, 1–11. [CrossRef]
45. Nizar, R.; Adel, M.A. Inverse Kinematics using Particle Swarm Optimization, a Statistical Analysis. *Procedia Eng.* **2013**, *64*, 1602–1611. [CrossRef]
46. Sarosh, P.; Tarek, S. Task Based Synthesis of Serial Manipulators. *J. Adv. Res.* **2015**, *6*, 479–492. [CrossRef]
47. Sarosh, P.; Tarek, S. Using Task Descriptions for Designing Optimal Task Specific Manipulators. In Proceedings of the IEEE/RSJ International Conference on Intelligent Robots and Systems (IROS), Hamburg, Germany, 29 September–3 October 2015; pp. 3544–3551. [CrossRef]
48. Mizuno, N.; Nguyen, C.H. Parameters Identification of Robot Manipulator based on Particle Swarm. In Proceedings of the 13th IEEE International Conference on Control and Automation, ICCA, Ohrid, Macedonia, 3–6 July 2017; pp. 307–312. [CrossRef]
49. Fang, L.; Dang, P. A step identification method of joint parameters of robots based on the measured pose of end-effector. *Proc. Inst. Mech. Eng. Part C J. Mech. Eng. Sci.* **2015**, *229*, 3218–3233. [CrossRef]
50. Chatterjee, A.; Siarry, P. Nonlinear Inertia Weight Variation for Dynamic Adaption in Particle Swarm Optimization. *Comput. Oper. Res.* **2006**, *33*, 859–871. [CrossRef]
51. Xin, J.; Chen, G.; Hai, Y. A Particle Swarm Optimizer with Multi-Stage Linearly-Decreasing Inertia Weight. In Proceedings of the International Joint Conference on Computational Sciences and Optimization, Sanya, China, 24–26 April 2009; Volume 1, pp. 505–508. [CrossRef]
52. Wang, H.; Qian, F. Improved PSO-based Multi-Objective Optimization using Inertia Weight and Acceleration Coefficients Dynamic Changing, Crowding and Mutation. In Proceedings of the 7th World Congress on Intelligent Control and Automation, Chongqing, China, 25–27 June 2008; pp. 4473–4478. [CrossRef]
53. Kononova, A.V.; Corne, D.W.; Wilde, P.D.; Shneer, V.; Caraffini, F. Structural Bias in Population-based Algorithms. *Inf. Sci.* **2015**, *298*, 468–490. [CrossRef]

Article

An Improved Bytewise Approximate Matching Algorithm Suitable for Files of Dissimilar Sizes

Víctor Gayoso Martínez, Fernando Hernández-Álvarez and Luis Hernández Encinas *

Institute of Physical and Information Technologies (ITEFI), Spanish National Research Council (CSIC), Serrano 144, 28034 Madrid, Spain; victor.gayoso@iec.csic.es (V.G.M.); fernando.hernandez@iec.csic.es (F.H.-Á.)

* Correspondence: luis@iec.csic.es; Tel.: +34-91-561-88-06

Received: 21 February 2020; Accepted: 30 March 2020; Published: 2 April 2020

Abstract: The goal of digital forensics is to recover and investigate pieces of data found on digital devices, analysing in the process their relationship with other fragments of data from the same device or from different ones. Approximate matching functions, also called similarity preserving or fuzzy hashing functions, try to achieve that goal by comparing files and determining their resemblance. In this regard, ssdeep, sdhash, and LZJD are nowadays some of the best-known functions dealing with this problem. However, even though those applications are useful and trustworthy, they also have important limitations (mainly, the inability to compare files of very different sizes in the case of ssdeep and LZJD, the excessive size of sdhash and LZJD signatures, and the occasional scarce relationship between the comparison score obtained and the actual content of the files when using the three applications). In this article, we propose a new signature generation procedure and an algorithm for comparing two files through their digital signatures. Although our design is based on ssdeep, it improves some of its limitations and satisfies the requirements that approximate matching applications should fulfil. Through a set of ad-hoc and standard tests based on the FRASH framework, it is possible to state that the proposed algorithm presents remarkable overall detection strengths and is suitable for comparing files of very different sizes. A full description of the multi-thread implementation of the algorithm is included, along with all the tests employed for comparing this proposal with ssdeep, sdhash, and LZJD.

Keywords: approximate matching; context-triggered piecewise hashing; edit distance; fuzzy hashing; LZJD; multi-thread programming; sdhash; signatures; similarity detection; ssdeep

---

1. Introduction

Digital forensics is the branch of Mathematics and Computer Science in charge of identifying, recovering, analysing, and providing conclusions about digital evidence found on electronic devices. When inspecting the content of a computer or a mobile phone, the first step for the digital forensics expert is to reduce all the data available, extracting the important information that can be analysed more efficiently [1].

An initial strategy to obtain that reduction consists in using cryptographic hashing functions such as MD5 or SHA-1. Even though those algorithms are not recommended for cryptographic purposes, they are still valid for determining if two files are the same, considering that the probability for two files to have the same hash value is negligible. Precisely due to this reason, NIST (National Institute of Standards and Technology) developed a database in the early 2000s, called NSRL (National Software Reference Library), which contains hash values of sets of files of several trusted operating systems [2]. Nevertheless, these cryptographic hashing functions face a common problem when comparing files. If one of the files is modified even just in one byte, the final outcome of the comparison is negative.

In contrast to cryptographic hashing functions, approximate matching functions [3], also known as similarity preserving hashing (SPH) or fuzzy hashing functions, try to detect the resemblance between two files by linking similar inputs to similar outputs, indistinctly called in this context similarity signatures, fingerprints or digests [3]. These functions, which analyse files at byte level, are useful to compare a large variety of data and detect similar texts and even embedded objects (e.g., an image in a Word or OpenDocument text file) or binary fragments (e.g., a virus inside a file, a specific data packet in a network connection or similar content in audio files). By using a diversity of methods (for example, computations with a sliding window, as it will be described in next sections), approximate matching functions are able to identify if even a single byte is changed.

In computer forensics, ssdeep is the best-known bytewise approximate matching application, and it is considered by some researchers as the de facto standard in some cybersecurity areas [4]. The implementation of ssdeep allows the generation of the signature of files (where the generated signature depends on the actual content of the file) and to compare two signature files or one signature file with a data file. The results in both cases are the same, using one method or the other depends on the data available to the user. However, even though ssdeep represented an important achievement in similarity detection techniques, and it is relatively up to date (at the time of writing this text, the latest version is 2.14.1, released in November 2017 [5]), during the last years some limitations have been highlighted by several researchers, publishing different enhancements or alternative theoretical approaches (see, for example, References [6–13]).

In this article we have delved on the most important limitations of ssdeep, sdhash, and LZJD, which are mainly the impossibility of comparing files of very different sizes in the case of ssdeep and LZJD, the scarce relationship between the comparison score obtained and the actual content of the files when using the three applications in some cases, and the excessive size of sdhash and LZJD signatures. As a result of our research, we are presenting in this contribution a signature generation procedure based on the one implemented in ssdeep had but that overcomes its design limitations, together with an easy-to-implement algorithm for comparing the content of two files through their digital signatures. We can confirm that, based on the list of requirements that, in our opinion, any similarity search function should fulfil, our algorithm provides results better adjusted to different situations than ssdeep and is able to compare any pair of files regardless of their respective size. Besides, our proposal can manage some signatures that represent certain special cases that include different content swapping schemes.

In order to obtain meaningful results, in addition to ssdeep we have also considered sdhash and LZJD in our tests. sdhash is also very popular and is representative of a completely different theoretical approach in approximate matching algorithms, as it uses Bloom filters in order to find the features that have the lowest empirical probability of being encountered by chance [14]. On the other hand, LZJD is a recent alternative that uses the Lempel-Ziv Jaccard distance [15]. As a result of the comparison with sdhash and LZJD, we can state that our algorithm provides better results regarding the recognition of the proportion of a file which is present in another file while generating significantly smaller signatures in almost all the cases where those two algorithms are used.

This article represents an improved version of the information presented as a PhD thesis by one of the authors in 2015 [16], where SiSe (Similarity Search) was first described (this thesis has not been published but it is available as open access at http://oa.upm.es/39099/). Compared to that work, this contribution features a different signature comparison algorithm and new tests, some of them using FRASH (Framework to test Algorithms of Similarity Hashing), a reputed tool designed for comparing this type of applications. As another novelty, this article includes the design details of the multi-threading C++ implementation of the proposed algorithm (allowing researchers to inspect the code of our implementation, uploaded to GitHub [17]), compares the application to the latest version of ssdeep (which in the last years has evolved and fine-tuned its capabilities), and broadens the comparative spectrum by adding LZJD to the test set.

The rest of this paper is organised as follows—Section 2 discusses the related work. Section 3 reviews the most significant features of ssdeep. In Section 4, we provide a complete description of our proposed algorithm. Section 5 describes the C++ implementation of SiSe, including the multi-thread design. Section 6 contains the ad-hoc tests that we have performed with a selected group of files in order to compare the similarity detection capabilities of our proposal to those of ssdeep, sdhash, and LZJD. Section 7 contains the tests performed with the FRASH framework and a well-known dataset. Finally, Section 8 summarises our conclusions about this topic.

2. Related Work

SPH functions can be divided into four categories [18]—Block-based hashing (BBH) functions, context-triggered piecewise hashing (CTPH) functions, statistically-improbable features (SIF) functions, and block-based rebuilding (BBR) functions, as described in the following paragraphs.

BBH functions generate and store cryptographic hashes for every block of a chosen fixed size (e.g., 512 bytes). In the next step, the block-level hashes from two different inputs are compared counting the number of common blocks and giving a measure of similarity between them. An implementation of this kind of fuzzy functions was developed by Nicholas Harbour, who created a program called dcfldd [19]. This function divides the input data into several blocks of a fixed length and calculates the corresponding cryptographic hash value for each of them. Even though this approach is very efficient from a computational point of view due to its simplicity, a single byte insertion or deletion at the beginning of a file could change all the block hashes, making this scheme too vulnerable.

The technique behind CTPH was originally proposed by Andrew Tridgell [20], who implemented spamsum, a context-triggered piecewise hashing application devoted to the identification of spam e-mails [21]. Its basic idea consists in locating content markers, called *contexts*, within a binary data object, calculating the hash of each fragment of the document delimited by the corresponding contexts, and storing the sequence of hashes. Thus, the boundaries of the fragments are based on the actual content of the object and not determined by an arbitrarily fixed block size. In 2006, based on spamsum, Jesse Kornblum developed ssdeep [22], one of the first programs for computing context-triggered piecewise signatures. This algorithm generates a matching score in the range 0–100. This score must be interpreted as a weighted measure of how similar these files are, where a higher result implies a greater similarity of the files [23].

The SIF approach is based on the idea of identifying a set of features in each of the objects under study and then comparing the features, where a feature in this context is a sequence of consecutive bits selected by some criteria from the file that stores the object. As a practical implementation of the concept, Vassil Roussev decided to use entropy in order to find statistically-improbable features [24]. With this idea, he proposed a new algorithm called sdhash, whose goal was to pick object features that are least likely to occur by chance in other data objects [25]. This algorithm produces a score between 0 and 100 that, according to its author, must be interpreted as a confidence value indicating how certain the tool is that the two data objects under comparison have non-trivial amounts of commonality [26].

A more recent example of this approach is LZJD, a proposal made by Edward Raff and Charles K. Nichola [15]. LZJD uses the same command-line arguments as sdhash but implements a different distance function, the Lempel-Ziv Jaccard distance [15]. According to its author, LZJD scores range from 0 to 100 and can be interpreted as the percentage of bytes shared by two files, which makes it comparable to the other applications considered in this contribution.

In turn, BBR functions use external data, which are blocks chosen randomly, uniformly or in a fixed way, in order to rebuild a file. The process compares the bytes of the original file to the chosen blocks and calculates the differences between them, using the Hamming distance or any other metric. Two of the best known implementations of this approach are bbHash [10] and SimHash [27].

It is important to mention that, while there are other high-level information retrieval techniques based on the semantic analysis of the content (e.g., see References [28–30]), SPH tools work at a lower level by directly analysing the value of the content as a byte sequence.

When analysing the similarity between two files, it is important to distinguish the concepts of resemblance and containment. While resemblance indicates how much an object looks like another one, containment indicates the presence of an object inside another one [31]. SPH applications adjust their results to either one of these concepts or both. As it will become clear later, our proposal uses a combination of both concepts.

Other interesting approach developed in recent years is TLSH (Trend Micro Locality Sensitive Hash) [32], a Locality Sensitive Hashing (LSH) application which is related to work done in the data mining area. LSH algorithms classify similar input items into the same groups with high probability, maximizing hash collisions [33]. However, while ssdeep, sdhash, and LZJD provide a similarity score between two digests in the range [0–100], in TLSH a distance score of 0 represents that the files are identical and higher scores represent a greater distance between the analysed documents. As TLSH does not imposes a limit on the score (according to the authors it can potentially go up to over 1000), it is not possible to directly compare its results to those of the other applications, so we decided not to include it in our tests.

3. Review of Ssdeep

The kernel of the signature generation algorithm in ssdeep is a rolling hash very similar to the one employed in spamsum [21] and rsync [34]. The rolling hash is used to identify a set of trigger points in the file that depend on the content of a sliding window of 7 bytes (i.e., the number and location of the trigger points totally depend on the content that is processed by the application, as every byte is taken into account in the calculations). The algorithm hits a trigger point every time the rolling hash (based on the Adler-32 function [35]), produces a value that matches a predefined condition. A second hashing function based on the FNV (Fowler-Noll-Vo) algorithm [36] is consequently used to calculate the hash values of the content located between two consecutive trigger points. Then, the last 6 bits of each hash value is translated into a Base64 character. The final signature is formed by concatenating the single characters generated at all the trigger points (with a maximum of 64 characters per signature).

The number of characters of the signature is strongly determined by the frequency of appearance of the trigger points. Therefore, the first step that must be completed by the algorithm is to estimate the value of the block size that would produce a final signature of 64 characters. More precisely, the value of the initial block size is the smaller number of the format $3 \cdot 2^n$ (where n is a positive integer value) which is not lower than the value obtained after dividing by 64 the input size in bytes. Another condition is that if the length of the signature generated does not reach 32 characters, ssdeep modifies the block size and executes the algorithm another time to produce a new signature. This process is done repeatedly until the signature obtained is of at least 32 characters

The signature generation algorithm produces two signatures in order to be able to compare a higher number of files. The reason for that is that sometimes similar files have slightly different block sizes. The two signatures are known as the leading signature, which uses a certain value for the block size (the leading block size), and the secondary signature, whose secondary block size value is twice the value of the leading block size.

The signature format produced by ssdeep consists of a header followed by one hash per line. The content of the header used in its latest versions is the following:

`ssdeep,1.1--blocksize:hash:hash,filename,`

where the element `ssdeep` identifies the file type, `1.1` informs about the version of the file format (not to be confused with the version of the program), `--` acts as a separator, and the remainder of the line identifies the elements displayed below the header (block size, primary hash, secondary hash, and filename).

In Information Theory, the edit distance between two strings of characters generally refers to the number of operations required to transform one string into the other. There are several ways to define the edit distance, depending on which operations are allowed. The ssdeep edit distance algorithm is based on the Damerau-Levenshtein distance between two strings [37,38].

That distance compares two strings and determines the minimum number of operations necessary to transform one string into the other. The only operations allowed in this distance comparison are insertions, deletions, substitutions of a single character, and transpositions of two adjacent characters [39,40].

In ssdeep, insertions and deletions have a weight of 1, substitutions a weight of 3, and transpositions a weight of 5. For instance, using this algorithm, the distance between the strings "Saturday" and "Sundays" is 5, as it can be seen in the following sequence, where the elements in bold format represent the characters that are added or removed at each step:

$$\text{S}\mathbf{a}\text{turday} \xrightarrow{\textit{deletion}} \text{S}\mathbf{t}\text{urday} \xrightarrow{\textit{deletion}} \text{Su}\mathbf{r}\text{day} \xrightarrow{\textit{deletion}} \text{Suday} \xrightarrow{\textit{insertion}} \text{Su}\mathbf{n}\text{day} \xrightarrow{\textit{insertion}} \text{Sunday}\mathbf{s}.$$

In practice, a consequence of assigning the value 3 to substitutions and 5 to transpositions is that the edit distance calculated only uses insertions and deletions. Therefore, substitutions and transpositions have a weight of 2 (a deletion followed by an insertion).

One of the limitations that this design has is that, for a specific string, a rotated version of it is assigned many insertion and deletion operations, when with regards to the content they are basically the same (i.e., the content is the same, although the order of the substrings is different). Consider for example the strings "abcd1234" and "1234abcd".

Finally, the score that ssdeep produces is normalised in the range $[0, 100]$, where 100 is associated to a perfect match and 0 to a complete mismatch. If the two signatures have different block sizes, then ssdeep automatically sets the score to 0 without performing further calculations.

In addition to the previous information, ssdeep defines the minimum length of the longest common substring as 7. If the longest common substring length detected during the procedure is less than that value, then ssdeep provides a score of 0.

The source code of ssdeep and implementations for both Windows and Linux are freely available [5], so interested readers can inspect the code and test the application.

4. Design of SiSe

4.1. Key Elements for Improvement

Frank Breitinger and Harald Baier presented in 2012 a list of four general properties for similarity preserving hashing functions [41], which they later extended to the following five characteristics [11]:

- *Compression*: The output must be several orders of magnitude smaller than the input for performance and storage reasons.
- *Ease of computation*: The processes of generating the hash value for a given file and making comparisons between files must be fast.

- *Similarity score*: The comparison function must provide a number representing a matching percentage value, so results for different files can be directly compared.
- *Coverage*: Every byte of the input data must be employed for calculating the similarity score.
- *Obfuscation resistance*: It must be difficult to obtain false negative and false positive results, even after manipulating the input data.

Once the analysis of the characteristics of ssdeep was completed, we were able to identify a list of additional practical features (which are closely related to ssdeep's limitations) that, in our consideration, any bytewise approximate matching function should provide (see Reference [16]):

- *Generation of results consistent with the content of the compared files*: This requirement implies the existence of a simple and clear definition of the concept of similarity, and how to express that concept as a number. In this context, this contribution uses a definition that states that similarity is the degree to which the contents of two files look like or are the same. This definition encompasses the concepts of *resemblance* and *containment* mentioned in Reference [3]. In order to measure it, a value that represents the percentage of the larger file which is also contained in the smaller file is provided.
- *Creation of signatures whose length is not affected by the input file size*: In order to avoid problems when storing the signatures of large files, the length of the signatures should be independent of the input file size.
- *Detection of content swapping*: This concept refers to the situation in which some portion of the document is taken from its original location and positioned in a different part, so from a graphical point of view it could be seen as moving data blocks inside the document. For a file where content swapping is the only operation performed, the result should be close to 100, as the content of the original and the modified files is basically the same.
- *Comparison of files of different sizes without limit*: In some cases, it is necessary to compare files of dissimilar size. If the only material available to the user are the signatures, then the files will be compared only if those signatures have a common block size. Implementations that allow the comparison of a signature file and a content file need to use the block size associated to the signature when processing the data file, which does not always produce the desired results, specially when the value of that block size is too big and does not generate a valid signature for the data file. This situation can happen, for instance, when comparing files five or ten times larger, so implementations need to provide a means for comparing such files.

As the reader may note, the previous requirements can be divided into two groups associated to the procedures of signature generation and signature comparison. For the sake of clarity, we provide below (see Table 1) the two differentiated sets of requirements as they will be referred to in the rest of the document:

Table 1. Requirements that should be fulfilled by any approximate matching application.

Signature Generation Procedure	Signature Comparison Procedure
(G1) Signature compression. (G2) Ease of computation. (G3) Coverage. (G4) Independence regarding the input file size.	(C1) Generation of a similarity score consistent with the content of the compared files. (C2) Ease of computation. (C3) Obfuscation resistance. (C4) Content swapping detection. (C5) Comparison of files of different sizes.

We have established a similarity definition that considers an important problem either not completely taken into account or implemented in an unsatisfactory way by other functions: We believe that the similarity comparison should detect the inclusion of the content of a file in another file, but at the same time it should also indicate in some way which parts of the larger file are contained in the smaller file. This feature would be very useful in many contexts, such as plagiarism detection. For example, if we consider three files extracted from the same book (one with the first chapter, another with the first ten chapters, and the last one with the whole book), the comparison should work in the following way—the result of comparing the first and the second file should be different from the one obtained when comparing the first and the third file. This outcome is possible with our similarity definition, as it provides more granularity about the results.

4.2. SiSe Signature Generation Procedure

As mentioned before, the SiSe signature generation procedure is based on that implemented in ssdeep. For example, the generation of the initial block size in SiSe and in ssdeep are the same (i.e., the minimum length of the leading signature is half the initially expected length), using the same versions of the Adler-32 and FNV (Fowler-Noll-Vo) functions, which were also selected by spamsum because their performance was better compared to other cryptographic functions such as MD5 or the SHA family.

However, SiSe includes important modifications in order to be aligned with the requirements described above. One of the more obvious differences between SiSe and ssdeep is the generation of two characters per trigger point instead of only one, using the last 12 bits of the FNV output instead of the last 6 bits. With this decision, the main drawback is that SiSe signatures require twice the number of characters than ssdeep signatures for the same file and block size. Nevertheless, in order to keep a low percentage of false positives, from our perspective it is essential to increase the number of characters generated with the output of the FNV function.

Even though SiSe does not impose a limit on the maximum length for the signatures when generating them, for performance reasons we have imposed in its implementation an arbitrary limit of 2560 double characters when comparing signatures. That value has been selected so, in most signatures, it is not exceeded, as the expected number of double characters in SiSe (and of single characters in ssdeep) is 64.

Another difference with respect to ssdeep is that SiSe computes three signatures in each execution loop, with block sizes b, $b/2$ and $b/4$ (in Reference [12] there is a similar technique, but Chen and Wang use b, $2b$, $4b$, and $8b$ instead, so our algorithm implements the opposite approach). The reason behind this decision is that the experimental work described in Reference [9] (which employed 15,036 files of several types totalling 3.84 gigabytes) demonstrated that 67.3% of the tested files needed the main loop of the algorithm (i.e., the loop that scans the file at byte level and processes its full content) to be executed once, while 26.2% needed to run the main loop twice, and the rest of the files needed at least three loop executions. In the case of our design, it is possible to choose as leading and secondary signature the ones related to block sizes b and $b/2$, or to block sizes $b/2$, and $b/4$, after any execution of the main loop. With this, it is not necessary to perform any additional loop execution and, consequently, the running time will decrease significantly in some cases. Figure 1 illustrates the difference between ssdeep and SiSe regarding the number of main loop executions for each case.

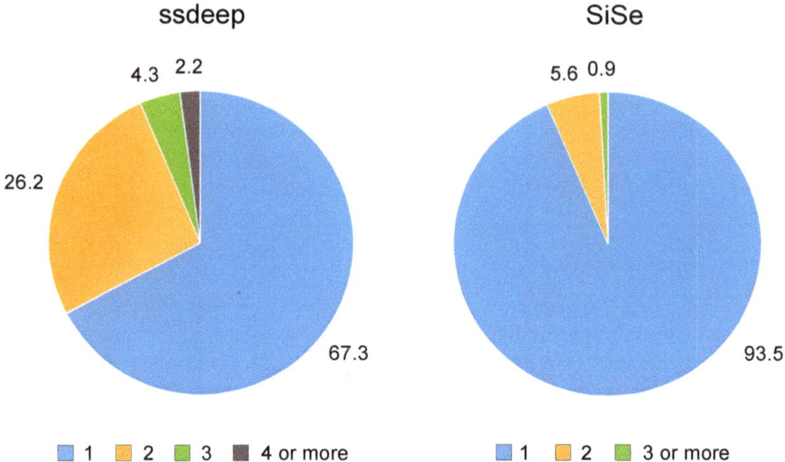

Figure 1. Percentage of files that need a different number of executions of the main loop.

While, in theory, calculating a third signature increments the complexity of the implementation and, in principle, would increment the running time, the difference is negligible in practice, since our algorithm checks if a certain byte is a trigger point when using block size b only if that byte is a trigger point for both $b/4$ and $b/2$. Besides, our approach allows the completion of only one pass most of the times (more precisely, in more than 90% of the cases) in comparison to ssdeep, where two or more passes are necessary in almost one third of the cases.

In our implementation we have decided to use b and $b/2$ (or $b/2$ and $b/4$) instead of b and $2b$ as the relationships between the block sizes of the leading and secondary signatures because, in this way, the length of both signatures surpasses the minimum length (32 double characters). This fact becomes important when using the secondary signature in a comparison: If, for the sake of precision, it is important for the length of the leading signature not to have less than 32 double characters, then the secondary signature should satisfy the same requirement. In our case, any signature selected for a comparison will surpass the minimum length, improving consequently the accuracy of the comparisons. It is obvious that the disadvantage of this design is that, in most files, the length of the secondary signature is larger than the length of the leading signature (the smaller the block size, the higher the number of trigger points that are likely to be detected), incrementing the size of the file containing both signatures.

Algorithm 1 presents the details of SiSe's signature generation procedure [16]. The meaning of the different functions included in the algorithm is as follows:

- `length(element)`: If the argument is a file, the function returns the length in bytes of the file. If it is a string, it returns the number of characters of the string.
- `floor(value)`: It computes the largest integer not greater than `value`.
- `subarray(array,start,end)`: It copies from `array` the subarray whose first byte is located at `start` and whose last byte is located just before the byte indicated by the position `end`.
- `updateWindow(byte)`: It updates the content of the sliding window taking into account the byte passed as the argument of the function.
- `Adler*(input)`: It computes the value associated to the byte array `input` using ssdeep's variant of the Adler-32 function.
- `FNV*(input)`: It computes the value associated to the byte array `input` using ssdeep's variant of the FNV function.

- `codify(value)`: It generates two Base64 characters (one using the last group of 6 bits of `value`, and the other one using the previous group of 6 bits).
- `append(string,chars)`: It appends two characters to `string`.

Algorithm 1 SiSe signature generation.

```
inputSize ← length(inputFile)
blockSize ← 3
dec ← 1
quotient ← floor(inputSize/64 )
while (blockSize < quotient) do
    blockSize ← 2·blockSize
end while
sig1Len, sig2Len ← 0
blockSize1 ← blockSize
blockSize2 ← blockSize1/2
blockSize4 ← blockSize2/2
while ((sig1Len < 64) and (sig2Len < 64)) do
    index ← 0
    last1, last2, last4 ← -1
    while (index < inputSize) do
        currentByte ← subArray(baInput,index,index+1)
        baWindow ← updateWindow(currentByte)
        resultAdler ← Adler*(baWindow)
        if ((resultAdler % blockSize4) = (blockSize4 - dec)) then
            baBlock4 ← subArray(baInput,last4+1,index+1)
            resultFNV ← FNV*(baBlock4)
            pairChars ← codify(resultFNV)
            sig4 ← append(sig4, pairChars)
            last4 ← index
            if ((resultAdler % blockSize2) = (blockSize2 - dec)) then
                baBlock2 ← subArray(baInput,last2+1,index+1)
                resultFNV ← FNV*(baBlock2)
                pairChars ← codify(resultFNV)
                sig2 ← append(sig2, pairChars)
                last2 ← index
                if ((resultAdler % blockSize1) = (blockSize1 - dec)) then
                    baBlock1 ← subArray(baInput,last1+1,index+1)
                    resultFNV ← FNV*(baBlock1)
                    pairChars ← codify(resultFNV)
                    sig1 ← append(sig1, pairChars)
                    last1 ← index
                end if
            end if
        end if
    end while
    sig1Len ← length(sig1)
    sig2Len ← length(sig2)
    if (sig1Len < 64) and (sig2Len < 64) then
        blockSize1 ← blockSize1/4
        blockSize2 ← blockSize1/2
        blockSize4 ← blockSize2/2
    else
        if (sig1Len < 64) then
            return sig1,sig2
        else
            return sig2,sig4
        end if
    end if
end while
```

4.3. SiSe Signature Comparison

In addition to the changes in the signature generation process explained above and the multi-threading design, in this contribution we have implemented a new signature comparison algorithm which compares two signature strings.

Our work is related to other edit distance algorithms that allow moves (e.g., References [42–45]). However, those contributions are mainly focused on the performance of such algorithms, and in many cases they impose important restrictions during the comparison process. For example, in Reference [42] the strings under comparison must be of equal size and must contain exactly the same characters; Reference [43] makes his study assuming that each letter occurs at most k times in the input strings; in Reference [44] the original position (p_1) and the final position (p_2) of the substring to be moved must satisfy a limiting rule (namely, if l is the length of the substring, it is necessary that p_2 is either smaller than p_1 or larger than $p_1 + l$). Regarding Reference [45], it uses a different technique based on the embedding of strings into a vector space and using a parsing tree.

Compared to the previously mentioned contributions, our algorithm manages text units comprised of two characters, while the rest of related algorithms, to the best of our knowledge, manage single characters, which allows us to improve the detection rate, as the probability of two pairs of characters being the same is $1/4096$ (the two Base64 characters of a pair use 12 bits) instead of $1/64$, as it is the case when using only one Base64 character (and the 6 bits associated to it).

Algorithm 2 shows all the calculations performed to calculate numerically the similarity of the strings `sig1` (with length `sig1Len`) and `sig2` (with length `sig2Len`), where the first step consists in creating a two-dimensional array of size (`sig1Len` + 1) × (`sig2Len` + 1). That array, called `arrComp`, initially contains the value 0 in all of its positions, but is updated whenever a double character of the first string matches a double character of the second string (only if both double characters are located in an even position, considering that the first position of a string is 0). The arrays identified as `arrx` and `array` mark if a character has already been taken into account during the computation of the score (a value `0` means that the character has not be considered for the score up to that point, whilst a value `1` means that the character belongs to a substring included in both input strings that has already been used during the calculation of the score).

Algorithm 2 SiSe signature comparison.

```
for all i such that 1 ≤ i ≤ sig1Len do
    for all j such that 1 ≤ j ≤ sig2Len do
        if sig1[i − 1] = sig2[j − 1] then
            if (i%2 = 1) and (j%2 = 1) then
                if sig1[i]=sig2[j] then
                    arrComp[i][j] = arrComp[i − 1][j − 1] + 1
                end if
            else
                if (i%2 = 0) and (j%2 = 0) then
                    if arrComp[i − 1][j − 1] > 0 then
                        arrComp[i][j] = arrComp[i − 1][j − 1] + 1
                        if i <sig1Len and j <sig2Len then
                            if sig1[i] ≠ sig1[j] then
                                list.add((arrComp[i][j],i,j))
                            end if
                        else
                            list.add((arrComp[i][j],i,j))
                        end if
                    end if
                end if
            end if
        end if
    end for
end for
index1 ← 0, index2 ← 0
for all item in list do
    if arrx[item.posRow]= 0 and arry[item.posCol]= 0 then
        points = points + item.val
        for all index1 such that (item.posRow - item.val + 1) ≤ index1 ≤ item.posRow do
            for all index2 such that (item.posCol - item.val + 1) ≤ index2 ≤ item.posCol do
                if ((arrx[index1]= 1) or (arry[index1]= 1)) then
                    points = points - 1
                end if
                arrx[index1] = 1
                arry[index2] = 1
            end for
        end for
    else
        dec = item.val;
        for all index1 such that item.posRow ≤ index1 ≥ item.posRow - item.val+1 do
            for all index2 such that item.posCol ≤ index2 ≥ item.posCol - item.val+1 do
                if arrx[index1]= 1 or arry[index2]= 1 then
                    dec ← dec - 1
                else
                    break
                end if
            end for
        end for
        if reduced>0 then
            points = points + dec
            for all index1 such that (item.posRow-item.val+1) ≤ index1 ≤ (item.posRow-removed) do
                for all index2 such that (item.posCol-item.val+1) ≤ index2 ← (item.posCol-(item.val-dec)) do
                    if arrx[index1]= 1 or arry[index2]= 1 then
                        points ← points - 1
                    end if
                    arrx[index1] ← 1
                    arry[index2] ← 1
                end for
            end for
        end if
    end if
end for
return (points*100)/max(sig1,sig2)
```

Table 2 shows an example of such an array when comparing the strings `A1B2C3D4F8` and `1A1BC3D4F7A1`. As it can be observed, the substring `1B` is not credited any point because in one string it starts at an odd position while in the other string it starts in an even position, so they do not produce a match.

Table 2. Comparison array example.

		A	1	B	2	C	3	D	4	F	8
		0	0	0	0	0	0	0	0	0	0
1	0	0	0	0	0	0	0	0	0	0	0
A	0	0	0	0	0	0	0	0	0	0	0
1	0	0	0	0	0	0	0	0	0	0	0
B	0	0	0	0	0	0	0	0	0	0	0
C	0	0	0	0	0	1	0	0	0	0	0
3	0	0	0	0	0	0	2	0	0	0	0
D	0	0	0	0	0	0	0	3	0	0	0
4	0	0	0	0	0	0	0	0	**4**	0	0
F	0	0	0	0	0	0	0	0	0	0	0
7	0	0	0	0	0	0	0	0	0	0	0
A	0	1	0	0	0	0	0	0	0	0	0
1	0	0	**2**	0	0	0	0	0	0	0	0

Once all the characters are compared and an array such as the one shown in Table 2 is completed, a list is created with elements that include three values: The highest value assigned to a series (displayed in bold in the table), and the row and column associated to that value. In the example, two elements would be added to the list, $(4, 8, 8)$ and $(2, 12, 2)$. Those elements, representing all the sequences of double characters (located at valid positions) are then processed by the second part of the algorithm in order to increase the number of points assigned to the comparison and remove those substrings from the comparison, so they are not taken into account more than once.

In the example shown, the final score would be 50, as 50% of string `1A1BC3D4F7A1` can be found in string `A1B2C3D4F8` (more specifically, the double characters `A1`, `C3`, and `D4` are shared by both strings, while the two strings contain three additional double characters that are not common).

5. SiSe Implementation

5.1. Interface

We have implemented our design of SiSe as a C++ multi-threaded command-line application that uses the features of the C++17 standard [46] and the OpenMP library [47]. The application implements a dual input command format: One adapted to the FRASH tests, and a proprietary one. Regarding the input format compatible with FRASH, SiSe can be used without modifiers with one or several files separated by spaces. This indicates SiSe that it must treat those files as content data and generate their signatures individually. In turn, if modifier `-x` is used along with the name of a file, it indicates the application that the file contains multiple signatures and that it must compare all signatures against the rest. Finally, if modifier `-r` is employed together with the path of a directory, SiSe will compute the signature of all the files contained in that folder.

In comparison, the proprietary input command format used by SiSe for creating a new signature is `sise -i input_file`, which allows the following additional optional elements:

`-o`: Output file name.
`-b`: Block size (if not indicated, the block size is computed by the application as in ssdeep).
`-d`: Decremental value, that is, the amount to be decremented to the block size in the trigger point computations (the default value is 1).
`-t`: Number of threads used by the application (if not indicated, the value employed is decided by SiSe based on the size of the input file).

Independently of the input command format used, the signature procedure creates a text string containing a header and two signatures (the leading and the secondary signatures). The header of the initial version is `SiSe-1.0--7:1--blocksize:hash,filename`, where the elements `7` and `1` are replaced by other numbers when non-default values selected by the user are employed for the window size and the decremental value, respectively.

The first signature that follows the header is the leading signature. Both signatures start with the block size and the value used in the modulus comparison to identify the trigger points (i.e., the block size minus the decremental value), the last one located between parenthesis and followed by a colon. The next item that appears is the actual signature string, corresponding to the previously stated block size. A double colon `::` separates both signatures. Finally, the complete path of the file that has been processed (written between quotation marks) appears separated from the previous information by a comma.

In order to illustrate the signature generation process, Table 3 shows the output provided by ssdeep and SiSe when processing the plain text file containing Robert Louis Stevenson's Treasure Island [48]. As it can be noted, the character that appears in the second position of each pair in the secondary signature of SiSe (the one with a block size of 3072) matches the corresponding character of the leading signature of ssdeep. This fact happens as the block size, windows size, and decremental value is the same for those two signatures. The coincidence in the series of characters ends at the 63th character of the ssdeep signature, as at that point ssdeep computes its final character using the rest of the document, independently of the length of that remaining part and the number of trigger points included in it. In contrast, SiSe continues computing all the characters derived from the presence of additional trigger points.

Table 3. ssdeep and SiSe signature examples.

```
> ssdeep pg27780.txt

ssdeep,1.1--blocksize:hash:hash,filename
3072:7ElbU+TnglXXzpfu2/8Fbo0ANtD3xKdeS/GVc0/bLe/sVcdJqZ/7y003FFh1g+Mp:27YpKwhKdeoD/JVjQT1J
+r0Sfitp4pc7,"/media/victor/USB/pg27780.txt"
```

```
> sise pg27780.txt

SiSe-1.0--7:1--blocksize(modvalue):hash,filename
6144(6143):k2C71YMp6KJw/hiKodceZoVDD//JfVrj0QVTT1GJA+Ir80+SrfGiEtMpq4UpucU7::3072(3071):Q7
BEU1UbaUe+1TInKgvluXnXkzMpDfLu12J/j8ZFfbwou0YAxNXtXDU3NxiKodceLSf/gGxVCcv0i/PbSLDeD/EsZVuc
zd+Jyq/ZC/M7fyH0a0G3JFaF1hc1Kgh+VMZVbrXJR/HrT1GJA+xYR6cvz7a580IfvpEBjAVAA+x3Tu0B+gTTW0Etn3
pfQ2wLkqx91FhovSu4uc1ePk,"/media/victor/USB/pg27780.txt"
```

The proprietary format of the command for comparing signatures and/or files is `sise -c input_file_1 input_file_2`, which allows these combinations:

- *Two signature files*: The comparison between signatures can be done only if they have at least one block size value in common. Additionally, the decremental value and the window size must be equal in both signature files. If the leading signatures use the same block size value, and the secondary signatures also employ the same block size value (but different from the block size of the leading signatures) at the same time, SiSe is able to apply the edit distance algorithm (see Algorithm 2 in Section 4.3) in both cases, and returns the highest score calculated.

- *A signature file and a data file*: The block size, decremental value, and window size are obtained from the signature file and used in the processing of the signature related to the data file.
- *Two data files*: In a first step, the leading and secondary signatures of both files are independently computed using the default parameters established for the block size, the decremental value, and the window size. Then, SiSe identifies the smallest block size value of the two leading signatures and processes again the signature of the other file with the new block size value.

5.2. Multi-Thread Design

The C++ implementation of SiSe uses the OpenMP library [47] both during the signature creation and the signature comparison procedures. Multi-threading is employed in the comparison phase when comparing digests using the -x option, that is, when requesting SiSe to compare all the signatures included in a file with the rest of signatures of that file. In that scenario, SiSe distributes the work between a number of threads that equals the number of logical cores of the processor, where the number of logical cores is the number of physical cores multiplied by the number of threads that can run on each core through the use of hyperthreading.

Multi-threading is also used during the signature generation process in two different ways. For any file larger than 1 megabyte, the strategy consists in dividing the file into several parts so each thread is assigned a portion of the file. Given the nature of the operations that each thread must perform, the multi-thread design tries to minimize the amount of bytes that are processed twice. Nevertheless, the double processing of some portions of the file cannot be avoided, as it will become clear with the description included below and Figure 2. Besides, when a command requests to process more than one file, SiSe creates a list of files whose size is less than 1 megabyte and assigns one thread per file up to the thread limit.

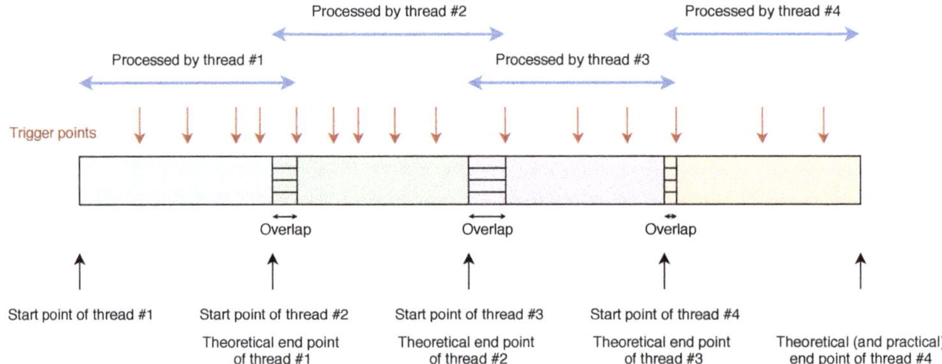

Figure 2. Representation of the multi-thread design.

Coming back to the multi-threading scheme for a file larger than 1 megabyte, when using n threads that file is divided into n parts, so each thread starts processing bytes at a different offset inside the file. If the working thread is thread #1, the first triggered point found by the thread provokes the computation of the first pair of characters of the signature. However, if the working thread is not thread #1, when the thread detects the first trigger point it adds no pair of characters to the signature, as at that point the thread lacks the knowledge of where the previous trigger point was located. From that moment on, all the threads work as expected and, whenever they find a trigger point, they compute the corresponding two-character element of the signature. Threads continue to work in that way and, except the last thread (which finishes its execution when it reaches the end of the file, computing at that moment the pair of characters associated to the end of the signature), they only stop when finding the first trigger point located after the byte that marks the theoretical end point of their respective file partitions.

This scheme brings as a consequence a certain overlapping of parts of the file processed both by threads $i-1$ and i, where the overlapped portion is read by thread $i-1$ at the end of its execution and by thread i at the beginning of its execution. The extent of the overlap depends on the content of the file, its size, and the number of threads. Figure 2 provides an example that illustrates the multi-thread scheme when four threads are used.

As the detection of trigger points depends on the value of the byte being processed and the value of the previous 7 bytes, before processing the first byte each thread fills in the sliding window array with the proper 7 bytes, so no false trigger points appear and thus the signatures are guaranteed to be the same independently of the number of threads used during the process.

Regarding the optimal number of threads to be employed by the application, after several tests it was decided to use one thread with files whose size is less than 1 megabyte and a number of threads that equal the number of logical cores in the rest of cases. The reason for doing this is that, for small files, the cost of setting up the threads and aggregating the results provided by them does not compensate the benefit of a reduced workload per thread. As for the buffer size used when reading the files, during the tests a buffer of one megabyte proved to be the best option.

6. Ad-Hoc Tests

This section shows the results of the comparisons performed with SiSe, ssdeep, sdhash, and LZJD in different scenarios designed by ourselves to verify the features associated to the characteristics described in Section 4.1.

We have classified the tests in four categories: Similarity tests, dissimilarity tests, special signatures tests, and suitability tests. For the first two groups, we have tested plain text files, Word documents, and BMP images. The tests with special signatures are intended to check the behaviour of the four algorithms in some special cases. The suitability tests try to determine the practicability of the applications in real-world scenarios by measuring the performance and signature size using files whose length span from less than 1 megabyte to several gigabytes.

As mentioned in Section 4, after performing several tests modifying the values of the block sizes, window sizes, and decremental values, we concluded that the results did not show appreciable differences. For this reason, we decided to use the default parameters (sliding window of 7 bytes, block size computed by the algorithm, and decremental value equal to 1) in all the tests that follow. Moreover, as they are the same values that ssdeep use, it is easier to derive conclusions from the results.

It is important to notice that, in all the tables related to tests in which two files are employed, the identifier linked to the row represents the first input argument and the identifier linked to the column represents the second input argument.

Tables associated to SiSe include four values in each cell. Those values are obtained when comparing two hash files, a content file and a hash file (in that order), a hash file and a content file (in that order), and two content files, respectively.

The values associated with the tests with ssdeep can be obtained either by comparing two signature files by means of the command `ssdeep -ak sigfile1 sigfile2` or by comparing a signature file and a content file (in that order) using the command `ssdeep -am sigfile contentfile`. As in all the tests both methods provide the same numerical value, in the tables related to ssdeep we have included each value once, even though all the tests have been performed separately with both commands.

Results can be obtained with sdhash and LZJD through two procedures. In the first one, a signature file with the hashes of all the files has been generated with the command `sdhash * > sigfile.sdbf` and, as a second step, that file has been used as input for the command `sdhash -c sigfile.sdbf` that compares all the hashes included in the file. In the second case, the results have been directly generated by using the command `sdhash -g *`. When using LZJD, the command `sdhash` must be replaced with the command `LZJD.jar`, as the author of LZJD recommends using the Java version of its algorithm [49]. In the tables with the sdhash results only one value appears in each cell, as the result obtained with the two procedures is exactly the same. In comparison, LZJD provides different values when using the two procedures with the same files, so in that case we have preferred to include both values in the corresponding table.

It is interesting to note that, while the ssdeep matrices are always symmetric, that is not always the case with the matrices related to SiSe. The reason for that fact is that the second score of each cell represents a test where a signature file (pertaining to the file identified by the row of the table) is compared against the content of a data file (whose identifier is at the top of the column that contains the cell with the score). According to its predefined behaviour, SiSe imposes the leading block size of the file associated to the row when processing the signature of the file associated to the column. If the leading block size of the file associated to the row is smaller than the theoretical leading block size of the file associated to the column, a signature (which theoretically is longer than the original signature that would have been generated with SiSe) is produced by the application. On the other hand, when the leading block size of the file associated to the row is larger than the theoretical leading block size of the file associated to the column, it could happen that, when that block size is forced on the file associated to the column, no signature of the minimum length could be generated. In this case, SiSe is not able to perform the comparison. In the rest of the cases there might be small differences in the outcome produced by SiSe for the second test (i.e., the second number displayed in every cell) in both comparisons (i.e., file A compared to file B, and file B compared to file A), depending on the leading block size used in each case. The same situation appears in the tests associated to the third score. However, when using two signature files or two data files (i.e., the first and fourth scenarios) the results are always symmetric for each comparison.

As a final comment, for better legibility in the tables, we have discarded the results related to the comparison of any file with itself.

6.1. Similarity Tests

6.1.1. Plain Text Documents

In this test, plain text files with the first 20 chapters of Miguel de Cervantes' Don Quijote, in the version offered by The Project Gutenberg [50], have been used. These files are the following: `Q01.txt` (10,887 bytes, Quijote's chapter 1), `Q02.txt` (23,877 bytes, chapters 1, 2), `Q03.txt` (37,342 bytes, chapters 1–3), `Q04.txt` (51,374 bytes, chapters 1–4), `Q05.txt` (60,526 bytes, chapters 1–5), `Q10.txt` (125,217 bytes, chapters 1–10), `Q15.txt` (204,198 bytes, chapters 1–15), and `Q20.txt` (305,527 bytes, chapters 1–20). Table 4 shows the percentage of the smaller file as a part of the larger file using the byte size as the comparison value, as in this test the smaller files are totally contained in the larger ones.

Table 4. Percentage of the larger file representing the smaller file with content in plain text format.

	Q01	Q02	Q03	Q04	Q05	Q10	Q15	Q20
Q01	-	45.6%	29.2%	21.2%	18.0%	8.7%	5.3%	3.6%
Q02	45.6%	-	63.9%	46.5%	39.4%	19.1%	11.7%	7.8%
Q03	29.2%	63.9%	-	72.7%	61.7%	29.8%	18.3%	12.2%
Q04	21.2%	46.5%	72.7%	-	84.9%	41.0%	25.2%	16.8%
Q05	18.0%	39.4%	61.7%	84.9%	-	48.3%	29.6%	19.8%
Q10	8.7%	19.1%	29.8%	41.0%	48.3%	-	61.3%	41.0%
Q15	5.3%	11.7%	18.3%	25.2%	26.6%	61.3%	-	66.8%
Q20	3.6%	7.8%	12.2%	16.8%	19.8%	41.0%	66.8%	-

Tables 5–9 present the outcomes of the tests using SiSe, ssdeep, sdhash, and LZJD. As it can be seen, SiSe is the application that offers results closer to the percentages included in Table 4, providing values better adapted to the similarity definition presented in Section 4.1. In addition to that, it is important to note that SiSe provides results (with two of its four operation modes) in several cases where ssdeep and LZJD return a value of 0, as for example in the comparison between `Q01.txt` and `Q04.txt`, the comparison between `Q02.txt` and `Q10.txt` or the comparison between `Q05.txt` and `Q15.txt`. The values generated by sdhash in those cases are 95, 99, and 100, respectively, which inform that the smaller file is definitely included in the larger file, but do not provide details about how much content of the larger file is replicated in the smaller file.

Table 5. Test results for similar plain text files with SiSe, part 1 (hash vs. hash/file vs. hash/hash vs. file/file vs. file).

	Q01	Q02	Q03	Q04
Q01	-	40/42/43/43	0/26/28/28	0/20/21/21
Q02	40/43/42/43	-	61/61/63/63	46/46/47/47
Q03	0/28/26/28	61/63/61/63	-	74/74/74/74
Q04	0/21/20/21	46/47/46/47	74/74/74/74	-
Q05	0/18/17/18	40/40/40/40	64/64/64/64	85/85/86/86
Q10	0/8/0/8	0/22/20/22	32/35/32/35	44/46/44/47
Q15	0/5/0/5	0/14/0/14	0/23/0/23	0/30/0/31
Q20	0/4/0/4	0/9/0/9	0/15/0/15	0/20/0/20

Table 6. Test results for similar plain text files with SiSe, part 2 (hash vs. hash/file vs. hash/hash vs. file/file vs. file).

	Q05	Q10	Q15	Q20
Q01	0/17/18/18	0/0/8/8	0/0/5/5	0/0/4/4
Q02	40/40/40/40	0/20/22/22	0/0/14/14	0/0/9/9
Q03	64/64/64/64	32/32/35/35	0/0/23/23	0/0/15/15
Q04	85/86/85/86	44/44/46/47	0/0/30/31	0/0/20/20
Q05	-	54/54/54/54	0/0/36/36	0/0/24/24
Q10	54/54/54/54	-	56/56/66/66	0/0/44/44
Q15	0/36/0/36	56/66/56/66	-	58/58/62/62
Q20	0/24/0/24	0/44/0/44	58/62/58/62	-

Table 7. Test results for similar plain text files with ssdeep.

	Q01	Q02	Q03	Q04	Q05	Q10	Q15	Q20
Q01	-	66	0	0	0	0	0	0
Q02	66	-	75	66	66	0	0	0
Q03	0	75	-	90	90	47	0	0
Q04	0	66	90	-	99	55	0	0
Q05	0	66	90	99	-	65	0	0
Q10	0	0	47	55	65	-	60	41
Q15	0	0	0	0	0	60	-	77
Q20	0	0	0	0	0	41	77	-

Table 8. Test results for similar plain text files with sdhash.

	Q01	Q02	Q03	Q04	Q05	Q10	Q15	Q20
Q01	-	95	95	95	95	95	95	95
Q02	95	-	99	99	99	99	99	99
Q03	95	99	-	100	100	100	100	100
Q04	95	99	100	-	100	100	100	100
Q05	95	99	100	100	-	100	100	100
Q10	95	99	100	100	100	-	100	100
Q15	95	99	100	100	100	100	-	100
Q20	95	99	100	100	100	100	100	-

Table 9. Test results for similar plain text files with LZJD (signature comparison/file comparison).

	Q01	Q02	Q03	Q04	Q05	Q10	Q15	Q20
Q01	-	25/35	20/22	0/0	0/0	0/0	0/0	0/0
Q02	25/35	-	59/54	45/38	37/31	0/0	0/0	0/0
Q03	20/22	59/54	-	66/63	53/50	20/21	0/0	0/0
Q04	0/0	45/38	66/63	-	77/77	25/28	0/0	0/0
Q05	0/0	37/31	53/50	77/77	-	34/34	0/0	0/0
Q10	0/0	0/0	20/21	25/25	34/34	-	46/49	29/29
Q15	0/0	0/0	0/0	0/0	0/0	46/49	-	53/53
Q20	0/0	0/0	0/0	0/0	0/0	29/29	53/53	-

There are two situations in which SiSe does not provide a comparison result. The first one happens when comparing two files with incompatible signatures. The second situation appears when comparing a small content file with the hash of a large file, as in that case the block size used in the comparison is too large for generating a valid signature in the smaller file. When the two other situations appear (comparing two content files or a large content file with the hash of a small file), valid results are always produced.

6.1.2. Word Documents

In this second test, the same book as in the previous test was used, although the chapters of Don Quijote have been saved as Microsoft Word 2016 using the identifiers `Q01.docx` (17,637 bytes), `Q02.docx` (24,140 bytes), `Q03.docx` (30,596 bytes), `Q04.docx` (37,422 bytes), `Q05.docx` (41,895 bytes), `Q10.docx` (68,691 bytes), `Q15.docx` (105,228 bytes), and `Q20.docx` (150,969 bytes).

Table 10 shows the percentage of the smaller file as a part of the larger file using the byte size as the comparison value, as in this test the smaller files are totally contained in the larger ones. The values included in this table are slightly higher than the values presented in Table 4 due to the internal format used by Microsoft Word.

Table 10. Percentage of the larger file representing the smaller file with content in Microsoft Word format.

	Q01	Q02	Q03	Q04	Q05	Q10	Q15	Q20
Q01	-	73.1%	57.6%	47.1%	42.1%	25.7%	16.8%	11.7%
Q02	73.1%	-	78.9%	64.5%	57.6%	35.1%	23.0%	16.0%
Q03	57.6%	78.9%	-	81.8%	73.0%	44.5%	29.1%	20.3%
Q04	47.1%	64.5%	81.8%	-	89.3%	54.5%	35.6%	24.8%
Q05	42.1%	57.6%	73.0%	89.3%	-	61.0%	39.8%	27.8%
Q10	25.7%	35.1%	44.5%	54.5%	61.0%	-	65.3%	45.5%
Q15	16.8%	23.0%	29.1%	35.6%	39.8%	65.3%	-	69.7%
Q20	11.7%	16.0%	20.3%	24.8%	27.8%	45.5%	69.7%	-

Tables 11–15 present the results obtained with SiSe, ssdeep, sdhash, and LZJD, respectively. With the information included in those tables, it is obvious that SiSe can compare more files than ssdeep. The values provided by SiSe in this test are lower than the values generated in the previous test, which is once again due to the internal structure of Word documents and the metadata that is contained in each file.

In the case of sdhash, it can be said that the values shown in Table 14 do not reflect the trend that can be visualised with SiSe where, for a specific file (e.g., `Q01.docx`) the comparison score decreases as the other file used in the comparison contains a larger portion of the book (`Q02.docx`, `Q05.docx`, `Q10.docx`, etc.), as the content of the initial file is diluted in the larger files.

Finally, it is worth mentioning that LZJD fails to produce results above zero in many cases.

Table 11. Test results for similar Word documents with SiSe, part 1 (hash vs. hash/file vs. hash/hash vs. file/file vs. file).

	Q01	Q02	Q03	Q04
Q01	-	32/32/32/32	29/29/30/30	23/23/23/23
Q02	32/32/32/32	-	32/32/32/32	24/24/24/24
Q03	29/30/29/30	32/32/32/32	-	52/53/52/53
Q04	23/23/23/23	24/24/24/24	52/52/53/53	-
Q05	21/21/21/21	22/22/22/22	47/47/48/48	49/49/49/49
Q10	0/14/0/14	0/15/12/15	11/15/11/15	11/15/11/15
Q15	0/9/0/9	0/10/8/10	6/9/6/10	6/10/6/10
Q20	0/7/0/7	0/8/0/8	0/7/0/7	0/7/0/7

Table 12. Test results for similar Word documents with SiSe, part 2 (hash vs. hash/file vs. hash/hash vs. file/file vs. file).

	Q05	Q10	Q15	Q20
Q01	21/21/21/21	0/0/14/14	00/9/9	0/0/7/7
Q02	22/22/22/22	0/12/15/15	0/8/10/10	0/0/8/8
Q03	47/48/47/48	11/11/15/15	6/6/9/10	0/0/7/7
Q04	49/49/49/49	11/11/15/15	6/6/10/10	0/0/7/7
Q05	-	11/11/16/16	6/6/9/9	0/0/7/7
Q10	11/16/11/16	-	45/45/45/45	27/27/32/32
Q15	6/9/6/9	45/45/45/45	-	65/65/65/65
Q20	0/7/0/7	27/32/27/32	65/65/65/65	-

Table 13. Test results for similar Word documents with ssdeep.

	Q01	Q02	Q03	Q04	Q05	Q10	Q15	Q20
Q01	-	40	0	0	0	0	0	0
Q02	40	-	0	0	0	0	0	0
Q03	0	0	-	63	49	0	0	0
Q04	0	0	63	-	50	0	0	0
Q05	0	0	49	50	-	0	0	0
Q10	0	0	0	0	0	-	52	32
Q15	0	0	0	0	0	52	-	68
Q20	0	0	0	0	0	32	68	-

Table 14. Test results for similar Word documents with sdhash.

	Q01	Q02	Q03	Q04	Q05	Q10	Q15	Q20
Q01	-	44	36	34	31	35	33	32
Q02	44	-	38	37	22	36	37	36
Q03	36	38	-	71	66	18	21	19
Q04	34	37	71	-	51	17	15	16
Q05	31	22	66	51	-	13	13	13
Q10	35	36	18	17	13	-	77	76
Q15	33	37	21	15	13	77	-	93
Q20	32	36	19	16	13	76	93	-

Table 15. Test results for similar Word documents with LZJD (signature comparison/file comparison).

	Q01	Q02	Q03	Q04	Q05	Q10	Q15	Q20
Q01	-	0/0	0/0	0/0	0/0	0/0	0/0	0/0
Q02	0/0	-	0/0	0/0	0/0	0/0	0/0	0/0
Q03	0/0	0/0	-	48/43	32/35	0/0	0/0	0/0
Q04	0/0	0/0	48/43	-	34/37	0/0	0/0	0/0
Q05	0/0	0/0	32/35	34/37	-	0/0	0/0	0/0
Q10	0/0	0/0	0/0	0/0	0/0	-	41/33	26/25
Q15	0/0	0/0	0/0	0/0	0/0	41/33	-	45/48
Q20	0/0	0/0	0/0	0/0	0/0	26/25	45/48	-

6.1.3. BMP Images

This test uses the three images shown in Figure 3. The first image is the Lenna's classic greyscale test image [51]. For the second and third images, although the image is the same, we have rearranged half of it horizontally and vertically, respectively. The three images are bitmaps of the same resolution, therefore their size is the same (786,486 bytes). Readers should note that, in those individual BMP images, each pixel is represented by 24 bits, and that the pixel content is stored sequentially in the file. Pixels are read processing rows from top to bottom and, for any given row, from left to right (which means that the first pixel stored in the file is the one located at the upper left corner of the image, and the last one is the pixel located at the bottom right corner of the file).

Figure 3. BMP test images.

The results obtained with the four applications are displayed in Tables 16–19, where we have used the identifiers BMP 1, BMP 2, and BMP 3 for the three images.

Table 16. Test results for similar BMP images with SiSe (hash vs. hash/file vs. hash/hash vs. file/file vs. file).

	BMP 1	BMP 2	BMP 3
BMP 1	100/100/100/100	98/98/98/98	20/20/20/20
BMP 2	98/98/98/98	100/100/100/100	20/20/20/20
BMP 3	20/20/20/20	20/20/20/20	100/100/100/100/

Table 17. Test results for similar BMP images with ssdeep.

	BMP 1	BMP 2	BMP 3
BMP 1	100/100	52/52	0/0
BMP 2	52/52	100/100	0/0
BMP 3	0/0	0/0	100/100

Table 18. Test results for similar BMP images with sdhash.

	BMP 1	BMP 2	BMP 3
BMP 1	-	69/69	75/75
BMP 2	69/69	-	50/50
BMP 3	75/75	50/50	-

Table 19. Test results for similar BMP images with LZJD (signature comparison/file comparison).

	BMP 1	**BMP 2**	**BMP 3**
BMP 1	-	37/44	43/50
BMP 2	37/44	-	36/43
BMP 3	43/50	36/43	-

Using the results obtained, we can conclude that ssdeep is not able to match the image vertically rearranged with the two other images. In contrast, SiSe obtains results in both cases. Additionally, SiSe assigns a larger similarity percentage when comparing the first and second images (98%) than ssdeep (52%) which, in our opinion, is closer to reality given our similarity definition and the content of the files. On the other hand, the results obtained when processing the third image with sdhash are better adapted to our similarity definition, but sdhash fails to identify the first two images as having basically the same content. Regarding LZJD, its results follow the trend of sdhash but with scores significantly lower that those of sdhash. The reason for SiSe providing a lower result when comparing the third file to the other ones than when comparing the second file to the third one, is that due to the way a BMP file stores its pixels, a lot more trigger points are shared by the first and second images.

6.2. Dissimilarity Tests

The aim of this group of tests is to detect undesirable false positives by comparing files of the same format but with different content. Given the results obtained, no file is interpreted as strongly related to the other files included in the same test, as the maximum value obtained through those tests is 10 (produced by sdhash).

6.2.1. Plain Text Documents

The plan text files used in this test contain the following books as offered by The Project Gutenberg: H. G. Wells' The Time Machine [52] (201,875 bytes), Miguel de Cervantes' Don Quijote [50] (2,198,927 bytes), Robert Louis Stevenson's Treasure Island [48] (397,415 bytes), and Jules Verne's Voyage au Centre de la Terre [53] (460,559 bytes).

In this test ssdeep and LZJD did not provide any positive results when comparing the different files. The results obtained with SiSe and sdhash are offered in Tables 20 and 21.

Table 20. Test results for dissimilar plain text files with SiSe (hash vs. hash/file vs. hash/hash vs. file/file vs. file).

	Machine	**Quijote**	**Treasure**	**Voyage**
Machine	-	0/0/2/2	5/5/5/5	1/1/4/4
Quijote	0/2/0/2	-	0/3/0/3	0/3/0/3
Treasure	5/5/5/5	0/0/3/3	-	3/3/3/3
Voyage	1/4/1/4	0/0/3/3	3/3/3/3	-

Table 21. Test results for dissimilar plain text files with sdhash.

	Machine	Quijote	Treasure	Voyage
Machine	-	10/10	9/9	7/7
Quijote	10/10	-	5/5	5/5
Treasure	9/9	5/5	-	5/5
Voyage	7/7	5/5	5/5	-

SiSe obtains a maximum value for dissimilar files of 5, which implies that in order to discard false positives a threshold could be established. In the case of ssdeep, it returns a value of 0 for every comparison without processing its edit distance algorithm. The reason for that is that the corresponding signatures do not have a substring with at least 7 characters common, and hence it directly returns a value of 0. The comparisons performed by sdhash show values that are slightly higher than the ones provided by SiSe, but without a significant difference.

6.2.2. Word Documents

For this test, new Microsoft Word 2016 (Microsoft Corporation, Redmond, WA, USA) files have been created with the same content of the books used in the previous test. The size in bytes of each file is 118,169 (`Machine.docx`), 1,918,754 (`Quijote.docx`), 372,941 (`Treasure.docx`), and 263,316 (`Voyage.docx`).

While ssdeep and LZJD did not provide any positive results when comparing the different files, SiSe and sdhash returned the residual values included in Tables 22 and 23.

Table 22. Test results for dissimilar Word files with SiSe (hash vs. hash/file vs. hash/hash vs. file/file vs. file).

	Machine	Quijote	Treasure	Voyage
Machine	-	0/0/1/1	2/2/2/2	1/1/2/2
Quijote	0/1/0/1	-	0/2/0/2	0/2/0/2
Treasure	2/2/2/2	0/0/2/2	-	1/1/1/1
Voyage	1/2/1/2	0/0/2/2	1/1/1/1	-

Table 23. Test results for dissimilar Word files with sdhash.

	Machine	Quijote	Treasure	Voyage
Machine	-	2/2	2/2	5/5
Quijote	2/2	-	3/3	2/2
Treasure	2/2	3/3	-	1/1
Voyage	5/5	2/2	1/1	-

As in the case of dissimilar plaint text files, the values returned by both SiSe and sdhash are very low and could be avoided by establishing a small threshold.

6.2.3. BMP Images

In this test, three well-known test colour images obtained from Reference [54] have been used. The first one is Lenna's portrait (`lenna.bmp`), the second one is the photograph of a combat jet (`airplane.bmp`), and the third one displays several vegetables (pepper.bmp) as it can be seen in Figure 4. The resolution of the three files is 512 × 512 pixels, and they use 24 bits per pixel, so their size is 786,486 bytes.

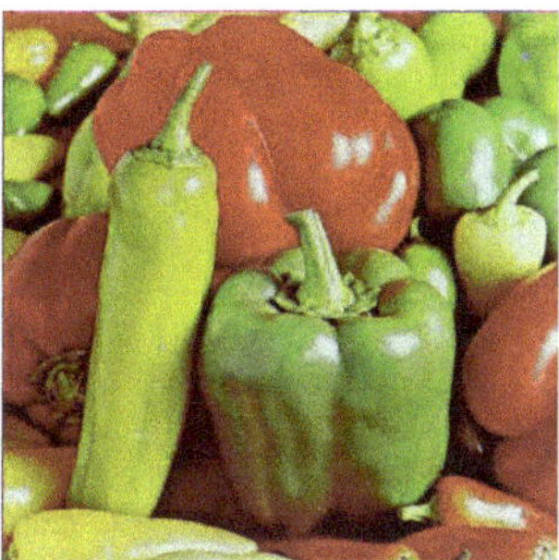

Figure 4. Color BMP test images.

While ssdeep and LZJD did not provide any positive results, sdhash returned a value of 1 when comparing pepper.bmp to the other two files and SiSe also returned a value of 1, but only when comparing lenna.bmp and pepper.bmp. Thus, the results are aligned with the expected behaviour, as the content of those files is clearly different.

6.3. Special Signatures

The aim of this test is to verify the behaviour of SiSe when it has to process some special signatures which represent cases that cannot be directly translated to regular input data files. Even so, we believe that it would be worthwhile to test SiSe and ssdeep in this scenario, as it represents different degrees of content modification. This test was not carried out with sdhash and LZJD as it was impossible to verify if the forged signatures truly reflected the nature of the tests, since their signature generation process is completely different.

Once again, it is important to point out that these special signatures do not represent actual files, instead they have been developed ad-hoc taking into account that the minimum length had to be 32 characters so they could be used with ssdeep. The special signatures for ssdeep are as follows:

```
S01: ABCDEFGHIJKLMNOPQRSTUVWXYZabcdefghijklmnopqrstuvwxyz
S02: ABCDEFGHIJKLMNOPQRSTUVWXYZABCDEFGHIJKLMNOPQRSTUVWXYZ
S03: abcdefghijklmnopqrstuvwxyzABCDEFGHIJKLMNOPQRSTUVWXYZ
S04: 12345678901234567890123456ABCDEFGHIJKLMNOPQRSTUVWXYZ
S05: BADCFEHGJILKNMPORQTSVUXWZYbadcfehgjilknmporqtsvuxwzy
S06: CDABGHEFKLIJOPMNSTQRWXUVabYZefcdijghmnklqropuvstyzwx
S07: EFGHABCDMNOPIJKLUVWXQRSTcdefYZabklmnghijstuvopqrwxyz
S08: IJKLMNOPABCDEFGHYZabcdefQRSTUVWXopqrstuvghijklmnwxyz
S09: QRSTUVWXYZabcdefABCDEFGHIJKLMNOPwxyzghijklmnopqrstuv
S10: ghijklmnopqrstuvwxyzABCDEFGHIJKLMNOPQRSTUVWXYZabcdef
```

The base element for this set is the first signature, S01. The second signature, S02, duplicates the first half of S01 in the second half. Signature S03 swaps the two blocks of S01. Additionally to this change, signature S04 replaces the first half of S03 with a string of digits. The remaining signatures, from S05 to

S10, have been created taking S01 and then applying transpositions of blocks whose sizes in characters are is 1, 2, 4, 8, 16, and 32, respectively.

Bearing in mind that SiSe signatures use two characters per trigger point, the special signatures of ssdeep have been adapted accordingly. In that sense, each character have been doubled (e.g., A and a have been transformed into AA and aa, respectively) in the signatures employed with SiSe.

The results obtained when comparing these special signatures are included in Tables 24 and 25.

Table 24. Test results with ad-hoc signatures and SiSe.

	S01	S02	S03	S04	S05	S06	S07	S08	S09	S10
S01	100	50	100	50	100	100	100	100	100	100
S02	50	100	50	50	50	50	50	50	50	50
S03	100	50	100	50	100	100	100	100	100	100
S04	50	50	50	100	50	50	50	50	50	50
S05	100	50	100	50	100	100	100	100	100	100
S06	100	50	100	50	100	100	100	100	100	100
S07	100	50	100	50	100	100	100	100	100	100
S08	100	50	100	50	100	100	100	100	100	100
S09	100	50	100	50	100	100	100	100	100	100
S10	100	50	100	50	100	100	100	100	100	100

Table 25. Test results with ad-hoc signatures and ssdeep.

	S01	S02	S03	S04	S05	S06	S07	S08	S09	S10
S01	100	50	50	50	0	0	0	55	63	63
S02	50	100	50	50	0	0	0	47	50	50
S03	50	50	100	50	0	0	0	36	44	90
S04	50	50	50	100	0	0	0	32	32	50
S05	0	0	0	0	100	0	0	0	0	0
S06	0	0	0	0	0	100	0	0	0	0
S07	0	0	0	0	0	0	100	0	0	0
S08	55	47	36	32	0	0	0	100	32	32
S09	63	50	44	32	0	0	0	32	100	32
S10	63	50	90	50	0	0	0	32	32	100

From the results obtained, it can be concluded that SiSe provides meaningful results in all the comparisons, while this is not the case for ssdeep. For instance, the comparison between S01 and S07, which provides a score of 0 in ssdeep, is considered to have a similarity degree of 100% by SiSe. A similar situation occurs when comparing for example S02 and S06, where the results provided by SiSe and ssdeep are 0 and 50, respectively.

Besides, the results provided by SiSe are significantly better adapted to our similarity definition. For instance, when comparing S01 to S03 and S04, it is clear that S03 is almost the same string as S01, whilst S04 only shares with S01 half of its content. However, ssdeep is not able to detect that difference and assigns a value of 50% in both cases. In comparison, SiSe identifies the similarity degree as 100% and 50%, respectively.

6.4. Suitability

The goal of this test is to compare the running time and the signature size of the four applications when processing the following files:

- `lshort.pdf` (613.131 bytes, File 1): The Not so short introduction to LaTeXversion 6.3 [55].
- `bctls-jdk15on-160.zip` (11,933,996 bytes, File 2): One of The Legion of Bouncy Castle's files [56].
- `inkscape-0.92.4-x64.exe` (66,983,821 bytes, File 3): An Inkscape 0.92.4 installer [57].
- `incubating-netbeans-10.0-source.zip` (113,256,313 bytes, File 4): The NetBeans version 10 source code [58].
- `VirtualBox-6.0.4-128413-Win.exe` (219,650,560 bytes, File 5): A Virtual Box installation file [59].
- `HD_Earth_Views.mov` (594,546,566 bytes, File 6): A NASA video taken from their digital archive [60].
- `ubuntu-17.10-desktop-amd64.iso` (1,502,576,640 bytes, File 7): One of the Ubuntu 17.10 distribution files [61].
- `debian-9.6.0-amd64-DVD-1.iso` (3,626,434,560 bytes, File 8): One of the Debian 9.6.0 distribution files [62].

The previous files were selected in order to cover a wide range of file size values, ranging from less than a megabyte to several gigabytes. Table 26 shows the running time in seconds when computing the signature of those files in a desktop with Ubuntu 18.04 operating system, an Intel i7-4790 processor (Intel Corporation, Santa Clara, CA, USA) at 3.60 GHz, and 16 gigabytes of RAM memory. All the applications have been executed using the Linux command `time`, and the output has been redirected to a file in all the cases in order not to penalize sdhash, as its signatures are very long and would provoke an important delay if they were to be printed on the screen.

Table 26. Signature generation time in seconds with the set of eight files tested.

	SiSe	ssdeep	sdhash	LZJD
File 1	0.012	0.005	0.025	0.499
File 2	0.078	0.087	0.326	1.275
File 3	0.394	0.464	0.352	6.066
File 4	0.723	0.821	0.671	10.535
File 5	1.333	1.573	1.158	20.869
File 6	2.680	3.992	3.054	-
File 7	6.529	9.280	8.042	-
File 8	16.048	24.791	18.657	-

As it can be observed in Table 26, SiSe is only slightly slower than ssdeep with very small files, and clearly faster than the other applications with medium and large-sized files. It was not possible to process the larger files with LZJD, so it seems that that application is not well-optimized for files of that size.

Regarding the study on the signature length, Table 27 shows the size of the files that store the signatures, where in each signature file only the hashes for that file have been included.

Table 27. Signature size in bytes with the set of eight files tested.

	SiSe	ssdeep	sdhash	LZJD
File 1	405	154	21,553	5476
File 2	457	172	405,902	5476
File 3	373	150	1,423,030	5476
File 4	283	170	2,405,783	5476
File 5	313	178	4,665,767	5476
File 6	367	152	12,628,901	-
File 7	577	179	31,915,871	-
File 8	343	136	77,028,165	-

It is clear that ssdeep provides the shortest signatures, followed by SiSe. Both applications generate signatures of less than 1 kilobyte. On the other hand, in most of the cases, the size of the signature files of sdhash is several orders of magnitude larger, which could lead to a storage problem when processing volumes with large files. This fact is due to the proportional relationship between the signature length and the input size in sdhash [41], something that does not happen in the case of ssdeep and SiSe. In a test performed by Breitinger and Baier, they found that the signature size when using sdhash is in average 3.3% the size of the original file. Regarding LZJD, the signatures generated by that application always have the same size, but the application was not able to process the three larger files.

Finally, it should also be noted that the complete file path is included in the signatures of SiSe and ssdeep, while sdhash and LZJD only include the file name in their signatures.

7. FRASH Tests

As mentioned in Reference [4], one of the major difficulties when comparing approximate matching algorithms is the diversity of approaches and files tested by each proposal. With that problem in mind, Breitinger et al. designed and implemented a test framework, called FRASH, that could be used for comparing different proposals.

In order to complement the results offered previously with another set of tests used by other authors, we have included in this section the results obtained with FRASH and the t5 dataset [63], a well-known set of 4457 files (1.8 gigabytes) derived from the GovDocs corpus designed to help test approximate matching algorithms [13].

The FRASH framework is implemented in Ruby 1.9.3 and requires a Linux environment to be run. In our evaluation of SiSe, ssdeep, sdhash, and LZJD we have used the same testing environment as in Section 6 (a desktop with Ubuntu 18.04 operating system, an Intel i7-4790 processor at 3.60 GHz, and 16 gigabytes of RAM memory). Instructions for installing the FRASH framework and compiling the four applications are included in SiSe's GitHub page [17].

FRASH allows us to perform five types of tests: efficiency tests, single-common-block correlation tests, fragment detection tests, alignment robustness tests, and random-noise-resistance tests. Information regarding how to lauch those tests can be found in References [4,17].

FRASH includes the Ruby files associated to ssdeep and sdhash. In order to include SiSe and LZJD in the tests, it is necessary to extend FRASH so that it supports both applications. The necessary files for that task are available in Reference [17]. Those files include the headers used by SiSe and LZJD as well as the methods for identifying the signature strings and retrieving the numerical result of a comparison.

Of all the tests implemented by FRASH, it has only been possible to perform (with the four applications) the tests related to efficiency, alignment, and fragmentation exactly as they are provided. The obfuscation test could only be completed with SiSe, ssdeep, and sdhash, due to the excessive time taken by LZJD (more than 3 days for processing just the first file). In comparison, the single-common-block test could not be completed with any application except ssdeep, as the other algorithms got stuck for days processing the first files of the batch being tested. For that reason, we have included the limited comparison regarding the obfuscation test, but have decided not to include the single-common-block test in our results.

7.1. Efficiency Tests

This test is composed of three parts suitably called runtime efficiency, fingerprint comparison, and compression [4]. Runtime efficiency measures the time needed by each application for reading the input files and generating the corresponding fingerprints. Those fingerprints are stored by FRASH in a temporary file so they can be used in the fingerprint comparison, which measures the time needed by the algorithms to complete an all-against-all comparison of the fingerprints. Finally, compression measures the ratio between input and output of each algorithm and returns a percentage value.

Table 28 shows the summary of the efficiency test when using all the files of the t5 corpus as input to the test (the screenshot and text file with the results of the test can be found in Reference [17]). This table contains the digest generation and all-pairs comparison time as well as the average digest length.

Table 28. Efficiency test results.

	SiSe	ssdeep	sdhash	LZJD
Digest generation (s)	8.83	13.50	11.07	22.87
All-pairs comparison (s)	45.17	7.19	137.38	2.26
Average digest length (bytes)	298	57	10,368	4098
Average digest ratio (%)	0.069	0.013	2.417	0.955

As the data included in the table shows, SiSe is the fastest application for digest generation thanks to its double multi-threading capabilities. As expected due to its design (usage of two-character elements, secondary digest larger than the primary digest and lack of digest limits), the average length for the digests generated by SiSe is approximately six times the length of ssdeep digests. It must be noted that FRASH does not compute the size of the digest as the size of the file generated: It removes both the header and the path of the file, and applies a factor of 3/4 to the Base64 string which forms each signature (which is a different approach from the one used in the tests described in Section 6.4, where the total length of the file was taken into account).

Finally, the all-pairs-comparison shows that LZJD clearly outperforms the other algorithms, especially sdhash and SiSe. In the case of SiSe, the fingerprint comparison is slow because of the complex method used in the comparison of digests and its implementation, though it is approximately three times faster than sdhash.

7.2. Alignment Tests

This test analyses the impact of inserting byte sequences at the beginning of an input by means of fixed and percentage blocks [4]. Regarding the interpretation of the results, in the case of percentual addition, a test associated to the value 100% means that new content of equal size to the original size has been added to the file, so if comparing the amount of the larger file (i.e., the modified file) that contains data included in the smaller file (i.e., the original file), in the case of that specific test the result of the comparison should be around 50 (following our similarity definition). In another example, adding 300% of new content means that the result of the comparison should be roughly 25% using the same definition of similarity.

Table 29 includes an extract of the alignment test performed by FRASH over a random selection of 100 files taken from the t5 corpus (the list of files tested and the screenshots and text file with the results of the test can be found in Reference [17]). In each cell, the first element represents the average score and the second one the number of matches (where the maximum is 100, the number of files tested).

Table 29. Alignment robustness test with percentage blocks showing the score (first value) and the number of matches (second value) when testing 100 files taken from the t5 corpus.

	SiSe	ssdeep	sdhash	LZJD
20%	78.41 100	85.67 97	80.08 100	33.50 100
40%	65.55 100	75.14 95	76.75 100	24.75 100
60%	56.69 100	68.66 91	79.12 100	19.98 100
80%	50.07 100	63.48 86	78.58 100	16.86 100
100%	44.83 100	60.68 71	78.76 100	14.74 100
200%	29.08 100	48.21 24	75.68 100	8.91 100
300%	21.79 100	31.00 5	78.68 100	6.52 100
400%	17.53 100	22.00 1	77.52 100	5.14 100
500%	14.45 100	- -	77.69 100	4.35 100

While sdhash is able to provide a match in all the cases, its results are located in the narrow band 70–80%, which does not permit to differentiate between the different testing scenarios. In comparison, LZJD is also able to match all the cases, but it produces values significantly lower than expected in all the tests (e.g., in the 100% test, the result is 14.74%). Both SiSe and ssdeep provide results better adapted to those expected according to our definition of similarity, but SiSe clearly outperforms ssdeep in the number of matches.

7.3. Fragment Detection Tests

Fragment detection identifies the minimum correlation between an input and a fragment (i.e., it determines what is the smallest fragment for which the similarity tool reliably correlates the fragment and the original file). It sequentially cuts a certain percentage (which varies along the test) of the original input length and generates the matching score. This test includes two modes: Random cutting (the framework randomly decides whether to start cutting at the beginning or at the end of an input and then continues randomly) and end side cutting (it only cuts blocks at the end of an input).

Table 30 includes an extract of the fragmentation tests performed by FRASH over the same random selection of 100 files belonging to the t5 corpus (the complete results can be found in Reference [17]). While SiSe and LZJD are able to match all the files, the scores of LZJD deviate significantly more from what is expected in several tests. In comparison to SiSe, both ssdeep and sdhash provide higher outputs, which from our point of view makes SiSe the best option for providing a value according to our similarity definition.

Table 30. Fragment detection test with two cutting modes (random start/end only) and number of matches in each case.

File Size	SiSe	ssdeep	sdhash	LZJD
95%	94.52/94.96 100/100	96.63/97.36 100/100	90.92/99.78 100/100	73.21/92.51 100/100
75%	76.54/79.97 100/100	83.49/88.54 99/100	75.96/99.58 100/100	40.12/65.97 100/100
50%	52.44/54.79 100/100	66.18/71.34 90/90	76.79/99.67 100/100	25.51/39.06 100/100
25%	26.69/29.35 100/100	48.36/60.44 11/25	81.81/99.39 100/98	13.57/19.02 100/100
5%	5.18/5.67 100/100	-/66.00 -/1	68.92/86.75 85/88	3.97/4.60 100/100

7.4. Single-Common-Block and Obfuscation Tests

Unfortunately, it has not been possible to complete the single common block and obfuscation tests in their original form with the four applications. Due to the design of the tests and the characteristics of the four algorithms evaluated, FRASH gets in a loop when processing most of the files used in the previous sections.

Both tests evaluate the algorithms for each of the files entered as an argument and change the files so the score gradually decreases until it reaches the value 0 (which is the halting condition). However, even with files whose size is less than 1 megabyte, LZJD gets active in the loop for several days with just one file, and the same happens (though to a lesser extent) with sdhash and SiSe. The result is that, after several days, there was no progress in the tests.

As an alternative, we were able to perform the obfuscation test with SiSe, ssdeep, and sdhash using a selection of the seven files (each one of a different file type) with minimum size from the t5 batch. The obfuscation test tries to determine what is the maximum number of changes if the match score is requested not to be lower than a certain value. This allows an estimation of how many bytes need to be changed all over the input to receive a non-match. The results of that test are offered in Table 31.

Table 31. Obfuscation resistance test showing the score (first value) and the number of matches (second value) when testing 7 files.

	SiSe	**ssdeep**	**sdhash**
≥90	8.89 7	6.86 7	5.71 7
≥70	33.71 7	26.00 7	21.86 7
≥50	84.57 7	40.50 5	38.86 7
≥30	183.14 7	- -	73.00 7
≥10	526.57 7	- -	182.71 7

The results of this test show that sdhash is the most resistant of the three algorithms. While SiSe and ssdeep offer similar values for the first three cases displayed in Table 31, ssdeep does not produce any output for the rest of the cases and SiSe is able to provide a match in all the cases.

8. Conclusions

In this contribution, a signature generation procedure based on ssdeep and a new algorithm for comparing two file signatures have been presented. We have implemented our tool, SiSe, to fulfil the requirements that any bytewise approximate matching application should satisfy, as indicated in Section 4.1:

(G1) Signature compression: The signature obtained is much smaller than the data file used as input. In this aspect ssdeep provides shorter signatures than SiSe, with both applications clearly outperforming sdhash and LZJD in the vast majority of cases (see Sections 6.4 and 7.1).

(G2) Ease of computation: The signature generation procedure is fast due to the use of the Adler-32 and FNV hashing functions, and the multi-threading design that allows distribution of the workload among the available processor logical cores (see Sections 4.2 and 5.2). As a result, SiSe outperforms ssdeep, sdhash, and LZJD in most of the signature generation tests, specially those with large files.

(G3) Coverage: Every byte of the input file is used by SiSe to process the hash value (see Sections 4.2 and 5.2). As there is no signature maximum length established in our design, it is guaranteed that all the data file is analysed under the same conditions. In this regard, it can be concluded that SiSe outperforms ssdeep, since if ssdeep reaches the maximum number of trigger points (63), the remaining content of the file is managed as a single trigger point, which consequently leads to potentially distinctive information loss.

(G4) Independence regarding the input file size: Thanks to the proposed signature generation procedure, the signature length does not depend on the file size. In this way, it allows us to obtain signatures of less than 1 kilobyte in the majority of cases (see Section 6).

(C1) Generation of a similarity score consistent with the content of the compared files: The score returned by SiSe is ranged between 0 and 100, representing the percentage of a file contained in another file (see Sections 4.2 and 6.1). Based on the set of tests performed, we have shown that, according to the definition of similarity managed in this contribution, SiSe generates results better adapted to that similarity definition than ssdeep, sdhash, and LZJD for most of the tests (see Sections 6 and 7). The results obtained with SiSe are adequate and show low values when comparing data files with different content, reducing the risk of false positives.

(C2) Ease of computation: Due to the design of the complex signature comparison procedure used in SiSe, its performance is not as good as the performance of ssdeep and LZJD, though SiSe clearly outperforms sdhash in tests with multiple files (see Sections 4.2 and 5.2).

(C3) Obfuscation resistance: The possibility (available to the user) to change parameters such as the decremental value, the length of the sliding window or the block size prevents attackers from figuring out which byte string would generate a trigger point. Additionally, as there is no limit in the size of the signature generated by SiSe, it is irrelevant if attackers insert at the beginning of the file 63 (or more) trigger points (see Sections 4.2 and 5.2).

(C4) Content swapping detection: In case data blocks are moved within the file, SiSe readily detects those movements, as shown in the ad-hoc signature tests (see Section 4.3).

(C5) Comparison of files of different sizes: Thanks to the modes made available in the implementation (comparison of two content files, two signature files, and one content file and one signature file), SiSe can compare files of very different sizes (see Sections 6 and 7). However, it must be noted that, given that the score provided by SiSe ranges from 0 to 100, in practice comparing files where the length proportion is greater that 100 (e.g., 1 megabyte and 100 megabytes) will in most cases produce a comparison score of 0, as the content of the larger file also found in the smaller file is expected to be less than 1%.

In the implementation of SiSe we have limited the maximum length to 2560 double characters for performance reasons, though given that the algorithm computes the initial block size with the goal to obtain a leading signature of 64 double characters, the number of files potentially affected by that limitation is practicable negligible.

Our algorithm, which can be applied to forensic operations in different types of digital media, shows results appropriately adapted to the similarity definition that we have used along this contribution. Our implementation is only slower than ssdeep when managing very small files, and faster than the other three algorithms in the rest of cases, specially when computing the signature of large files. Additionally, the signatures generated by our algorithm are much smaller than the ones of sdhash and, to a lesser extent, LZJD, making it more suitable for processing sets of large files.

From a practical standpoint, SiSe is able to compare files without imposing any limitation to their respective size, which consequently allows us to use it in many different situations (e.g., detecting plagiarism, searching for a list of items inside a much larger document, looking for malware hidden in executable files, etc.). Besides, it is able to detect the resemblance in some special cases where the other algorithms considered in this contribution fail.

In our opinion, according to the previously mentioned features, our proposal is a good alternative to the most common tools used nowadays (i.e., ssdeep, sdhash, and LZJD) for evaluating files, producing a clear value representing the similarity degree, specially when the goal is to obtain accurate information about the percentage of a file included in another file.

Author Contributions: Elaboration, validation, reviewing, edition, and resource allocation, V.G.M., F.H.-Á., L.H.E. Funding acquisition, V.G.M., L.H.E. All authors have read and agreed on the published version of the manuscript.

Funding: This work was supported in part by the Ministerio de Economía, Industria y Competitividad (MINECO), in part by the Agencia Estatal de Investigación (AEI), in part by the Fondo Europeo de Desarrollo Regional (FEDER, UE) under Project COPCIS, Grant TIN2017-84844-C2-1-R, and in part by the Comunidad de Madrid (Spain) under Project reference P2018/TCS-4566-CM (CYNAMON), also cofunded by European Union FEDER funds. Víctor Gayoso Martínez would like to thank CSIC Project CASP2/201850E114 for its support.

Conflicts of Interest: The authors declare no conflict of interest.

Abbreviations

The following abbreviations are used in this manuscript:

BBH	Block-Based Hashing
BBR	Block-Based Rebuilding
BMP	Bitmap
CTPH	Context-Triggered Hashing
FNV	Fowler-Noll-Vo
FRASH	Framework to test Algorithms Of Similarity Hashing
LSH	Locality Sensitive Hashing
LZJD	Lempel-Ziv Jaccard Distance
NIST	National Institute of Standards and Technology
NSRL	National Software Reference Library
SIF	Statistically-Improbable Features
SISE	Similarity Search
SPH	Similarity Preserving Hashing
TLSH	Trend Micro Locality Sensitive Hash

References

1. Stamm, M.C.; Wu, M.; Liu, K.J.R. Information forensics: An overview of the first decade. *IEEE Access* **2013**, *1*, 167–200. [CrossRef]
2. NIST. National Software Reference Library. Available online: http://www.nsrl.nist.gov (accessed on 21 March 2020).
3. NIST. Approximate Matching: Definition and Terminology, SP 800-168. 2014. Available online: https://csrc.nist.gov/publications/detail/sp/800-168/final (accessed on 21 March 2020).
4. Breitinger, F.; Stivaktakis, G.; Baier, H. Frash: A framework to test algorithms of similarity hashing. *Digit. Investig.* **2013**, *10*, 50–58. [CrossRef]
5. Jesse Kornblum. Ssdeep Project. 2017. Available online: https://ssdeep-project.github.io/ssdeep/index.html (accessed on 21 March 2020).
6. Astebøl, K. mvHash—A New Approach for Fuzzy Hashing. Master's Thesis, Gjøvik University College, Gjøvik, Norway, 2012.
7. Baier, H.; Breitinger, F. Security aspects of piecewise hashing in computer forensics. In *Sixth International Conference on IT Security Incident Management and IT Forensics (IMF 2001)*; Morgenstern, H., Ehlert, R., Frings, S., Goebel, O., Guenther, D., Kiltz, S., Nedon, J., Schadt, D., Eds.; IEEE Computer Society: Los Alamitos, CA, USA, 2011; pp. 21–36.
8. Breitinger, F.; Astebøl, K.; Baier, H.; Busch, C. mvHash-B—A new approach for similarity preserving hashing. In *Seventh International Conference on IT Security Incident Management and IT Forensics (IMF 2013)*; IEEE Computer Society: Washington, DC, USA, 2013; pp. 33–44.
9. Breitinger, F.; Baier, H. Performance issues about context-triggered piecewise hashing. In *Third ICST Conference on Digital Forensics & Cyber Crime (ICDF2C)*; Gladyshev, P., Rogers, M., Eds.; Springer: Berlin/Heidelberg, Germany, 2011; pp. 141–155.
10. Breitinger, F.; Baier, H. A fuzzy hashing approach based on random sequences and Hamming distance. In *7th Annual Conference on Digital Forensics, Security and Law (ADFSL 2012)*; Dardick, G., Ed.; Association of Digital Forensics, Security and Law: Ponce Inlet, FL, USA, 2012; pp. 89–100.
11. Breitinger, F.; Baier, H. Similarity preserving hashing: Eligible properties and a new algorithm mrsh-v2. In *Digital Forensics and Cyber Crime*; Rogers, M., Seigfried-Spellar, K., Eds.; Springer: Berlin/Heidelberg, Germany, 2013; pp. 167–182.
12. Chen, L.; Wang, G. An efficient piecewise hashing method for computer forensics. In *First International Workshop on Knowledge Discovery and Data Mining (WKDD 2008)*; IEEE Computer Society: Los Alamitos, CA, USA, 2008; pp. 635–638.

13. Roussev, V. An evaluation of forensic similarity hashes. *Digit. Investig.* **2011**, *8*, 34–41. [CrossRef]
14. Roussev, V. Sdhash Home. 2013. Available online: http://roussev.net/sdhash/sdhash.html (accessed on 21 March 2020).
15. Raff, E.; Nicholas, C. Lempel-ziv jaccard distance, an effective alternative to ssdeep and sdhash. *Digit. Investig.* **2018**, *24*, 34–49. [CrossRef]
16. Hernández Álvarez, F. Biometric Authentication fo Users through Iris by Using Key Binding and Similarity Preserving Hash Functions. Ph.D. Thesis, Universidad Politécnica de Madrid, Madrid, Spain, 2015.
17. Gayoso, V. SiSe (Similarity Search). 2019. Available online: https://github.com/victor-gayoso/sise (accessed on 21 March 2020).
18. Gayoso Martínez, V.; Hernández Álvarez, F.; Hernández Encinas, L. State of the art in similarity preserving hashing functions. In *Proceedings of WorldComp 2014—International Conference on Security & Management—SAM'14*; Daimi, K., Arabnia, H., Eds.; CSREA Press: Athens, GA, USA, 2014; pp. 139–145.
19. Harbour, N. dcfldd—Defense Computer Forensics Lab. 2002. Available online: http://dcfldd.sourceforge.net (accessed on 21 March 2020).
20. Tridgell, A. Efficient Algorithms for Sorting and Synchronization. Master's Thesis, The Australian National University, Canberra, Australia, 1999.
21. Tridgell, A. Spamsum Readme. 1999. Available online: http://samba.org/ftp/unpacked/junkcode/spamsum/README (accessed on 21 March 2020).
22. Kornblum, J. Identifying almost identical files using context triggered piecewise hashing. *Digit. Investig.* **2006**, *3*, 91–97. [CrossRef]
23. Kornblum, J. Ssdeep Usage. 2017. Available online: https://ssdeep-project.github.io/ssdeep/usage.html (accessed on 21 March 2020).
24. Roussev, V. Building a better similarity drap with statistically improbable features. In *Proceedings of the 42nd Hawaii International Conference on System Science*; IEEE Computer Society: Washington, DC, USA, 2009; pp. 1–10.
25. Roussev, V. Data fingerprinting with similarity digests. In *Advances in Digital Forensics VI*; Chow, K.-P., Shenoi, S., Eds.; Springer: Berlin/Heidelberg, Germany, 2010; pp. 207–226.
26. Roussev, V. Sdhash Tutorial. 2013. Available online: http://roussev.net/sdhash/tutorial/03-quick.html (accessed on 21 March 2020).
27. Sadowsky, C.; Levin, G. Simhash: Hash-Based Similarity Detection. 2007. Available online: http://citeseerx.ist.psu.edu/viewdoc/download?doi=10.1.1.473.7179&rep=rep1&type=pdf (accessed on 21 March 2020).
28. Blanco, R.; Lioma, C. Graph-based term weighting for information retrieval. *Inf. Retr. J.* **2012**, *15*, 54–92. [CrossRef]
29. Brosseau-Villeneuve, B.; Nie, J.-Y.; Kando, N. Latent word context model for information retrieval. *Inf. Retr. J.* **2014**, *17*, 21–51. [CrossRef]
30. Goyal, P.; Behera, L.; McGinnity, T.M. A novel neighborhood based document smoothing model for information retrieval. *Inf. Retr. J.* **2013**, *16*, 391–425. [CrossRef]
31. Broder, A.Z. On the resemblance and containment of documents. In *Proceedings of the Compression and Complexity of Sequences 1997 (SEQUENCES '97)*; IEEE Computer Society: Washington, DC, USA, 1997; pp. 21–29.
32. Oliver, J.; Cheng, C.; Chen, Y. TLSH—A locality sensitive hash. In *Proceedings of the 2013 Fourth Cybercrime and Trustworthy Computing Workshop*; IEEE Computer Society: Piscataway, NJ, USA, 2013; pp. 7–13.
33. Leskovec, J.; Rajaraman, A.; Ullman, J.D. *Mining of Massive Datasets*, 3rd ed.; Cambridge University Press: Cambridge, UK, 2020.
34. Davison, W. rsync. 1996. Available online: http://rsync.samba.org/ (accessed on 21 March 2020).
35. Deutsch, P.; Gailly, J.-L. Zlib Compressed Data Format Specification Version 3.3. IETF RFC 1950. 1996. Available online: http://tools.ietf.org/html/rfc1950 (accessed on 21 March 2020).
36. Fowler, G.; Noll, L.C.; Vo, K.-P.; Eastlake, D. The FNV Non-Cryptographic Hash Algorithm. IETF Internet Draft. 2012. Available online: http://tools.ietf.org/html/draft-eastlake-fnv-03 (accessed on 21 March 2020).
37. Damerau, F.J. A technique for computer detection and correction of spelling errors. *Commun. ACM* **1964**, *7*, 171–176. [CrossRef]

38. Levenshtein, V.I. Binary codes capable of correcting deletions, insertions, and reversals. *Sov. Phys. Dokl.* **1966**, *10*, 707–710.
39. Karpinski, M. On approximate string matching. *Lect. Notes Comput. Sci.* **1983**, *158*, 487–495.
40. Wagner, R.A.; Fischer, M.J. The string-to-string correction problem. *J. ACM* **1974**, *21*, 168–173. [CrossRef]
41. Breitinger, F.; Baier, H. Properties of a similarity preserving hash function and their realization in sdhash. In *Information Security for South Africa (ISSA 2012)*; Institute of Electrical and Electronics Engineers: Piscataway, NJ, USA, 2012; pp. 1–8.
42. Shapira, D.; Storer, J.A. Edit distance with move operations. *Lect. Notes Comput. Sci.* **2002**, *2373*, 85–98.
43. Chrobak, M.; Kolman, P.; Sgall, J. The greedy algorithm for the minimum common string partition problem. *Lect. Notes Comput. Sci.* **2004**, *3122*, 84–95.
44. Kaplan, H.; Shafrir, N. The greedy algorithm for edit distance with moves. *Inf. Process. Lett.* **2006**, *97*, 23–27. [CrossRef]
45. Cormode, G.; Muthukrishnan, S. The string edit distance matching problem with moves. *ACM Trans. Algorithms* **2007**, *3*, 85–98. [CrossRef]
46. ISO. Programming Languages—C++, ISO Std. 14 882:2017. 2017. Available online: https://www.iso.org/standard/68564.html (accessed on 21 March 2020).
47. OpenMP. The OpenMP API Specification for Parallel Programming. 2018. Available online: https://www.openmp.org/ (accessed on 21 March 2020).
48. Stevenson, R.L. Treasure Island. 2009. Available online: http://www.gutenberg.org/ebooks/27780 (accessed on 21 March 2020).
49. Raff, E. Java Implementation of the Lempel-Ziv Jaccard Distance. 2019. Available online: https://github.com/EdwardRaff/jLZJD (accessed on 21 March 2020).
50. Cervantes Saavedra, M. Don Quijote. 2010. Available online: http://www.gutenberg.org/ebooks/2000 (accessed on 21 March 2020).
51. Wakin, M. Standard Test Images—Lenna. 2003. Available online: https://www.ece.rice.edu/~wakin/images/ (accessed on 21 March 2020).
52. Wells, H.G. The Time Machine. 2011. Available online: http://www.gutenberg.org/ebooks/35 (accessed on 21 March 2020).
53. Verne, J. Voyage au Centre de la Terre. 2011. Available online: http://www.gutenberg.org/ebooks/4791 (accessed on 21 March 2020).
54. Levkin, H. Set of Classic Test Still Images. 2013. Available online: http://www.hlevkin.com/TestImages/classic.htm (accessed on 21 March 2020).
55. Oetiker, T.; Partl, H.; Hyna, I.; Schlegl, E. The Not so Short Introduction to LaTeX Version 6.3. 2018. Available online: http://tobi.oetiker.ch/lshort/lshort.pdf (accessed on 21 March 2020).
56. Legion of the the Bouncy Castle Inc. JCE with Provider and Lightweight API. 2018. Available online: https://www.bouncycastle.org/download/lcrypto-jdk15on-160.zip (accessed on 21 March 2020).
57. Inkscape. Inkscape 0.92.4. 2019. Available online: https://inkscape.org/gallery/item/13318/inkscape-0.92.4-x64.exe (accessed on 21 March 2020).
58. The Apache Software Foundation. Apache NetBeans 10.0 Source Code. 2018. Available online: https://archive.apache.org/dist/incubator/netbeans/incubating-netbeans/incubating-10.0/incubating-netbeans-10.0-source.zip (accessed on 21 March 2020).
59. Oracle, Index of Virtual Box 6.0.4 Downloads. 2019. Available online: http://download.virtualbox.org/virtualbox/6.0.4/VirtualBox-6.0.4-128413-Win.exe (accessed on 21 March 2020).
60. NASA. Earth in HD. 2013. Available online: http://s3.amazonaws.com/akamai.netstorage/HD_downloads/HD_Earth_Views.mov (accessed on 21 March 2020).
61. Canonical Ltd. Ubuntu 17.10 (Artful Aardvark). 2017. Available online: http://old-releases.ubuntu.com/releases/17.10/ (accessed on 21 March 2020).

62. Software in the Public Interest. Debian 9.6. 2018. Available online: https://cdimage.debian.org/mirror/cdimage/archive/9.6.0/amd64/iso-dvd/debian-9.6.0-amd64-DVD-1.iso (accessed on 21 March 2020).
63. Roussev, V. The t5 Corpus. 2011. Available online: http://roussev.net/t5/t5.html (accessed on 21 March 2020).

Article

Gene-Similarity Normalization in a Genetic Algorithm for the Maximum k-Coverage Problem

Yourim Yoon [1] and Yong-Hyuk Kim [2,*]

[1] Department of Computer Engineering, College of Information Technology, Gachon University, 1342 Seongnamdaero, Sujeong-gu, Seongnam-si, Gyeonggi-do 13120, Korea; yryoon@gachon.ac.kr

[2] School of Software, Kwangwoon University, 20 Kwangwoon-ro, Nowon-gu, Seoul 01897, Korea

* Correspondence: yhdfly@kw.ac.kr

Received: 28 February 2020; Accepted: 30 March 2020; Published: 2 April 2020

Abstract: The maximum k-coverage problem (MKCP) is a generalized covering problem which can be solved by genetic algorithms, but their operation is impeded by redundancy in the representation of solutions to MKCP. We introduce a normalization step for candidate solutions based on distance between genes which ensures that a standard crossover such as uniform and n-point crossovers produces a feasible solution and improves the solution quality. We present results from experiments in which this normalization was applied to a single crossover operation, and also results for example MKCPs.

Keywords: maximum k-coverage; redundant representation; normalization; genetic algorithm

---

1. Introduction

The maximum k-coverage problem (MKCP) is regarded as a generalization of several covering problems. The problem has a range of applications in combinatorial optimization such as scheduling, circuit layout design, packing, facility location, and covering graphs by subgraphs [1,2]. Recently, besides some theoretical approaches [3,4], it has been extended to many real-world applications such as blog-watch [5], seed selection in a Web crawler [6], map-reduce [7], influence maximization in social networks [8], recommendation in e-commerce [9], sensor deployment in wireless sensor networks [10], multi-depot train driver scheduling [11], cloud computing [12], and location problem [13].

Let $A = (a_{ij})$ be an $m \times n$ 0-1 matrix, and let w_i be a weight applied to each row of A. The objective of MKCP is to choose k columns so as to maximize the sum of the weights of the rows that contain '1' and are also located in one of these k columns.

This problem can be represented formally as follows:

$$\begin{aligned} \text{maximize} \quad & \sum_{i=1}^{m} w_i \cdot I\left(\sum_{j=1}^{n} a_{ij}x_j \geq 1\right) \\ \text{subject to} \quad & \sum_{j=1}^{n} x_j = k \\ & x_j \in \{0,1\}, \quad j = 1,2,\ldots,n, \end{aligned}$$

where $I(\cdot)$ is an indicator function ($I(false) = 0$ and $I(true) = 1$).

We are concerned with the case that $w_i = 1$ for all i, when we say that a row is covered by the selected k columns if it contains '1' which is also located in one of these columns. In this case, MKCP is to find k columns that cover as many rows as possible.

MKCP has many real-world applications. For example, MKCP can be applied to the following practical applications about museum touring [14]: Suppose that a museum can operate only k guided tour programs of visiting a set of exhibits among n possible programs due to some constraints such as operating costs. If there are m visitors and it is possible to predict whether or not each visitor i is satisfied with the experience of each program j (meaning a_{ij}), the problem of choosing exactly k programs to satisfy as many visitors as possible is exactly formulated as MKCP. Among the visitors, if it is preferred to satisfy specific visitors who have more importance such as VIPs, the problem can be modeled more elaborately by giving different weight (w_i) to each visitor i. Similarly, MKCP can be applied to other practical applications such as public safety networks and systems [15,16]. For example, we can consider a disaster management system in which n agencies are involved and m positions for Unmanned Aerial Vehicles (UAVs) are available to enable the communication between the agencies. In this situation, it can be easily investigated whether or not each agency i is covered when a UAV is placed in each position j (meaning a_{ij}). If only k UAVs are available for resource management, the problem of choosing exactly k positions of UAVs to cover as many agencies in the disaster area as possible can also be formulated as MKCP.

The NP-hardness of MKCP can easily be deduced from the NP-hardness of the minimum set covering problem (MSCP) [17]. Many meta-heuristics have been applied to the MSCP such as tabu search [18], genetic algorithms [19,20], particle swarm optimization [21], ant colony optimization [22], etc., but MKCP has been scarcely addressed. Some naïve greedy heuristics [1,7,23–25] and an improved local search [26], which do not scale well to large datasets, have been studied. To the best of our knowledge, there is only one meta-heuristic [27] that uses particle swarm optimization, applied to MKCP, and genetic algorithms have not been applied. The present authors previously conducted an initial investigation of MKCP [28], which we now extend. Because MKCP selects a fixed number of columns, the representation of solutions is simpler than it is in other covering problems, and this favors the adoption of a genetic algorithm (GA). When we apply a GA to the MKCP, it also has an interesting inherent property that each gene of a chromosome is not just a number but actually a column of a matrix, which we analyze and relate to the characteristic of its solution space. We go on to present a problem-specific normalization for use in a GA which takes account of the feasibility of solutions and improves the solution quality.

The remaining parts are organized as follows. We analyze the solution space of MKCP, and the representation of solutions in Section 2. Based on this analysis, in Section 3, we propose a normalization method for producing feasible and improved solutions. In Section 4, we present experimental results to assess the effectiveness of our method. Finally, we draw conclusions in Section 5.

2. Representation and Space of Solution to MKCP

When we solve a problem with GAs, the representation of chromosomes is an important issue. Representation depends on the problem and reflects the properties of the problem. One representation of a solution to MKCP is a vector of k integers representing column indices; another is a binary vector of length n, in which each element indicates whether or not the corresponding column is selected. Thus, when $k = 2$ and $n = 4$, the integer representation $(1, 4)$ is equivalent to the binary vector representation $(1, 0, 0, 1)$. We focus on the integer vector representation.

Using the integer vector representation, the encodings $(1, 4)$ and $(4, 1)$ both represent the same solution. The coverage, which is the value of the objective function, is also independent of this ordering. Since for the same solution there may be more than one different encoding, encoding space (genotypes) is different from solution space (phenotypes). We can consider permutations and combinations as genotypes and phenotypes, respectively. In the integer vector representation, it is natural to encode combinations using permutation encoding. However, in this case, many encodings correspond to the same combination. That is, this representation is redundant. Let G be the space of encodings in the integer vector representation, and let two encodings x and y in G be $(x_1, x_2, \ldots, x_n)$ and $(y_1, y_2, \ldots, y_n)$, respectively. Now, the relation $\sim$ is defined on G:

Definition 1. *$x \sim y$ if and only if we have a permutation $\sigma \in \Sigma_k$ such that $\sigma(x_1, x_2, \ldots, x_k) = (y_1, y_2, \ldots, y_k)$, where Σ_k is the set of permutations of size k, and $\sigma(x)$ represents x permuted by σ.*

Proposition 1. *The relation* $\sim$ *becomes an equivalence relation.*

Proof. For each $i \leq k$, $\langle \Sigma_i, \circ \rangle$ forms a symmetric group S_i, where $\circ$ is the function composition operator [29]. Then, the direct product $P := \Pi_{i=1}^{k} S_k$ is also a group [29], and therefore has an identity meaning that the relation $\sim$ is reflexive. When $\sigma \in P$, its inverse $\sigma^{-1} \in P$ exists, i.e., the relation $\sim$ is symmetric. The group P is closed under the operator $\circ$: It means if $\sigma^1, \sigma^2 \in P$, then $\sigma^1 \circ \sigma^2 \in P$. That is, the relation $\sim$ is also transitive. Taken together, the relation $\sim$ becomes an equivalence relation. □

For example, the solutions $(1,3,5,7)$ and $(3,7,5,1)$ are equivalent. The number of vectors equivalent to a vector $g \in G$ is $k!$.

The equivalence relation of Definition 1 allows us to consider the real solution space (phenotype space) as the set of equivalence classes of elements in G, i.e., the quotient space $G/\sim$.

We can measure the similarity of two vectors x and y in G, the space of MKCP solution encodings, using a distance metric D, which we obtain by summing values of a subsidiary metric d which measures the difference between two columns of the matrix A which expresses the MKCP, as follows:

$$D(x,y) := \sum_{i=1}^{n} d(x_i, y_i). \tag{1}$$

Here, the genes of two chromosomes, x_i and y_i, represent chosen column indices. If we regard the indices as just labels, not column vectors, the discrete metric, which becomes one if the two indices are the same, and zero otherwise, can be used as a metric d. This satisfies all the conditions for a distance metric including the triangular inequality.

An alternative metric is Hamming distance, which is a measure of the dissimilarity of two binary vectors. If we consider each column of the matrix A as a binary vector (the column vector), not just an integer, we can find the Hamming distance between any two column vectors, and hence between the corresponding genes x_i and y_i in two chromosomes from G. These distances can then be summed as shown in Equation (1).

Figure 1 shows two distances in G calculated using discrete metric and Hamming distance. As shown in Figure 1c, the discrete metric simply compares the column indices of x_i and y_i, and ignores the contents of the corresponding columns of A. In Figure 1d, we see that the distance between x_1 and y_1 is the Hamming distance between the first and the second column vectors of A.

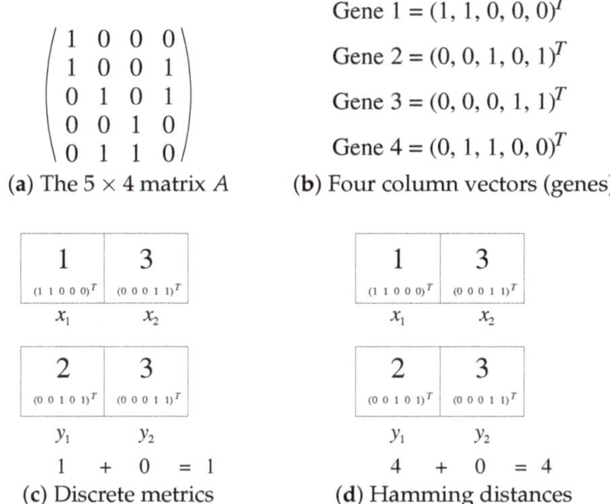

Figure 1. Discrete metric vs. Hamming distance, for the chromosomes (1,3) and (2,3).

Now, we establish a metric in the quotient space $G/\sim$ by the following proposition:

Proposition 2. *Let $x = (x_1, x_2, \ldots, x_n)$ and $y = (y_1, y_2, \ldots, y_n)$ be in G, and let a metric D on G be defined by Equation (1). Then,*

$$\bar{D}(\bar{x}, \bar{y}) := \min_{\sigma \in \Sigma_k} \sum_{i=1}^{k} d(x_i, \sigma_i(y)) \quad (2)$$

becomes a metric on $G/\sim$, where $\sigma_i(y)$ represents the ith element of permuted y by σ.

Proof. Let σ be in the group $\Pi_{i=1}^{k} S_k$. The computation $D(x, y) := \sum_{i=1}^{n} d(x_i, y_i)$, is unaffected by summation order, and thus $D(x, y) = D(\sigma(x), \sigma(y))$. Hence, σ is an isometry on G, and thus $\Pi_{i=1}^{k} S_k$ is an isometry subgroup. The relation $\sim$ is an equivalent relation obtained from $\Pi_{i=1}^{k} S_k$. Hence from [30,31] $\bar{D}(\bar{x}, \bar{y})$ is a metric on $G/\sim$. □

3. Normalization in MKCP

As shown above, redundancy in the integer representation of solutions to MKCP means that the encoding (genotype) space G is unnecessarily larger than the true solution (phenotype) space, which is the quotient space $G/\sim$. Redundant representations can be expected to reduce the performance of genetic algorithms significantly, which, in particular, undermines the effectiveness of standard crossovers defined using masks [32]. The problem of redundant representations have been addressed by a number of methods such as adaptive crossover [33–36], and among which the normalization technique [37] is representative. Normalization changes a parent genotype into a different genotype with the same phenotype which is similar to the genotype of the other parent before a standard crossover is performed. It is based on adaptive crossovers [34,35] and many variants have appeared [31,38,39].

3.1. Preserving Feasibility

A representation is infeasible if it does not meet the requirements of a solution. Figure 2a shows an integer representation which is infeasible because it contains duplicate column indices, and Figure 2b shows a binary representation which is infeasible because it contains the wrong number of '1's. Figure 2 also shows how these infeasible solutions are likely to be created by standard crossover operators such as uniform and n-point crossovers.

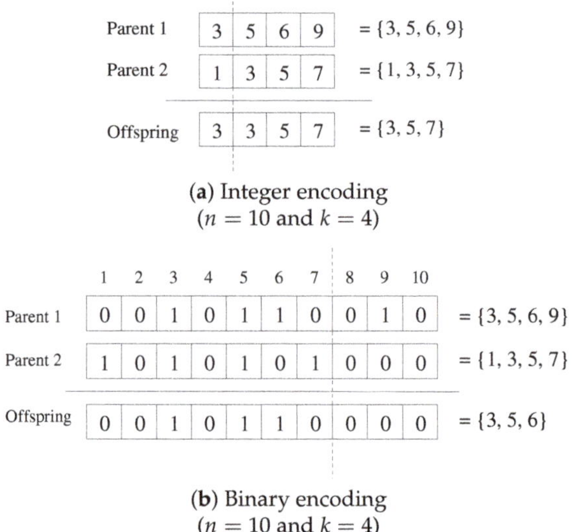

Figure 2. How infeasible solutions can be created by crossover operations.

A repairing step can be performed to restore feasibility, but this has the effect of a mutation, and may garble gene sequences inherited from parents. Using integer encoding, the problem can be avoided by rearranging the parents so that any shared column indices are in the same position. Such rearrangement naturally makes offspring preserve feasibility and no special repairing step is required. This form of normalization is easily implemented, as shown in the pseudocode in Figure 3, and takes $O(k^2)$ time.

```
FP_normalization(x, y)
{
    for i ← 0 to k
        for j ← 0 to k
            if x_i = y_j, then
                Swap values of y_i and y_j;
}
```

Figure 3. Pseudocode of normalization to preserve feasibility.

In Figure 4, this normalization to preserve feasibility for the example in Figure 2a is shown.

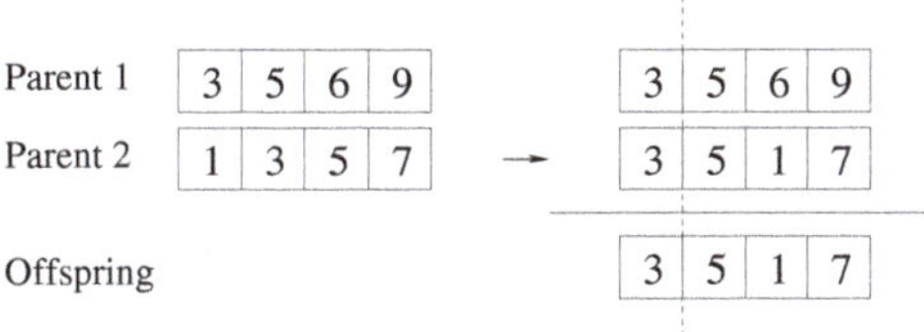

Figure 4. Normalization to preserve feasibility in the example from Figure 2a.

3.2. *Normalization for Improving Solution Quality*

A good solution to MKCP will consist of dissimilar columns. For example, in Figure 1, $(1,3)$ is a better solution than $(1,4)$ because the '1's in Columns 1 and 3 all occur in different rows, while Columns 1 and 4 share a '1' in Row 2.

In the context of a genetic algorithm, we would expect chromosomes with genes corresponding to dissimilar columns to be most effective. Looking again at Figure 1, suppose that we apply a standard one-point crossover to the parents $(1,2)$ and $(3,4)$. If the cutting line lies between the first gene and the second one, then the offspring will be $(1,4)$, and this offspring covers the rows $\{1,2,3\}$ of A. However, if the second parent is rearranged to $(4,3)$, the offspring becomes $(1,3)$, which covers $\{1,2,4,5\}$. This example is illustrated in Figure 5. In general, we want the offspring to have genes that correspond to columns that are as dissimilar as possible. Thus, it is helpful to rearrange chromosomes so that genes corresponding to similar columns are in the same positions. The goal of this rearrangement can be formulated in terms of distance in the phenotype space $G/\sim$. Consider two parents x and y in the genotype space G. Rearranging genes so that those corresponding to the most similar columns are located in the same positions is equivalent to the search for a permutation σ^* such that the distance between the equivalence class of x and that of y is equal to the distance between x and $\sigma^*(y)$. This is also equivalent to finding the σ which minimizes the distance between x and the permuted y in Equation (2).

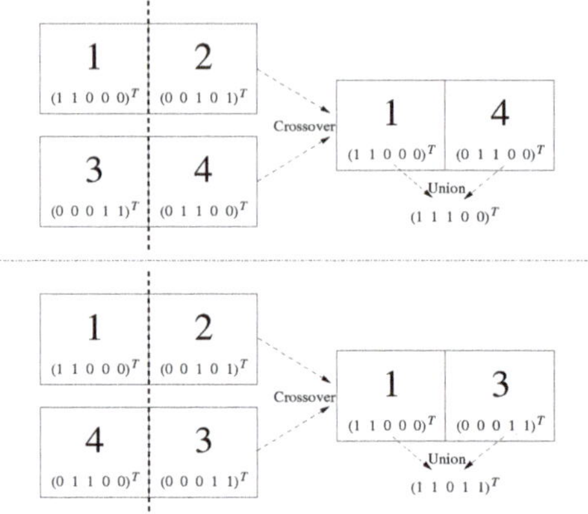

Figure 5. Effect of rearranging genes to match similar columns in the example from Figure 1.

The optimal rearrangement is achieved by considering all of the permutations of genes in the second parent, and choosing the permutation that minimizes the distance sum between the column vectors corresponding to gene pairs in the same locations. If Hamming distance H is used to get the dissimilarity between the column vectors corresponding to two genes, we will choose the permutation σ^* such that

$$\sigma^* = \underset{\sigma \in \Sigma_k}{\operatorname{argmin}} \sum_{i=1}^{k} H(x_i, \sigma_i(y)), \tag{3}$$

where Σ_k is the set of all the permutations of length k and $\sigma_i(y)$ denotes the ith element of permuted y by σ.

We give an example case in Figure 6, in which the chromosomes $(1,2)$ and $(3,4)$ are both parents, and now we normalize $(3,4)$. Because k is 2, the number of all the permutations is just 2. We compute $\sum_{i=1}^{k} H(x_i, \sigma_i(y))$ for each permutation. If $\sigma^1 = \binom{1\,2}{1\,2}$, $\sigma^1(y) = (3,4)$. Then, $H(x_1, \sigma_1^1(y)) = 4$, $H(x_2, \sigma_2^1(y)) = 2$, and their sum is 6. For the second permutation $\sigma^2 = \binom{1\,2}{2\,1}$, $\sigma^2(y) = (4,3)$. Then, $H(x_1, \sigma_1^2(y)) = 2$, $H(x_2, \sigma_2^2(y)) = 2$, and their sum is 4. Since this is smaller than 6, $\sigma^2 = \binom{1\,2}{2\,1}$ is the optimal permutation.

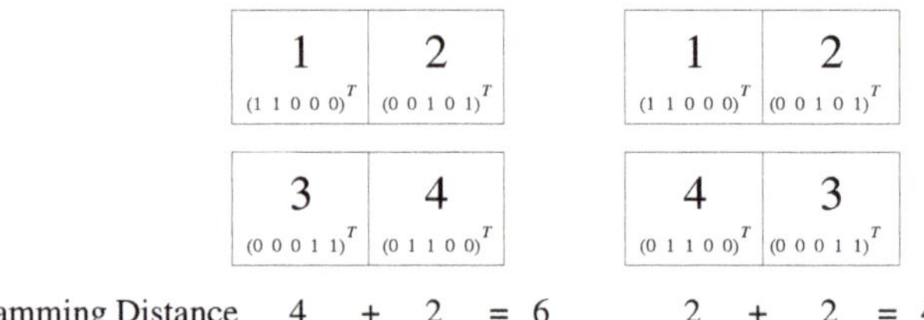

Figure 6. Normalization with Hamming distance in the example from Figure 1.

Enumerating all $k!$ permutations is intractable for a large k, and thus this procedure is intractable. However, the problem can be solved using the Hungarian method [40], which is a network-flow-based technique. It provides an optimal result, and runs in $O(k^3)$ time [41]. Alternatively, we can use a fast heuristic [42], which runs in $O(k^2)$ and produces results very close to the optimum. Either of these methods can be treated as a function that accepts a 2D array, in which the elements are the distances between two genes, and returns the permutation with the minimum total distance between chromosomes. This normalization is shown in the pseudocode of Figure 7.

```
OPT_normalization(x, y)
{
    for i ← 0 to k
        for j ← 0 to k
            D[x_i, y_j] ← H(x_i, y_j);
    σ* ← argmin_{σ∈Σ_k} Σ_{i=1}^{k} H(x_i, σ_i(y)); // using the Hungarian method or its fast variant
    y ← σ*(y);
}
```

Figure 7. Pseudocode of normalization by the optimal rearrangement.

After we rearrange the second parent according to the permutation σ^*, a standard crossover is applied.

Now, we investigate the relation between this optimal rearrangement and the feasibility. There may be more than one optimal rearrangement satisfying Equation (3), but we can show that one of them is always feasible. Feasibility is preserved by an optimal permutation σ^* which locates all common indices in the same positions by σ^*, as we now prove:

Proposition 3. *If $x = (x_1, x_2, \ldots, x_k)$ and $y = (y_1, y_2, \ldots, y_k)$ are two chromosomes and $x_p = y_q$, then there exists $\sigma^* \in \Sigma_k$ such that $\sigma^*_p(y) = y_q$ and $\sum_{i=1}^{k} d(x_i, \sigma^*_i(y)) \leq \sum_{i=1}^{k} d(x_i, \sigma_i(y))$ for all $\sigma \in \Sigma_k$, where d is a distance metric.*

Proof. Let σ' be $\mathrm{argmin}_{\sigma\in\Sigma k} \sum_{i=1}^{k} d(x_i, \sigma_i(y))$. There is an index r satisfying $\sigma'_r(y) = y_q$. Let σ'' be the same permutation as σ', except that $\sigma'(y)_p$ and $\sigma'_r(y)$ are exchanged:

$$\sigma''_i(y) = \begin{cases} \sigma'_r(y) & \text{if } i = p, \\ \sigma'_p(y) & \text{if } i = r, \\ \sigma'_i(y) & \text{otherwise.} \end{cases}$$

Then,

$$\begin{aligned}
\sum_{i=1}^{k} d(x_i, \sigma''_i(y)) &= d(x_p, \sigma''_p(y)) + d(x_r, \sigma''_r(y)) + \sum_{i\neq p, i\neq r} d(x_i, \sigma''_i(y)) \\
&= d(x_p, \sigma'_r(y)) + d(x_r, \sigma'_p(y)) + \sum_{i\neq p, i\neq r} d(x_i, \sigma'_i(y)) \\
&= d(x_p, y_q) + d(x_r, \sigma'_p(y)) + \sum_{i\neq p, i\neq r} d(x_i, \sigma'_i(y)) \\
&= d(x_r, \sigma'_p(y)) + \sum_{i\neq p, i\neq r} d(x_i, \sigma'_i(y)) \quad (\because d(x_p, y_q) = 0 \text{ by assumption}) \\
&\leq d(x_r, x_p) + d(x_p, \sigma'_p(y)) + \sum_{i\neq p, i\neq r} d(x_i, \sigma'_i(y)) \quad (\because \text{triangular inequality}) \\
&= d(x_r, y_q) + d(x_p, \sigma'_p(y)) + \sum_{i\neq p, i\neq r} d(x_i, \sigma'_i(y)) \\
&= d(x_r, \sigma'_r(y)) + d(x_p, \sigma'_p(y)) + \sum_{i\neq p, i\neq r} d(x_i, \sigma'_i(y)) \\
&= \sum_{i=1}^{k} d(x_i, \sigma'_i(y)) \\
&\leq \sum_{i=1}^{k} d(x_i, \sigma_i(y)) \text{ for all } \sigma \in \Sigma_k.
\end{aligned}$$

Hence, $\sum_{i=1}^{k} d(x_i, \sigma_i''(y)) \le \sum_{i=1}^{k} d(x_i, \sigma_i(y))$ for all $\sigma \in \Sigma_k$ and $\sigma_p''(y) = y_q$. □

This proof relies on the triangular inequality, which is a key property of any valid distance. Thus, Proposition 3 holds for distances other than Hamming distance. We could use discrete metric, introduced in Section 2. This distance provides a very rough comparison of two solutions, but Proposition 3 still holds. Equation (3) can be rewritten using discrete metric ρ instead of Hamming distance H:

$$\sigma^* = \operatorname*{argmin}_{\sigma \in \Sigma_k} \sum_{i=1}^{k} \rho(x_i, \sigma_i(y)). \tag{4}$$

In particular, in this case, feasibility is preserved by any optimal permutation σ^* in Equation (4). Its proof is quite similar to the proof of Proposition 3.

Proposition 4. *If $x = (x_1, x_2, \ldots, x_k)$ and $y = (y_1, y_2, \ldots, y_k)$ are two chromosomes and $x_p = y_q$, then $\sigma_p^*(y) = y_q$, where σ^* is a permutation such that $\sum_{i=1}^{k} \rho(x_i, \sigma_i^*(y)) \le \sum_{i=1}^{k} \rho(x_i, \sigma_i(y))$ for all $\sigma \in \Sigma_k$.*

Proof. We assume that $\sigma_p^*(y) \neq y_q$. There is an index r satisfying $\sigma_r^*(y) = y_q$. Let σ' be the same permutation as σ^*, except that $\sigma^*(y)_p$ and $\sigma_r^*(y)$ are exchanged:

$$\sigma_i'(y) = \begin{cases} \sigma_r^*(y) & \text{if } i = p, \\ \sigma_p^*(y) & \text{if } i = r, \\ \sigma_i^*(y) & \text{otherwise.} \end{cases}$$

Then,

$$\begin{aligned}
\sum_{i=1}^{k} \rho(x_i, \sigma_i'(y)) &= \rho(x_p, \sigma_p'(y)) + \rho(x_r, \sigma_r'(y)) + \sum_{i \neq p, i \neq r} \rho(x_i, \sigma_i'(y)) \\
&= \rho(x_p, \sigma_r^*(y)) + \rho(x_r, \sigma_p^*(y)) + \sum_{i \neq p, i \neq r} \rho(x_i, \sigma_i^*(y)) \\
&= \rho(x_p, y_q) + \rho(x_r, \sigma_p^*(y)) + \sum_{i \neq p, i \neq r} \rho(x_i, \sigma_i^*(y)) \\
&= \rho(x_r, \sigma_p^*(y)) + \sum_{i \neq p, i \neq r} d(x_i, \sigma_i^*(y)) \quad (\because x_p = y_q \text{ by assumption}) \\
&< 1 + 1 + \sum_{i \neq p, i \neq r} d(x_i, \sigma_i^*(y)) \\
&= \rho(x_p, \sigma_p^*(y)) + \rho(x_r, x_p) + \sum_{i \neq p, i \neq r} \rho(x_i, \sigma_i^*(y)) \quad (\because x_p \neq \sigma_p^*(y) \text{ and } x_r \neq x_p) \\
&= \rho(x_p, \sigma_p^*(y)) + \rho(x_r, y_q) + \sum_{i \neq p, i \neq r} \rho(x_i, \sigma_i^*(y)) \\
&= \rho(x_p, \sigma_p^*(y)) + \rho(x_r, \sigma_r^*(y)) + \sum_{i \neq p, i \neq r} \rho(x_i, \sigma_i^*(y)) \\
&= \sum_{i=1}^{k} \rho(x_i, \sigma_i^*(y)).
\end{aligned}$$

This contradicts the assumption that $\sum_{i=1}^{k} \rho(x_i, \sigma_i^*(y)) \le \sum_{i=1}^{k} \rho(x_i, \sigma_i(y))$ for all $\sigma \in \Sigma_k$. □

Using discrete metric will only cause identical indices to be rearranged into the same positions. In fact, normalization by discrete metric is exactly the same as rearranging for preserving feasibility introduced in Section 3.1.

4. Experiments

4.1. Test Sets and Test Environments

Our experiments were conducted on 65 instances of 11 set cover problems with various size and densities, from the OR-library [43]. Although these benchmark data were designed as set cover problems, the data can also be considered as maximum k-coverage problems, as in [27]. Some details of these problems are presented in Table 1, where m and n are the numbers of rows and columns, respectively, and density is the percentage of '1's in the MKCP matrix A. The present authors previously experimented with problems with fixed values of k of 10 and 20 [28]. However, in this study, we varied k with the tightness ratio α, which is the product of k and the density of a problem. The higher is the tightness ratio, the larger is the value of the object function, which is the coverage, that we are likely to achieve. If the tightness ratio is 1, the optimum coverage is likely to be very close to n. We used tightness ratios of 0.8, 0.6, and 0.4.

We implemented all our tested algorithms in C language using *gcc* version 5.4.0, and ran them on Ubuntu 16.04.6.

Table 1. Problem set.

Problem Set	m	n	Density	#Instances Tested
I-4	200	1000	2%	10
I-5	200	2000	2%	10
I-6	200	1000	5%	5
I-A	300	3000	2%	5
I-B	300	3000	5%	5
I-C	400	4000	2%	5
I-D	400	4000	5%	5
I-E	500	5000	10%	5
I-F	500	5000	20%	5
I-G	1000	10,000	2%	5
I-H	1000	10,000	5%	5

4.2. Effect of Normalization on a Crossover

To see whether or not normalization is effective at crossover, we performed experiments with three methods of rearranging the second of two parents before a crossover:

- REPAIR: The second parent is not rearranged before the crossover, but infeasible offspring are repaired to restore feasibility after the crossover.
- FP: The second parent is rearranged to produce only feasible offspring using the normalization in Figure 3.
- OPT: The second parent is rearranged by the normalization in Figure 7, to minimize the sum of the distances using the permutation σ^* of Equation (3).

REPAIR produces feasible offspring by the method of replacing duplicate column indices with a randomly chosen one among the indices that are not contained in the offspring. OPT was implemented using the Hungarian method [40], which runs in $O(k^3)$ time.

To determine the most effective method, we performed the following steps:

1. N parent chromosomes were randomly generated. (N was set to 100.)
2. The chromosomes were randomly paired, making $N/2$ couples.
3. The second parent in each pair was rearranged using the methods described above.
4. A uniform crossover was applied to each couple.
5. We computed the mean and the standard deviation for the coverage of each of the $N/2$ offspring.

This procedure was applied to a single instance of each problem listed in Table 1 using REPAIR, FP, and OPT. Table 2 shows the results for each of these 11 instances. We see that OPT normalization outperforms the others, and can therefore be expected to improve the performance of a GA. Moreover, the results of the one-tailed t-test show that the objective values of offspring produced using OPT normalization were better than those of their parents. On the contrary, the values produced using REPAIR and FP were similar to those of their parents. This suggests that, compared to other methods, OPT normalization would strongly support the function of the crossover operator in searching the solution space in a promising direction, without replacement strategy.

Table 2. Effect of normalization on a single crossover, after the REPAIR, FP, and OPT procedures.

Tightness			Parents		REPAIR		FP		OPT	
Ratio	Instance	k	Ave	SD	Ave	SD	Ave	SD	Ave	SD
	scp41	40	110.27	6.33	110.30	5.89	110.00	5.46	**113.80**	6.12
	scp51	40	110.72	6.62	110.00	6.95	111.82	6.90	**115.02**	6.77
	scp61	16	111.33	6.43	111.24	6.64	111.20	6.73	**114.20**	6.24
	scpa1	40	167.69	7.76	166.80	7.83	167.68	8.47	**173.66**	8.93
	scpb1	16	168.31	9.27	170.80	9.79	168.90	9.11	**171.36**	9.25
0.8	scpc1	40	220.40	10.53	221.04	10.83	220.42	9.76	**225.84**	9.91
	scpd1	16	223.74	9.22	222.34	10.15	224.36	8.69	**228.56**	9.48
	scpnre1	8	284.43	10.23	285.08	10.93	285.76	11.08	**287.14**	9.27
	scpnrf1	4	295.44	10.76	294.94	9.94	294.44	9.50	**297.34**	9.75
	scpnrg1	40	553.30	14.49	553.96	14.56	553.82	16.10	**561.28**	12.16
	scpnrh1	16	560.94	14.49	562.12	16.57	563.14	16.24	**567.46**	17.09
	scp41	30	90.40	6.41	90.80	6.34	89.98	4.82	**95.34**	6.14
	scp51	30	90.77	6.74	91.92	6.92	91.44	6.50	**94.20**	6.64
	scp61	12	90.96	6.54	90.56	6.17	90.94	5.86	**91.88**	7.75
	scpa1	30	137.54	8.09	136.74	8.95	137.60	7.63	**143.74**	8.01
	scpb1	12	138.05	8.31	138.44	8.36	139.68	9.08	**141.68**	8.28
0.6	scpc1	30	181.59	10.64	178.52	10.55	180.24	11.30	**186.06**	10.09
	scpd1	12	185.27	9.08	184.80	9.06	185.34	9.01	**188.30**	7.72
	scpnre1	6	233.02	11.34	231.60	12.83	234.04	12.41	**234.12**	11.04
	scpnrf1	3	244.40	11.37	245.04	10.48	244.24	10.28	**245.68**	9.57
	scpnrg1	30	452.91	14.48	452.70	14.52	452.58	14.07	**461.34**	13.42
	scpnrh1	12	461.87	13.66	460.80	16.68	462.22	16.58	**462.40**	15.61
	scp41	20	66.44	5.52	66.84	5.59	67.30	6.09	**68.06**	5.19
	scp51	20	66.41	5.54	67.44	6.61	66.84	6.44	**68.98**	6.71
	scp61	8	66.55	6.80	67.68	6.05	66.72	5.91	**67.74**	5.71
	scpa1	20	100.82	8.40	99.80	7.62	100.00	7.63	**103.54**	6.80
	scpb1	8	101.25	7.43	101.74	8.91	101.06	7.38	**104.06**	8.43
0.4	scpc1	20	132.47	10.42	131.68	8.78	133.14	9.07	**136.14**	9.92
	scpd1	8	135.48	10.49	136.32	8.49	134.72	9.46	**136.94**	9.96
	scpnre1	4	171.43	10.32	172.04	10.14	172.94	12.05	**173.16**	11.10
	scpnrf1	2	179.95	10.83	180.04	9.85	179.94	11.46	**182.90**	11.22
	scpnrg1	20	331.51	13.57	332.10	13.64	330.52	14.89	**337.30**	14.60
	scpnrh1	8	338.58	14.59	336.84	14.80	337.30	14.42	**343.44**	13.90
t-test p-value *			-		4.25×10^{-1}		1.09×10^{-1}		7.52×10^{-12}	

Ave and SD are the average and the standard deviation of the fitness of 100 parents (in the column of "Parents") or 50 offspring (in the remaining columns of REPAIR, FP, and OPT), respectively. REPAIR produces feasible offspring by random repair. FP rearranges the second parent to produce feasible offspring using the normalization in Figure 3. OPT is optimized normalization of the second parent. * The one-tailed t-test of the null hypothesis that the result of given method is equal to fitness of parents.

4.3. Performance of GAs with Normalization Methods

Our underlying evolutionary model is similar to the model of CHC [44], which was applied to many problems [45–50]. We paired a population of N chromosomes randomly, and then we applied crossover to each pair, generating a total of $N/2$ offspring. We ranked all parents and offspring and the fittest N individuals among them became the population in the next generation. We used 100 as the size of population in our experiments. We reinitialized the population, except for the best individual, if there were no changes over $kr(1-r)$ generations, where r is a divergence ratio that wes set to 0.25. This GA stopped after 500 generations and returned the best it has found. The pseudocode of our GA is given in Figure 8.

```
Initialize a population P of N individuals;
for i ← to maximum generations
{
    Randomly pair a population P of N individuals;  {N/2 pairs}
    for each pair (p1, p2) ∈ P   {N/2 iterations}
    {
        Normalize p2 to make close to p1;  {optionally applied}
        o ← crossover(p1, p2);  {make offspring from parents}
    }
    P ← the best N individuals among N parents and N/2 offspring;
    if there are no changes in P over kr(1 − r) generations, then
        Reinitialize a population P, except for the best individual;
}
return the best individual;
```

Figure 8. Pseudocode of our genetic algorithm.

In the following experiments, we changed the normalization method in a single GA. We compared the output of our GA with the best result that we found in this study, using the metric %-gap, which is $100 \times |best - output|/best$.

Thirty trials were performed for each method, and the averaged results are shown in Table 3. The results labeled RR-GA were produced without normalization: infeasible offspring produced were repaired randomly using the same method as REPAIR in Section 4.2. We also compared the results from the GA with a multi-start method using randomly generated solutions. In each run of this method, called Multi-Start, we sampled 10^6 random solutions and chose the best one. Even RR-GA performed significantly better than Multi-Start, suggesting that GAs are an appropriate mechanism for solving the MKCP.

The results labeled FP-GA in Table 3 were produced by rearranging the genes of the second parent to produce feasible offspring without the need for repair. FP-GA outperformed RR-GA for large values of k but not for small values. It seems that the effect of mutation by repair is rather effective when the solution space is small.

The results labeled OPT-GA in Table 3 were produced by the GA with the proposed normalization. Using the same GA as FP-GA, OPT-GA rearranges the genes of the second parent to minimize the sum of distances between genes (column vectors) before applying recombination. OPT-GA clearly outperforms Multi-Start and RR-GA; the results of one-tailed t-tests prove that OPT-GA also outperforms FP-GA significantly.

The results of the t-tests show that Multi-Start is clearly the worst technique, even though it is allowed 10^6 evaluations, while the GA-based methods produce relatively small 2.5×10^4 chromosomes. RR-GA and FP-GA have similar performance, and OPT-GA clearly does best.

Table 3. Results of three GAs compared with random solutions to MKCP.

Tightness	Instance		Multi-Start		RR-GA		t-Test [1]	FP-GA		t-Test [2]	OPT-GA		t-Test [3]
Ratio	Set	k	%-Gap	Ave	%-Gap	Ave	p-Value	%-Gap	Ave	p-Value	%-Gap	Ave	p-Value
0.8	I-4	40	28.17	141.35	7.08	182.86	6.62×10^{-16}	3.36	190.19	3.45×10^{-11}	**1.96**	192.95	2.56×10^{-10}
	I-5	40	28.67	141.52	6.84	184.82	2.92×10^{-19}	3.95	190.57	1.71×10^{-9}	**1.64**	195.14	5.16×10^{-10}
	I-6	16	20.56	143.77	4.22	173.35	8.21×10^{-10}	4.21	173.37	4.67×10^{-1}	**2.53**	176.42	6.86×10^{-5}
	I-A	40	27.62	207.57	6.16	269.13	1.09×10^{-9}	4.29	274.49	4.85×10^{-6}	**1.81**	281.61	5.45×10^{-6}
	I-B	16	20.56	208.93	4.52	251.10	2.69×10^{-7}	4.73	250.54	8.86×10^{-2}	**2.68**	255.94	1.05×10^{-4}
	I-C	40	26.81	269.33	6.12	345.48	2.89×10^{-8}	4.85	350.16	7.50×10^{-5}	**2.10**	360.28	2.11×10^{-5}
	I-D	16	18.78	271.43	3.91	321.14	8.71×10^{-8}	4.15	320.31	6.58×10^{-2}	**1.90**	327.84	2.65×10^{-6}
	I-E	8	12.65	336.81	3.63	371.60	7.01×10^{-8}	4.04	369.99	9.37×10^{-3}	**2.65**	375.37	1.31×10^{-4}
	I-F	4	6.23	346.58	2.25	361.27	2.18×10^{-6}	2.42	360.66	1.27×10^{-1}	**1.83**	362.82	1.78×10^{-3}
	I-G	40	21.61	630.13	4.73	765.78	3.64×10^{-10}	3.82	773.12	4.74×10^{-4}	**1.64**	790.58	1.47×10^{-5}
	I-H	16	14.79	634.61	3.31	720.14	1.81×10^{-8}	3.75	716.83	1.29×10^{-2}	**1.93**	730.41	5.89×10^{-5}
0.6	I-4	30	31.62	121.57	6.72	165.85	3.47×10^{-16}	4.75	169.35	8.86×10^{-9}	**2.69**	173.01	1.76×10^{-9}
	I-5	30	32.71	121.79	6.76	168.76	1.25×10^{-17}	5.69	170.70	1.25×10^{-5}	**2.53**	176.42	1.53×10^{-12}
	I-6	12	21.51	124.00	4.35	151.12	3.29×10^{-7}	4.21	151.35	1.62×10^{-1}	**2.56**	153.95	1.02×10^{-3}
	I-A	30	31.03	178.49	6.12	242.97	2.64×10^{-8}	5.30	245.09	6.19×10^{-3}	**2.36**	252.70	6.20×10^{-5}
	I-B	12	22.15	179.34	4.83	219.27	5.21×10^{-7}	5.22	218.36	1.65×10^{-2}	**3.00**	223.47	1.61×10^{-5}
	I-C	30	29.54	230.25	5.89	307.54	1.41×10^{-9}	5.19	309.85	1.43×10^{-3}	**2.63**	318.20	3.56×10^{-7}
	I-D	12	20.18	232.11	4.28	278.34	5.40×10^{-8}	4.73	277.03	1.78×10^{-2}	**2.71**	282.91	8.94×10^{-6}
	I-E	6	11.82	287.65	3.12	316.02	8.78×10^{-8}	3.48	314.85	2.63×10^{-2}	**2.25**	318.84	7.78×10^{-5}
	I-F	3	4.29	296.71	1.43	305.56	5.35×10^{-7}	1.91	304.09	1.05×10^{-3}	**1.13**	306.50	1.72×10^{-4}
	I-G	30	24.17	531.23	5.18	664.26	2.06×10^{-9}	5.00	665.56	1.31×10^{-1}	**2.42**	683.65	2.75×10^{-5}
	I-H	12	15.06	535.09	2.95	611.43	2.77×10^{-8}	3.34	608.92	9.63×10^{-3}	**1.80**	618.66	2.49×10^{-4}
0.4	I-4	20	32.30	95.98	5.56	133.91	1.71×10^{-14}	4.54	135.36	2.39×10^{-7}	**2.40**	138.39	2.27×10^{-9}
	I-5	20	34.33	96.27	6.20	137.51	5.07×10^{-16}	6.12	137.62	2.93×10^{-1}	**2.76**	142.55	8.79×10^{-10}
	I-6	8	19.63	98.21	3.14	118.37	3.63×10^{-6}	3.38	118.08	1.62×10^{-1}	**2.08**	119.66	6.08×10^{-4}
	I-A	20	32.95	140.65	5.88	197.46	1.94×10^{-7}	6.12	196.95	1.14×10^{-1}	**2.92**	203.66	1.14×10^{-5}
	I-B	8	21.41	141.44	4.02	172.76	5.61×10^{-7}	4.84	171.27	3.75×10^{-3}	**2.48**	175.53	1.48×10^{-4}
	I-C	20	31.50	179.73	5.50	247.96	2.99×10^{-8}	5.77	247.24	1.12×10^{-1}	**2.66**	255.41	5.49×10^{-7}
	I-D	8	19.61	180.87	3.74	216.59	1.92×10^{-6}	4.48	214.91	2.07×10^{-2}	**2.56**	219.23	2.34×10^{-6}
	I-E	4	9.00	222.76	2.14	239.55	1.89×10^{-8}	2.64	238.34	1.59×10^{-2}	**1.64**	240.78	1.21×10^{-6}
	I-F	2	1.87	229.63	1.72	229.97	1.22×10^{-1}	1.97	229.39	2.48×10^{-2}	**1.55**	230.37	8.79×10^{-3}
	I-G	20	25.35	405.93	4.45	519.61	5.35×10^{-9}	4.65	518.51	2.74×10^{-2}	**2.32**	531.16	2.91×10^{-5}
	I-H	8	14.64	408.88	3.15	463.92	2.75×10^{-9}	3.52	462.16	2.56×10^{-2}	**1.87**	470.06	4.82×10^{-5}

Averages from 30 runs. Multi-Start generates random solutions and chooses the best. RR-GA is a genetic algorithm with random repair. FP-GA is a genetic algorithm with FP normalization. OPT-GA is a genetic algorithm with OPT normalization. [1] The one-tailed t-test with the null hypothesis of RR-GA = Multi-Start. [2] The one-tailed t-test with the null hypothesis of FP-GA = RR-GA. [3] The one-tailed t-test with the null hypothesis of OPT-GA = FP-GA.

5. Conclusions

We present the maximum k-coverage problem (MKCP) and analyze its representation and solution space. If we apply a GA to the MKCP, then we immediately encounter the issue of redundancy in the genotype space, which is larger than the phenotype space that we characterize as a quotient space. We introduce a method of normalizing chromosomes that ensures a crossover produces feasible offspring with genes that are column vectors of the MKCP matrix and are as dissimilar as possible. This normalization was implemented using the Hungarian method [40]. We performed experiments which showed the effectiveness of this approach.

In this study, we adopted two locus-based metrics of the discrete metric and its extended version derived from Hamming distance between genes (column vectors). However, other metrics such as some variants of Cayley metric on permutations [51,52] may also be applied to the proposed theoretical framework. In the case of such non-locus-based metric, we should design a new crossover tailored to the metric. This investigation will be a promising work, which we leave for future study.

As mentioned in the Introduction, the proposed theoretical framework can be applied to real-world applications such as the cyber-physical social systems and public safety networks. We leave this applied work for future study. We also expect that this approach can be applied to other problems which have the same representation for their solutions. By expanding this technique to solution representations of variable length as in [53,54], we believe it could also be applied to the set cover problem.

Author Contributions: Conceptualization, Y.Y. and Y.-H.K.; methodology, Y.-H.K.; software, Y.Y.; validation, Y.Y.; formal analysis, Y.Y.; investigation, Y.Y.; resources, Y.Y.; data curation, Y.Y.; writing—original draft preparation, Y.Y.; writing—review and editing, Y.-H.K.; visualization, Y.Y.; supervision, Y.-H.K.; project administration, Y.-H.K.; and funding acquisition, Y.-H.K. All authors have read and agreed to the published version of the manuscript.

Funding: The present research was conducted by the Research Grant of Kwangwoon University in 2020. This research was a part of the project titled 'Marine Oil Spill Risk Assessment and Development of Response Support System through Big Data Analysis' funded by the Korea Coast Guard. This work was also supported by the National Research Foundation of Korea (NRF) grant funded by the Korea government (Ministry of Science, ICT & Future Planning) (No. 2017R1C1B1010768).

Acknowledgments: The authors thank Byung-Ro Moon for his valuable suggestions, which improved this paper.

Conflicts of Interest: The authors declare that they have no conflict of interest.

Disclosure: A preliminary version of this paper appeared in the Proceedings of the Genetic and Evolutionary Computation Conference, pp. 593–598, 2008. In comparison with the conference paper, this paper was newly rewritten with the following new materials: (i) new work: complete literature survey (Section 1) and complete theoretical work of the proposed normalization (Sections 2 and 3.2); and (ii) improved work: an improved genetic algorithm (through the improvement of normalization technique), and its largely-extended experiments together with their statistical verification (Section 4).

References

1. Hochbaum, D.S.; Pathria, A. Analysis of the greedy approach in problems of maximum k-coverage. *Nav. Res. Logist.* **1998**, *45*, 615–627. [CrossRef]
2. Caprara, A.; Fischetti, M.; Toth, P.; Vigo, D.; Guida, P.L. Algorithms for railway crew management. *Math. Program.* **1997**, *79*, 125–141. [CrossRef]
3. Indyk, P.; Mahabadi, S.; Mahdian, M.; Mirrokni, V.S. Composable Core-sets for Diversity and Coverage Maximization. In Proceedings of the 33rd ACM SIGMOD-SIGACT-SIGART Symposium on Principles of Database Systems, Snowbird, UT, USA, 22–27 June 2014; pp. 100–108.
4. Indyk, P.; Vakilian, A. Tight Trade-offs for the Maximum k-Coverage Problem in the General Streaming Model. In Proceedings of the 38th ACM SIGMOD-SIGACT-SIGAI Symposium on Principles of Database Systems, Amsterdam, The Netherlands, 30 June–5 July 2019; pp. 200–217.
5. Saha, B.; Getoor, L. On Maximum Coverage in the Streaming Model & Application to Multi-topic Blog-Watch. In Proceedings of the the SIAM International Conference on Data Mining, Sparks, NV, USA, 30 April–2 May 2009; pp. 697–708.

6. Zheng, S.; Dmitriev, P.; Giles, C.L. Graph Based Crawler Seed Selection. In Proceedings of the 18th International Conference on World Wide Web, WWW '09, Madrid, Spain, 20–24 April 2009; pp. 1089–1090.
7. Chierichetti, F.; Kumar, R.; Tomkins, A. Max-cover in Map-reduce. In Proceedings of the 19th International Conference on World Wide Web, Raleigh, NC, USA, 26–30 April 2010; pp. 231–240.
8. Li, F.H.; Li, C.T.; Shan, M.K. Labeled Influence Maximization in Social Networks for Target Marketing. In Proceedings of the IEEE International Conference on Privacy, Security, Risk, and Trust, and IEEE International Conference on Social Computing, Boston, MA, USA, 9–11 October 2011; pp. 560–563.
9. Hammar, M.; Karlsson, R.; Nilsson, B.J. Using Maximum Coverage to Optimize Recommendation Systems in e-Commerce. In Proceedings of the 7th ACM Conference on Recommender Systems, Hong Kong, China, 12–16 October 2013; pp. 265–272.
10. Yoon, Y.; Kim, Y.H. An Efficient Genetic Algorithm for Maximum Coverage Deployment in Wireless Sensor Networks. *IEEE Trans. Cybern.* **2013**, *43*, 1473–1483. [CrossRef] [PubMed]
11. Yaghini, M.; Karimi, M.; Rahbar, M. A set covering approach for multi-depot train driver scheduling. *J. Comb. Optim.* **2015**, *29*, 636–654. [CrossRef]
12. Liu, Q.; Cai, W.; Shen, J.; Fu, Z.; Liu, X.; Linge, N. A speculative approach to spatial-temporal efficiency with multi-objective optimization in a heterogeneous cloud environment. *Secur. Commun. Netw.* **2016**, *9*, 4002–4012. [CrossRef]
13. Máximo, V.R.; Nascimento, M.C.V.; Carvalho, A.C.P.L.F. Intelligent-guided adaptive search for the maximum covering location problem. *Comput. Oper. Res.* **2017**, *78*, 129–137. [CrossRef]
14. Tsiropoulou, E.E.; Thanou, A.; Papavassiliou, S. Quality of Experience-based museum touring: A human in the loop approach. *Soc. Netw. Anal. Min.* **2017**, *7*, 33. [CrossRef]
15. Sikeridis, D.; Tsiropoulou, E.E.; Devetsikiotis, M.; Papavassiliou, S. Socio-spatial resource management in wireless powered public safety networks. In Proceedings of the IEEE Military Communications Conference (MILCOM), Los Angeles, CA, USA, 29–31 October 2018, pp. 810–815.
16. Fragkos, G.; Tsiropoulou, E.E.; Papavassiliou, S. Disaster Management and Information Transmission Decision-Making in Public Safety Systems. In Proceedings of the IEEE Global Communications Conference (GLOBECOM), Waikoloa, HI, USA, 9–13 December 2019, pp. 1–6.
17. Garey, M.; Johnson, D.S. *Computers and Intractability: A Guide to the Theory of NP-Completeness*; Freeman: San Francisco, CA, USA, 1979.
18. Caserta, M.; Doerner, K.F. Tabu search-based metaheuristic algorithm for the large-scale set covering problem. In *Metaheuristics: Progress in Complex Systems Optimization*; Springer: New York, NY, USA, 2007; pp. 43–63.
19. Aickelin, U. An indirect genetic algorithm for set covering problems. *J. Oper. Res. Soc.* **2002**, *53*, 1118–1126. [CrossRef]
20. Beasley, J.E.; Chu, P.C. A Genetic Algorithm for the Set Covering Problem. *Eur. J. Oper. Res.* **1996**, *94*, 392–404. [CrossRef]
21. Balaji, S.; Revathi, N. A new approach for solving set covering problem using jumping particle swarm optimization method. *Nat. Comput.* **2016**, *15*, 503–517. [CrossRef]
22. Al-Shihabi, S.; Arafeh, M.; Barghash, M. An improved hybrid algorithm for the set covering problem. *Comput. Ind. Eng.* **2015**, *85*, 328–334. [CrossRef]
23. Ausiello, G.; Boria, N.; Giannakos, A.; Lucarelli, G.; Paschos, V. Online maximum *k*-coverage. *Discret. Appl. Math.* **2012**, *160*, 1901–1913. [CrossRef]
24. Yu, H.; Yuan, D. Set coverage problems in a one-pass data stream. In Proceedings of the the SIAM International Conference on Data Mining, Austin, TX, USA, 2–4 May 2013, pp. 758–766.
25. Chandu, D.P. Big Step Greedy Heuristic for Maximum Coverage Problem. *Int. J. Comput. Appl.* **2015**, *125*, 19–24.
26. Wang, Y.; Ouyang, D.; Yin, M.; Zhang, L.; Zhang, Y. A restart local search algorithm for solving maximum set *k*-covering problem. *Neural Comput. Appl.* **2018**, *29*, 755–765. [CrossRef]
27. Lin, G.; Guan, J. Solving maximum set *k*-covering problem by an adaptive binary particle swarm optimization method. *Knowl.-Based Syst.* **2018**, *142*, 95–107. [CrossRef]
28. Yoon, Y.; Kim, Y.H.; Moon, B.R. Feasibility-Preserving Crossover for Maximum *k*-Coverage Problem. In Proceedings of the Genetic and Evolutionary Computation Conference, Atlanta, GA, USA, 12–16 July 2008; pp. 593–598.

29. Fraleigh, J.B. *A First Course in Abstract Algebra*, 7th ed.; Addison Wesley: Boston, MA, USA, 2002.
30. Burago, D.; Burago, Y.; Ivanov, S.; Burago, I.D. *A Course in Metric Geometry*; American Mathematical Society: Providence, RI, USA, 2001.
31. Yoon, Y.; Kim, Y.H.; Moraglio, A.; Moon, B.R. Quotient geometric crossovers and redundant encodings. *Theor. Comput. Sci.* **2012**, *425*, 4–16. [CrossRef]
32. Choi, S.S.; Moon, B.R. Normalization in Genetic Algorithms. In Proceedings of the Genetic and Evolutionary Computation Conference, Chicago, IL, USA, 12–16 July 2003; pp. 862–873.
33. Dorne, R.; Hao, J.K. A New Genetic Local Search Algorithm for Graph Coloring. In Proceedings of the Fifth Conference on Parallel Problem Solving from Nature, Amsterdam, The Netherlands, 27–30 September 1998; pp. 745–754.
34. Laszewski, G. Intelligent Structural Operators for the k-way Graph Partitioning Problem. In Proceedings of the Fourth International Conference on Genetic Algorithms, San Diego, CA, USA, 13–16 July 1991; pp. 45–52.
35. Mühlenbein, H. Parallel genetic algorithms in combinatorial optimization. In *Computer Science and Operations Research: New Developments in Their Interfaces*; Pergamon: Oxford, MS, USA, 1992; pp. 441–456.
36. Van Hoyweghen, C.; Naudts, B.; Goldberg, D.E. Spin-flip symmetry and synchronization. *Evol. Comput.* **2002**, *10*, 317–344. [CrossRef]
37. Kang, S.J.; Moon, B.R. A Hybrid Genetic Algorithm for Multiway Graph Partitioning. In Proceedings of the Genetic and Evolutionary Computation Conference, Las Vegas, NV, USA, 8–12 July 2000, pp. 159–166.
38. Moraglio, A.; Kim, Y.H.; Yoon, Y.; Moon, B.R. Geometric Crossovers for Multiway Graph Partitioning. *Evol. Comput.* **2007**, *15*, 445–474. [CrossRef]
39. Choi, S.S.; Moon, B.R. Normalization for Genetic Algorithms with Nonsynonymously Redundant Encodings. *IEEE Trans. Evol. Comput.* **2008**, *12*, 604–616. [CrossRef]
40. Kuhn, H.W. The Hungarian Method for the assignment problem. *Nav. Res. Logist. Q.* **1955**, *2*, 83–97. [CrossRef]
41. Papadimitriou, C.H.; Steiglitz, K. *Combinatorial Optimization: Algorithms and Complexity*; Prentice-Hall: Englewood Cliffs, NJ, USA, 1955.
42. Avis, D. A survey of heuristics for the weighted matching problem. *Networks* **1983**, *13*, 475–493. [CrossRef]
43. Beasley, J.E. OR-Library: Distributing test problems by electronic mail. *J. Oper. Res. Soc.* **1990**, *41*, 1069–1072. [CrossRef]
44. Eshelman, L.J. The CHC adaptive search algorithm: How to have safe search when engaging in nontraditional genetic recombination. In *Foundations of Genetic Algorithms*; Morgan Kaufmann: Burlington, MA, USA, 1991; pp. 265–283.
45. Alba, E.; Luque, G.; Araujo, L. Natural language tagging with genetic algorithms. *Inf. Process. Lett.* **2006**, *100*, 173–182. [CrossRef]
46. Cordón, O.; Damasb, S.; Santamaría, J. Feature-based image registration by means of the CHC evolutionary algorithm. *Image Vis. Comput.* **2006**, *24*, 525–533. [CrossRef]
47. Guerra-Salcedo, C.; Whitley, D. Genetic Search for Feature Subset Selection: A Comparison Between CHC and GENESIS. In Proceedings of the Third Annual Conference on Genetic Programming; Morgan Kaufmann: Burlington, MA, USA, 1998; pp. 504–509.
48. Nebro, A.J.; Alba, E.; Molina, G.; Chicano, F.; Luna, F.; Durillo, J.J. Optimal antenna placement using a new multi-objective CHC algorithm. In Proceedings of the 9th Annual Conference on Genetic and Evolutionary Computation, London, UK, 7–11 July 2007; pp. 876–883.
49. Tsutsui, S.; Goldberg, D.E. Search space boundary extension method in real-coded genetic algorithms. *Inf. Sci.* **2001**, *133*, 229–247. [CrossRef]
50. Yoon, Y.; Kim, Y.H.; Moraglio, A.; Moon, B.R. A Theoretical and Empirical Study on Unbiased Boundary-extended Crossover for Real-valued Representation. *Inf. Sci.* **2012**, *183*, 48–65. [CrossRef]
51. Pinch, R.G.E. The distance of a permutation from a subgroup of S_n. *arXiv* **2005**, arXiv:math/0511501.
52. Kim, Y.H.; Moraglio, A.; Kattan, A.; Yoon, Y. Geometric generalisation of surrogate model-based optimisation to combinatorial and program spaces. *Math. Probl. Eng.* **2014**, *2014*, 184540. [CrossRef]

53. Nam, Y.W.; Kim, Y.H. Automatic jazz melody composition through a learning-based genetic algorithm. In Proceedings of the 8th International Conference on Computational Intelligence in Music, Sound, Art and Design (EvoMUSART), Leipzig, Germany, 24–26 April 2019; Lecture Notes in Computer Science. Springer: Berlin, Germany; Volume 11453, pp. 217–233.
54. Lee, J.; Kim, Y.H. Epistasis-based basis estimation method for simplifying the problem space of an evolutionary search in binary representation. *Complexity* **2019**, *2019*, 2095167. [CrossRef]

Article

Hybridization of Multi-Objective Deterministic Particle Swarm with Derivative-Free Local Searches

Riccardo Pellegrini [1], Andrea Serani [1,*], Giampaolo Liuzzi [2], Francesco Rinaldi [3], Stefano Lucidi [4] and Matteo Diez [1]

1 CNR-INM, National Research Council—Institute of Marine Engineering, 00139 Rome, Italy; riccardo.pellegrini@inm.cnr.it (R.P.); matteo.diez@cnr.it (M.D.)
2 CNR-IASI, National Research Council—Institute for Systems Analysis and Computer Science, 00185 Rome, Italy; giampaolo.liuzzi@iasi.cnr.it
3 Department of Mathematics, University of Padua, 35121 Padua, Italy; rinaldi@math.unipd.it
4 Department of Computer, Control, and Management Engineering "A. Ruberti", Sapienza University, 00185 Rome, Italy; lucidi@diag.uniroma1.it
* Correspondence: andrea.serani@cnr.it

Received: 5 March 2020; Accepted: 31 March 2020; Published: 7 April 2020

Abstract: The paper presents a multi-objective derivative-free and deterministic global/local hybrid algorithm for the efficient and effective solution of simulation-based design optimization (SBDO) problems. The objective is to show how the hybridization of two multi-objective derivative-free global and local algorithms achieves better performance than the separate use of the two algorithms in solving specific SBDO problems for hull-form design. The proposed method belongs to the class of memetic algorithms, where the global exploration capability of multi-objective deterministic particle swarm optimization is enriched by exploiting the local search accuracy of a derivative-free multi-objective line-search method. To the authors best knowledge, studies are still limited on memetic, multi-objective, deterministic, derivative-free, and evolutionary algorithms for an effective and efficient solution of SBDO for hull-form design. The proposed formulation manages global and local searches based on the hypervolume metric. The hybridization scheme uses two parameters to control the local search activation and the number of function calls used by the local algorithm. The most promising values of these parameters were identified using forty analytical tests representative of the SBDO problem of interest. The resulting hybrid algorithm was finally applied to two SBDO problems for hull-form design. For both analytical tests and SBDO problems, the hybrid method achieves better performance than its global and local counterparts.

Keywords: hybrid algorithms; memetic algorithms; particle swarm; multi-objective deterministic optimization, derivative-free; global/local optimization; simulation-based design optimization

---

1. Introduction

The research and development of new technologies, along with a reduction of design and production costs, are enabling factors to face the challenges imposed by the worldwide-market competition. Computer simulations have played an increasingly important role in the design process of engineering products whose efficiency is greatly affected by shape parameters, such as air-, ground-, and water-borne vehicles. For these reasons, addressing real-world complex industrial applications involves high-fidelity physics-based solvers, particularly where innovative products are pursued for which past experience is not available.

In this context, the simulation-based design (SBD) paradigm has demonstrated the ability to support the design decision-making process. The recent development of high-performance computing systems has led the SBD to integrate optimization algorithms and uncertainty quantification (UQ)

methods, shifting the SBD paradigm to automatic deterministic and stochastic SBD optimization (SBDO) [1] with potential global solutions to the design problem. The computational costs associated with the solution of the SBDO problem with high-fidelity solvers remain a limiting factor for the implementation of the SBDO in industries and, in particular, small- and medium-sized enterprises. The major critical issues related to computational cost and implementation complexity of the SBDO procedure can be summarized as: (a) physical solvers are computationally expensive to run; (b) they are often combined in a multidisciplinary analysis framework that increases the computational complexity; (c) solvers are often available only as a black box; (d) solutions are often affected by non-negligible residual noise; (e) assessing design performance in a variety of operating/environmental conditions requires multiple solvers to operate; (f) the search for improved performance often faces conflicting objectives; and (g) achieving optimum design through an optimization method requires a large number of evaluations, especially for high-dimensional design spaces and, if their global exploration is attempted, making high-fidelity SBDO a very complex theoretical, algorithmic, and technical task. Despite the unquestionable and significant achievements in the field, with the capability of solving high-fidelity multidisciplinary design optimization , the design-space exploration and exploitation in the quest for global optima remains a hard-to-reach objective, due to its almost-unaffordable computational cost [2], especially when dealing with multi-objective (constrained) problems.

Real-world applications are pervaded by almost conflicting objectives and can be formulated as multi-objective (constrained) problems (MOPs) [3]. For such problems, the optimum is defined by the Pareto definition, where a set of non-dominated solutions provides the trade-off among the objectives. From the optimization view point, the Pareto optimality definition separates the roles of *problem solver* and *decision maker*. The problem solver finds one or more non-dominated solutions. The decision maker, based on her/his preferences, selects one of the non-dominated solutions. For this reason, considering when the decision maker operates, the following classification of the optimization algorithms can be drawn up: (i) *without preferences*, decision maker preferences are indifferent and the problem is solved by finding a single non-dominated solution; (ii) *a priori*, the optimization procedure is driven by the knowledge of the decision maker preferences, making provision for the acceptance of the optimal solution by the decision maker; and (iii) *a posteriori*, decision maker preferences are taken into account at the end of the optimization. Even though there is interest in without preferences and a priori methods (capable of providing a single non-dominated solution), the a posteriori methods have a high relevance in the SBDO context, since they are able to approximate the whole set of Pareto solutions. For this purpose, evolutionary algorithms (EAs) [4] have long been used as a posteriori methods: by adapting a group of individuals, EAs naturally allow for the solution of MOPs. Nevertheless, EAs are in general stochastic and their testing and assessment requires statistically converged results (non-dominated solution set) from multiple optimization runs, which are almost unattainable for complex industrial applications. Therefore, deterministic methods have also been developed and assessed [5–7].

Historically, the aeronautics industry and the associated research have played a very important role in tackling and solving SBDO for complex multidisciplinary applications. For technological and computational-efficiency reasons, researchers have focused on local optimization methods, often relying on availability of derivatives from adjoint solvers [2]. On the one hand, local methods require the knowledge of derivatives, which may require intrusive approaches (e.g., adjoint methods, unfortunately not suitable for black-box solvers) and can get easily trapped in local minima. On the other hand, global optimization methods usually do not require derivatives, but are affected by the *curse of dimensionality*, where the complexity and associated computational cost to solve the problem increases dramatically with the dimension of the design space. Furthermore, local algorithms investigate accurately a limited domain region, also providing proof of convergence (generally not available for global methods), whereas global methods are designed to explore the whole design domain, providing approximate solutions to the decision problem. Since in the SBDO context objectives may be noisy and/or their derivatives are often not provided by the simulation tool, derivative-free optimization algorithms [8] are often a viable option. Examples of local and global derivative-free SBDO approaches to the design problem have been investigated in the ship hydrodynamic community [9], with focus on both civil and military applications and development of dual-use technologies. In the last years, several EAs have been proposed to solve design problems, such as dragonfly [10], salp swarm [11], polar bear [12], and dolphin pod optimization [13]. Among other derivative-free global methods, particle swarm optimization (PSO) [14] represents a suitable option for the solution of SBDO problems [15]. The numerical experiences reported in literature seem to indicate that it is quite efficient in identifying the regions of the feasible set where the globally optimal solutions are located [16]. Although the multi-objective extension of PSO results in an increased algorithmic complexity, the associated increased computational cost (overhead) is orders of magnitude lower than the CPU-time associated to the typical objectives evaluations in SBDO for ship hydrodynamics, and therefore not considered an issue.

In the last decades, the optimization community has explored the possibility of combining global and local approaches, with the aim of enriching global search capability with the local search accuracy in some hybrid formulations. This kind of algorithms falls in the class of *memetic algorithms* (MAs), where a population-based global method is coupled with an individual learning procedure capable of performing local refinements [17]. Among other MAs, PSO has been extended to single-objective hybrid global/local formulations by local line search in [18]. In the MOPs context, PSO has been combined with several local search methods, such as scatter search [19], synchronous particle local search [20], gradient-based [21], charged system search [22], modified local search [23], multi-objective dichotomy line search [24], and optimal particle search strategy [25]. Further examples of hybrid methods in the context of multi-objective optimization can be found in [26,27]. Most MAs and PSO methods ensure global exploration and solution diversity via use of random components. These require running multiple times to achieve statistically significant results, which may be not affordable in computationally-expensive SBDO problems. Therefore, deterministic PSO formulations are considered in this work.

At the present state of the art, to the authors best knowledge, studies are still limited on memetic multi-objective deterministic derivative-free EA formulations for an effective and efficient solution of SBDO for hull-form design. In earlier work [28], the authors proposed the hybridization of a multi-objective deterministic version of the PSO algorithm (MODPSO) [6] with local searches by a deterministic derivative-free multi-objective (DFMO) [29] line search method. The resulting multi-objective deterministic hybrid algorithm (MODHA) was applied to a limited number of test cases and algorithm setups. Criteria for the activation of local searches were proposed and discussed, based on the particles velocity and the hypervolume metric (HV) [30], which is a widely-used indicator for the assessment of convergence and distribution of multi-objective problem solutions [31]. Although the study was limited to only few cases/setups, HV was found the most promising criterion to drive the hybridization within MODHA. A more extensive study was recommended on the local activation and computational-cost breakdown between global and local searches.

The objective of the present work is to show how the use of a global/local hybrid algorithm based on MODPSO and DFMO achieves better performance than stand-alone global and local algorithms in solving multi-objective SBDO for hull-form design. Specifically, the problems of interest are characterized by a number of variables of the order of 10 and up to three objectives (for instance, resistance, seakeeping, and maneuverability). Specifically, the paper advances the study on MODHA presented in [28], with focus on the use of the HV metric and the identification of the proper setup for the hybridization scheme.

The proposed hybridization scheme is driven by two parameters: the first defines the threshold for the activation of local searches, whereas the second defines the "deepness" of the local search (i.e., how many function evaluations are performed at each call of the local algorithm). To identify reasonable values for these parameters, a full factorial combination of three activation and three deepness parameters is evaluated and the performance compared. Following the guidelines in [32], the assessment was targeted towards a representative benchmark of the specific SBDO problem of interest. Specifically, the benchmark was composed by 40 analytical test problems, with two and three objectives, 1 to 12 variables, and several features of variables and functions spaces. MODHA performance was assessed by the HV bounded by the non-dominated solution set. Finally, the most promising parameter setup was applied to two SBDO problems pertaining to the shape optimization of a high-speed catamaran in waves and of a small waterplane area twin hull (SWATH) model in calm water. Comparisons with stand-alone global [6] and local [29] methods are presented and discussed.

2. Multi-Objective Optimization Problem Formulation and Definitions

The general formulation of a multi-objective minimization problem reads

$$\begin{aligned} \text{minimize} \;\; & \mathbf{f}(\mathbf{x}) = \{f_m(\mathbf{x})\}, \;\; \text{with} \;\; m = 1, \ldots, M \\ \text{subject to} \;\; & z_i(\mathbf{x}) \leq 0, \;\; \text{with} \;\; i = 1, \ldots, I \\ \text{and to} \;\; & h_j(\mathbf{x}) = 0, \;\; \text{with} \;\; j = 1, \ldots, J \\ \text{and to} \;\; & \mathbf{l} \leq \mathbf{x} \leq \mathbf{u} \end{aligned} \tag{1}$$

where $\mathbf{x} \in \mathbb{R}^N$ is the variables vector with N the number of variables, M is the number of objectives f_m, z_i are the inequality constraints, h_j are the equality constraints, and $\mathbf{l}$ and $\mathbf{u}$ are the lower and upper bound for $\mathbf{x}$, respectively. Defining the feasible solution set as

$$\mathcal{X} = \{\mathbf{x} \in \mathbb{R}^N : \mathbf{l} \leq \mathbf{x} \leq \mathbf{u}, \;\; z_i(\mathbf{x}) \leq 0 \;\; \forall i \in I, \;\; h_j(\mathbf{x}) = 0 \;\; \forall j \in J\} \tag{2}$$

the theoretical solution of the MOP in Equation (1) is the locus of non-dominated feasible solutions given (both in the variable and objective function space) by the Pareto solution set $\mathcal{P}$ (see Figure 1)

$$\mathcal{P} = \{(\mathbf{x}, \mathbf{p}) | \mathbf{x} \in \mathcal{X}, \, \mathbf{p} = \mathbf{f}(\mathbf{x}), \, \nexists \mathbf{y} \in \mathcal{X} : \mathbf{f}(\mathbf{y}) \preceq \mathbf{f}(\mathbf{x})\} \tag{3}$$

where $\mathbf{f}(\mathbf{y}) \preceq \mathbf{f}(\mathbf{x})$ indicates that $\mathbf{f}(\mathbf{y})$ dominates $\mathbf{f}(\mathbf{x})$ (see, e.g. [33]).

In the following, the approximate and discrete solution set $\mathcal{S}^i$ (see Figure 1) achieved by the optimizer at the ith iteration and used for the algorithm performance assessment is defined similarly to Equation (3) as

$$\mathcal{S}^i = \{(\mathbf{x}, \mathbf{s}) | \mathbf{x} \in \mathcal{X}^i, \, \mathbf{s} = \mathbf{f}(\mathbf{x}), \, \nexists \mathbf{y} \in \mathcal{X}^i : \mathbf{f}(\mathbf{y}) \preceq \mathbf{f}(\mathbf{x})\} \tag{4}$$

where $\mathcal{X}^i$ is the subset of feasible points produced by the algorithm and available at the ith iteration.

Likewise, the approximate and discrete reference solution set $\mathcal{R}$ used for the performance analysis is defined as

$$\mathcal{R} = \{(\mathbf{x}, \mathbf{r}) \in \cup_{n=1}^{N_s} \mathcal{S}_n | \nexists (\mathbf{y}, \mathbf{s}) \in \cup_{n=1}^{N_s} \mathcal{S}_n : \mathbf{s} \preceq \mathbf{r}\} \tag{5}$$

where N_s is the number of final solution sets $\mathcal{S}_n$ provided by different algorithm formulations and/or setups.

2.1. Performance Metrics

The performance assessment of a multi-objective optimization algorithm can be established in terms of *convergence* and *diversity*: the former is related to the distance between $\mathcal{S}$ and $\mathcal{R}$, whereas the latter deals with the wideness and the spread of $\mathcal{S}$. A suitable convergence-diversity metric is the HV [30]. It provides the (hyper)volume dominated by the solution set $\mathcal{S}$, evaluated as follows

$$\text{HV}(\mathcal{S}, \mathcal{R}) = \text{volume}\left(\bigcup_{i=1}^{|\mathcal{S}|} v_i\right) \tag{6}$$

by using the anti-ideal point $\mathcal{A}$ of the reference solution set $\mathcal{R}$ [33] as a reference point, as shown in Figure 1.

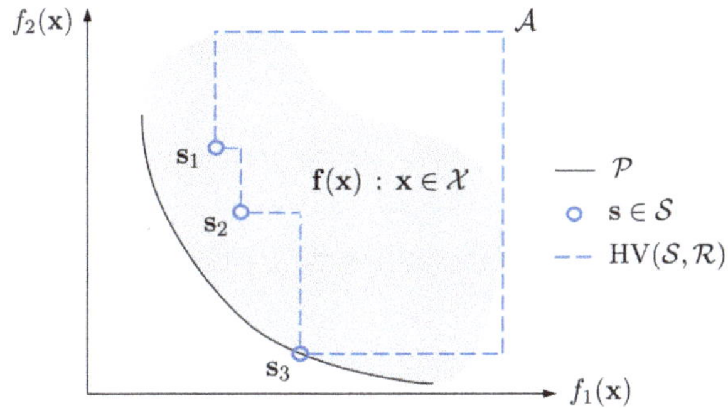

Figure 1. Hypervolume metric.

Herein, to assess and compare the performance of different algorithms, a normalized version of HV is used

$$\mathrm{NHV} = \frac{\mathrm{HV}(\mathcal{S},\mathcal{R})}{\mathrm{HV}(\mathcal{R},\mathcal{R})} \tag{7}$$

To further assess the impact of the kth setting parameter on the algorithm performance, the relative variability $\sigma^2_{r,k}$ [34] is additionally used. Considering the algorithm parameters vector as $\mathbf{t} \in \mathcal{T}$, the relative performance variability associated to its kth component is

$$\sigma^2_{r,k} = \frac{\sigma^2_k}{\sum_{k=1}^{|\mathcal{T}|} \sigma^2_k} \tag{8}$$

where

$$\sigma^2_k = \frac{1}{|\Gamma|} \sum_{\gamma \in \Gamma} [\hat{\mu}_k(\gamma)]^2 - \left[\frac{1}{|\Gamma|} \sum_{\gamma \in \Gamma} \hat{\mu}_k(\gamma) \right]^2 \tag{9}$$

with Γ containing the positions γ assumed by the parameter t_k,

$$\hat{\mu}_k(\gamma) = \frac{1}{|\mathcal{B}|} \sum_{b \in \mathcal{B}} \bar{\mu}(\mathbf{t}), \mathcal{B} = \{\mathbf{t} : t_k = \gamma\} \qquad \text{and} \qquad \bar{\mu}(\mathbf{t}) = \frac{1}{|\mathcal{Q}|} \sum_{q \in \mathcal{Q}} [\mu(\mathbf{t})]_q \tag{10}$$

where $[\mu(\mathbf{t})]_q$ is the value of the NHV given by the parameters $\mathbf{t}$, for the problem q.

3. Multi-Objective Deterministic Hybrid Algorithm: MODHA

The proposed global/local hybrid algorithm is based on a multi-objective deterministic version of the particle swarm optimization [6] and on the derivative-free multi-objective local searches algorithm [29]. The following subsections present the global and local algorithms and finally the hybridization proposed.

3.1. Multi-Objective Deterministic Particle Swarm Optimization (MODPSO)

The PSO algorithm [14] belongs to the class of metaheuristic algorithms for single-objective derivative-free global optimization. It is based on the social-behavior metaphor of a flock of birds or a swarm of bees searching for food. The original PSO formulation makes use of random coefficients, to enhance the swarm dynamics' variability. This approach might be too expensive in SBDO for real industrial applications, since statistically convergent results can be obtained only through extensive numerical campaigns. Therefore, a deterministic PSO is introduced in [35] and a discussion for its effective and efficient use in SBDO, with comparison to the original (stochastic) PSO, is presented in [36]. The extension of deterministic PSO to a multi-objective formulation is proposed in [37] as follows

$$\begin{cases} \mathbf{v}_j^{i+1} = \chi \left[\mathbf{v}_j^i + \phi_1 \left(\mathbf{h}_j - \mathbf{x}_j^i \right) + \phi_2 \left(\mathbf{g}_j - \mathbf{x}_j^i \right) \right] \\ \mathbf{x}_j^{i+1} = \mathbf{x}_j^i + \mathbf{v}_j^{i+1} \end{cases} \tag{11}$$

where $\mathbf{v}_j^i$ and $\mathbf{x}_j^i$ are the velocity and the position of the jth particle at the ith iteration, χ is the constriction factor, ϕ_1 and ϕ_2 are the cognitive (or personal) and social (or global) learning rates, and $\mathbf{h}_j$ and $\mathbf{g}_j$ are the cognitive and social (personal and global) attractors. Specifically, $\mathbf{h}_j$ is the personal minimizer of an aggregated objective function defined as

$$\mathbf{h}_j = \underset{\mathbf{x}_{h,j}}{\operatorname{argmin}} \sum_{m=1}^{M} f_m(\mathbf{x}_{h,j}) \qquad \text{with} \qquad \mathbf{x}_{h,j}^i \in \mathcal{S}_j^i \tag{12}$$

where $\mathbf{x}_{h,j}^i$ are the points of the personal non-dominated solution set $\mathcal{S}_j^i$ at the ith iteration, whereas $\mathbf{g}_j$ is the closest point to the jth particle of the solution set $\mathcal{S}^i$ defined as

$$\mathbf{g}_j = \underset{\mathbf{x}_j}{\operatorname{argmin}} \|\mathbf{x}_j^i - \mathbf{x}\| \qquad \text{with} \qquad \mathbf{x} \in \mathcal{S}^i \tag{13}$$

Here, a memory-based formulation of MODPSO is used, where all points $\mathbf{x}$ ever visited until the current iteration are assessed in Equation (4).

A discussion on MODPSO formulations and parameters setup for an effective and efficient use in SBDO is presented in [6]. The current MODPSO code is available at github.com/MAORG-CNR-INM/POT.

3.2. Derivative-Free Multi-Objective Local Searches (DFMO)

DFMO [29] is an a posteriori derivative-free algorithm for constrained (possibly) non-smooth multi-objective problems. The algorithm produces (or updates) a set of non-dominated solutions (rather than a single one, as it is common in the single-objective case), tending towards the $\mathcal{P}$ of the problem, as the iteration count grows.

At each iteration (see Figure 2), from each point $\mathbf{x}_j \in \mathcal{S}^i$, DFMO starts a line-search along a suitably generated direction $\hat{d}_k$. If such a direction is able to guarantee "sufficient" decrease, then a "sufficiently" large step λ along this direction is performed, allowing to (possibly) improve $\mathcal{S}$.

It is worth noting that DFMO belongs to a particular class of derivative-free methods proposed to tackle optimization problems where first order derivatives of the objective function and/or of the constraints can be neither calculated nor explicitly approximated [38]. In particular, DFMO does not perform any line-searches along approximations of gradient-related directions but rather along directions that are not affected by possible errors on the calculation of the objective function and constraints. Further details on DFMO formulation and implementation can be found in [29]. DFMO source code is available at www.iasi.cnr.it/liuzzi/DFL/index.php/home.

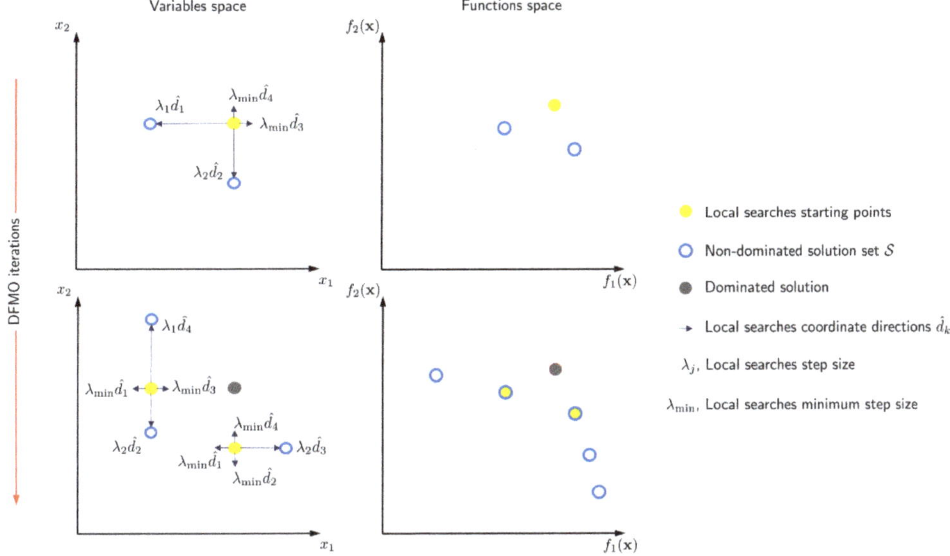

Figure 2. DFMO conceptual scheme.

3.3. Hybridization Scheme

When and where from the local search starts, represent one of the critical issues when combining global and local optimization methods. Herein, the proposed hybridization scheme uses the HV metric. Defining the local search activation parameter as

$$\alpha = \frac{\mathrm{HV}(\mathcal{S}^i, \mathcal{S}^i)}{\mathrm{HV}(\mathcal{S}^{i-1}, \mathcal{S}^i)} \tag{14}$$

where $\mathrm{HV}(\mathcal{S}^i, \mathcal{S}^i)$ is the hypervolume associated to $\mathcal{S}^i$ and $\mathrm{HV}(\mathcal{S}^{i-1}, \mathcal{S}^i)$ is the hypervolume associated to solution set at the previous algorithm iteration $\mathcal{S}^{i-1}$, the hybrid algorithm starts a local search from each point of the current solution set if $\alpha < \alpha^\star$, with $\alpha^\star$ a prescribed threshold value.

The number of problem evaluations performed at each call of the local algorithm ($\mathcal{L}$) is defined as $\mathcal{L} = \omega \mathcal{Z}$, with ω the local search deepness parameter and $\mathcal{Z}$ the number of particles. Hybridization scheme block-diagram and pseudo-code are shown in Figure 3 and Algorithm 1, respectively.

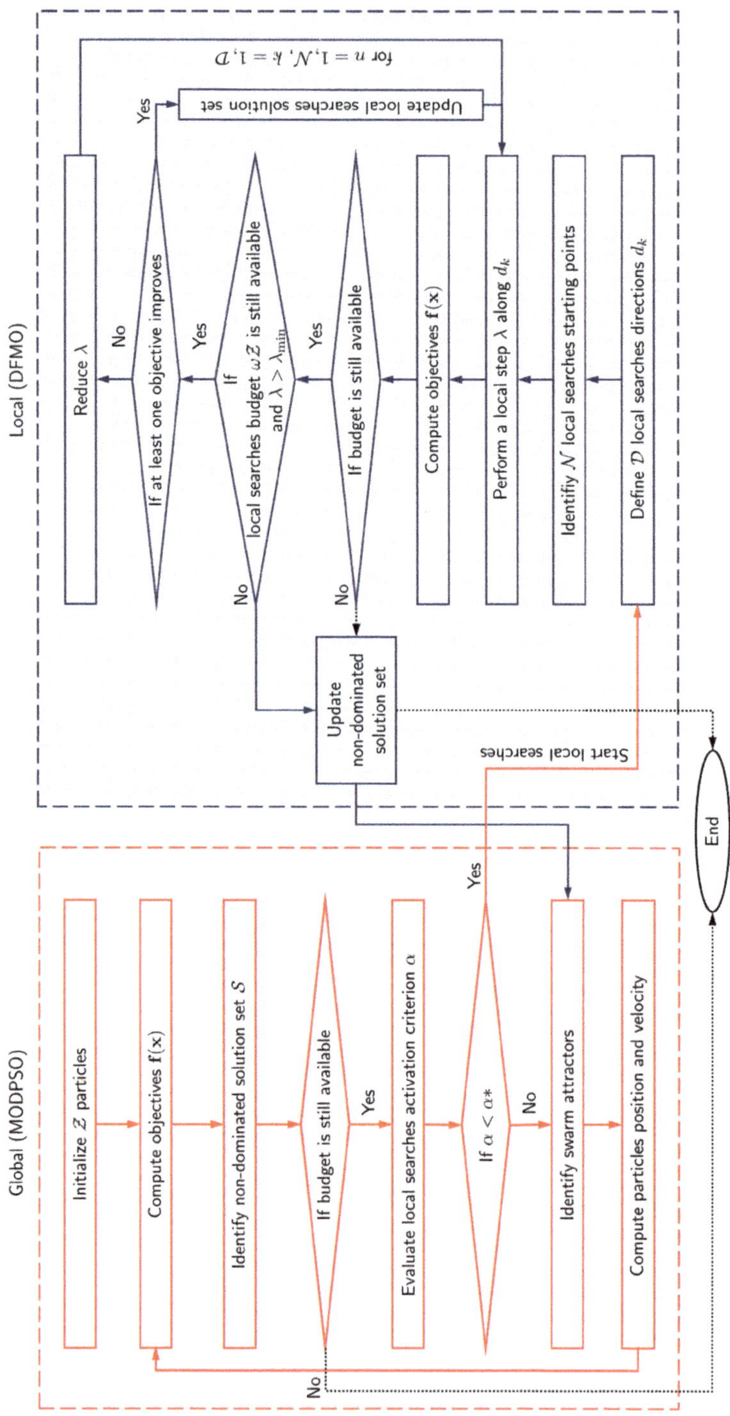

Figure 3. MODHA block diagram.

Algorithm 1 MODHA pseudo-code.

```
1: Initialize a swarm of $\mathcal{Z}$ particles
2: while $i <$ Maximum number of iterations (Maximum budget of problem evaluations) do
3:     for $j = 1, \mathcal{Z}$ do
4:         Evaluate the objective functions $\mathbf{f}$
5:     end for
6:     Compute the personal and global non-dominated solution sets
7:     for $j = 1, \mathcal{Z}$ do
8:         Identify cognitive $\mathbf{h}_j$ and social $\mathbf{g}_j$ attractors,
9:         Update particles velocity $\mathbf{v}_j^{i+1}$ and position $\mathbf{x}_j^{i+1}$
10:    end for
11:    Evaluate the local search activation parameter ($\alpha$)
12:    if Condition for performing local search is true ($\alpha < \alpha^\star$) then
13:        Define $\mathcal{D} = 2N$ coordinate directions $\hat{d}_k$
14:        Identify $\mathcal{N}$ starting points for the local search
15:        for $n = 1, \mathcal{N}$ do
16:            Set the number of local search $\mathcal{L}$ to zero
17:            for $k = 1, \mathcal{D}$ do
18:                while $\mathcal{L} < \omega\mathcal{Z}$ and $\lambda > \lambda_{\min}$ do
19:                    Perform one step equal to $\lambda$ along $\hat{d}_k$ from the $n$th starting point
20:                    Evaluate the objective functions $\mathbf{f}$
21:                    Set $\mathcal{L}$ to $\mathcal{L} + 1$
22:                    if At least one objective function improves then
23:                        Update local search solution set and go to step 18
24:                    else
25:                        Reduce $\lambda$
26:                    end if
27:                end while
28:            end for
29:        end for
30:        Update the non-dominated solution set $\mathcal{S}^i$ with local search solution set
31:    end if
32:    Update the number of problem evaluations performed
33: end while
34: Output the non-dominated solution set $\mathcal{S}^i$
```

4. Analytical Test Problems

The most promising activation and deepness parameter values are identified using a test set representative of a SBDO problem with up to three objectives and a number of variables of the order of 10. Specifically, 40 real-valued test problems were selected from the literature [39], with two and three objectives and 1 to 12 variables. The selected test problems include several features, such as convex/concave/mixed/disconnected Pareto fronts or some combination, separable/non-separable variables, Pareto one-to-one/many-to-one, and uni/multi-modality [40]. The multi-objective test problems are summarized in Table 1. Note that problems $q = 26, \ldots, 40$ are multi-objective combination of well-known single-objective test functions.

To provide a proper comparison between different problems with different function-space range, the non-dominated solution set $\mathcal{S}$ for each problem was normalized with the functions range, therefore each non-dominated solution $\mathbf{s}_j \in [0, 1]$ and the anti-ideal reference point for the HV is $\{1\}_{m=1}^{M}$ (see Figure 1). The code provided in [41] is used for the computation of the HV metric.

Table 1. Analytical test problems.

Problem q	Name	Reference	N	M
1	Deb 4.1	[42]	2	2
2	Deb 5.3	[42]	2	2
3	Deb 5.1.3	[42]	2	2
4	DTLZ1	[43]	7	3
5	DTLZ3	[43]	12	3
6	DTLZ3n2	[43]	2	2
7	DTLZ5	[43]	12	3
8	F2	[44]	2	2
9	Far1	[40]	2	2
10	FES2	[40]	10	3
11	I2	[40]	8	3
12	I5	[40]	8	3
13	IKK1	[40]	2	3
14	IM1	[40]	2	2
15	lovison4	[45]	2	2
16	lovison5	[45]	3	3
17	lovison6	[45]	3	3
18	MOP3	[40]	2	2
19	MOP4	[40]	3	2
20	MOP6	[40]	2	2
21	Sch1	[40]	1	2
22	TKLY1	[40]	4	2
23	VU2	[40]	2	2
24	WFG4	[40]	8	3
25	ZDT6	[46]	10	2
26	Freudenstein-Roth—Multi Modal	[34]	2	2
27	Freudenstein-Roth—Sphere	[34]	2	2
28	Freudenstein-Roth—Styblinski-Tang	[34]	2	2
29	Freudenstein-Roth—Three-Hump Camel Back	[34]	2	2
30	Levy 5 —Schubert	[34]	2	2
31	Levy 10—Griewank	[34]	2	2
32	Levy 15—Ackley	[34]	2	2
33	Schubert P1—Matyas	[34]	2	2
34	Schubert P2—Exponential	[34]	2	2
35	Sphere—Booth	[34]	2	2
36	Sphere—Schubert P1	[34]	2	2
37	Sphere—Six-Hump Camel Back	[34]	2	2
38	Test Tube Holder—Ackley	[34]	2	2
39	Test Tube Holder—Schubert	[34]	2	2
40	Test Tube Holder—Schubert P1	[34]	2	2

5. Simulation-Based Design Optimization Problems

Two SBDO problems were solved pertaining to hull-form optimization of a high-speed catamaran and a SWATH model. Details are provided in the following subsections.

5.1. Catamaran Problem

The stochastic (reliability-based and robust) design optimization of a high-speed catamaran in irregular head waves was solved [1]. Realistic environment conditions are associated to the North Pacific Ocean, whereas operating conditions include stochastic sea state and speed. A multi-objective problem ($M = 2$) was solved, aiming at the reduction of the expected value of the mean total resistance in irregular waves (f_1) and the increase of the ship operational effectiveness (operability) referring to a set of motion-related constraints (f_2). The SBDO problem reads

$$\begin{array}{rl} \text{minimize} & \{f_1(\mathbf{x}), -f_2(\mathbf{x})\}^\mathsf{T} \\ \text{subject to} & f_1 \leq 0;\ \ f_2 \geq 0 \\ \text{and to} & \mathbf{l} \leq \mathbf{x} \leq \mathbf{u} \end{array} \tag{15}$$

The problem was solved by means of stochastic radial-basis function (SRBF) interpolation [47] of high-fidelity unsteady Reynolds-averaged Navier–Stokes (URANS) simulations (CFDShip-Iowa v4.5 [48]). Further details on design variables, objective functions, constraints, and conditions can be found in [48]. A snapshot of the catamaran behavior in irregular head waves computed by the URANS solver is shown in Figure 4.

The inequalities in Equation (15) are handled by a penalty function, so that, if $f_1 > 0$ or $f_2 < 0$,

$$f_m = 10000 f_m \qquad \text{for} \qquad m = 1,2 \tag{16}$$

whereas if domain bounds violation occurs

$$f_m = f_m + 100 \sum_{j=1}^{N} \max(x_j - u_j, 0) + 100 \sum_{j=1}^{N} \max(l_j - x_j, 0) \qquad \text{for} \qquad m = 1,2 \tag{17}$$

Four design variables ($N = 4$) control the catamaran hull global modifications, based on the Karhunen–Loève expansion (KLE) of the shape modification vector [49] based on free-form deformation. Further details can be found in [1]. The reference non-dominated solution set $\mathcal{R}$ is taken from [6].

Figure 4. High-speed catamaran in irregular head waves by URANS simulations [50].

5.2. SWATH Problem

A multi-objective optimization problem ($M = 2$) was solved for calm water resistance (f_1) and displacement (f_2) at constant speed [51]. The SBDO problems reads

$$\begin{array}{rl} \text{minimize} & \{f_1(\mathbf{x}), -f_2(\mathbf{x})\}^\mathsf{T} \\ \text{and to} & \mathbf{l} \leq \mathbf{x} \leq \mathbf{u} \end{array} \tag{18}$$

The hydrodynamic resistance was evaluated by an adaptive multi-fidelity SRBF metamodel [52] trained with high-fidelity URANS (χnavis [53]) and low-fidelity potential flow (WARP [54]) solvers. Further details on design variables, objective functions, constraints, and conditions can be found in [51].

Figure 5 shows non-dimensional pressure distribution on the SWATH torpedoes and wave elevation for the original geometry by potential flow simulation.

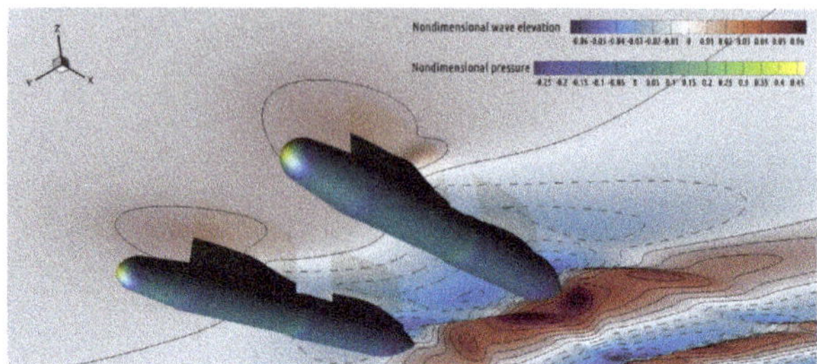

Figure 5. Original SWATH, non-dimensional pressure distribution and wave elevation by potential flow simulation.

Four design variables ($N = 4$) control SWATH hull global modifications, based on the KLE of the shape modification vector [49] based on a CAD parametric model developed with the CAESES® software from FRIENDSHIP SYSTEMS [51]. The reference non-dominated solution set $\mathcal{R}$ is taken from [55].

6. Numerical Results

The MODPSO setup was defined as in [6]: number of particles $\mathcal{Z}$ equal to $8MN$, initialized using a Hammersley sequence sampling [56] over variables domain and boundaries; coefficients $\chi = 0.721$ and $\phi_1 = \phi_2 = 1.655$ [57]; box constraints handled by a semi-elastic wall-type method [15].

Three threshold values for DFMO activation parameter (see Equations (14)) were set as $\alpha^\star = \{1.0, 1.1, 1.2\}$. Three deepness parameters (budget of local search evaluations for each call) were set as $\omega = \{1, 5, 10\}$. The line-search step λ was halved at each local search step until it reached a minimum step size $\lambda_{\min} = 1 \times 10^{-9}$. The starting local search step size was set equal to the 10% of the variables-space dimension.

The computational budget was assessed by νMN problem evaluations (one problem evaluation involves the evaluation of each objective function) where $\nu = 125 \times 2^c, c \in \mathbb{N}[0, 4]$ therefore ranges between $125MN$ and $2000MN$. The MODHA formulations under analysis performs $\omega\mathcal{Z}$ local search for each call, starting from the current solution set $\mathcal{S}^i$, and are activated by the HV threshold value α (see Equation (14)).

A preliminary study on the analytical test problems (see Table 1) was used to identify the most promising values of the activation and deepness parameters. The most promising parameters setup was finally applied to the two SBDO problems and compared with its global (MODPSO) and local (DFMO) counterparts, using a budget of $2000MN$ problem evaluations.

6.1. Analytical Benchmark Problems

Figure 6 shows the NHV metric versus the budget of problem evaluations. Average values are presented, conditional to the local search activation parameter α (Figure 6 left) and deepness parameter ω (Figure 6 right). It can be noted how MODHA effectiveness improves by using higher values of α, meaning that the hybrid formulation is pushed to exploit local searches capability and its effectiveness is enriched by global exploration only if a significant improvement of the HV is achieved between iterations (i.e., the global exploration improves the algorithm performance if the HV improves by 10–20% between MODHA iterations). Considering the deepness parameter ω, the use of a low number of local search ($\omega = 1$, at each DFMO call) is more effective when a low computational budget is available. On the contrary (using higher budgets), the "deeper" is the local search, the greater is the effectiveness of the hybrid formulation.

Figure 7 shows the relative variance σ_r^2 of the NHV retained by each of the aforementioned parameters. The hybrid formulation is mainly affected by the local searches activation parameter α rather than by the actual number of local searches (related to ω).

The MODHA performance is summarized in Table 2. For each budget of problem evaluations, the NHV mean values are provided along with standard deviations. Furthermore, budget-averaged performance is also provided. The hybridization scheme with $\alpha = 1.2$ and $\omega = 1$ is the most effective on average, providing the highest NHV with the lowest standard deviation, in accordance with results shown in Figure 6.

Finally, MODHA with the most promising parameters setup ($\alpha = 1.2, \omega = 1$) is compared with MODPSO and DFMO. Comparison results are shown in Figure 8 and summarized in Table 2. MODHA shows better average performance than its global and local counterparts. Furthermore, the results achieved from MODHA with the most promising parameters setup are found always very close to the results achieved with the "best" parameters setup (see Figure 8). Some illustrative examples of the non-dominated solution sets achieved by MODPSO, DFMO, and MODHA are shown in Figure 9. It can be noted how MODHA improves the quality and accuracy of the non-dominated solution sets (obtained by MODPSO and DFMO) in terms of both convergence towards the reference and diversity of the solutions.

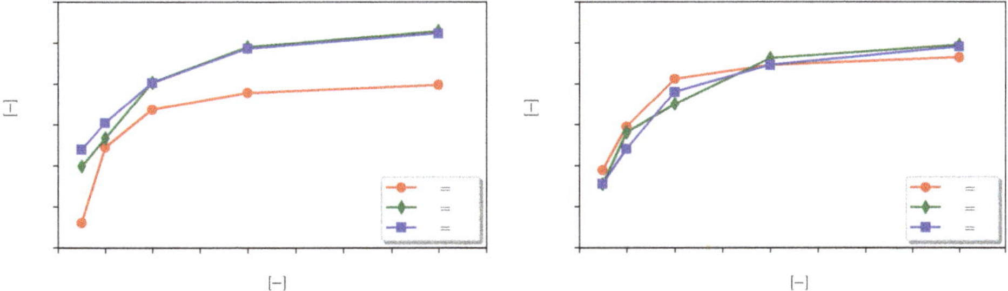

Figure 6. MODHA average performance for the analytical test problems, conditional to: the local search activation parameter α (**left**); and deepness parameter ω (**right**).

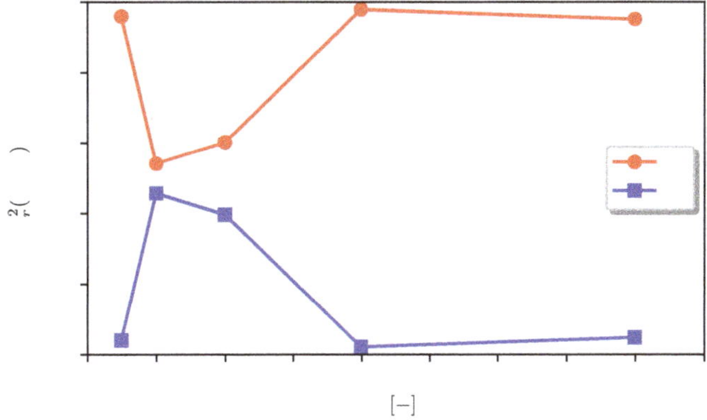

Figure 7. Normalized hypervolume relative variability of MODHA parameters (α and ω) for the analytical test problems.

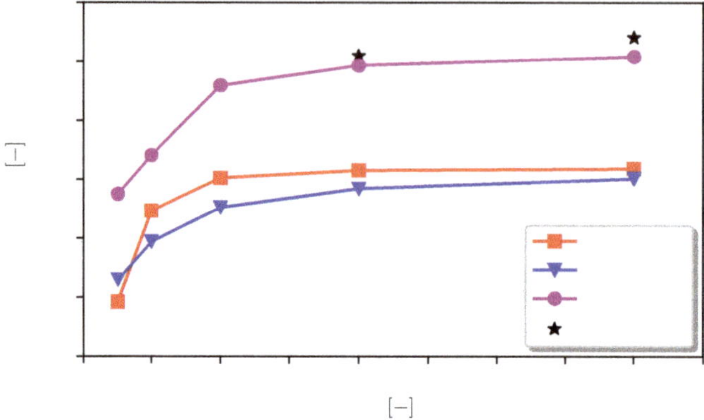

Figure 8. Comparison of MODHA average performance (on the analytical test problems) to MODPSO and DFMO.

Table 2. MODHA average performance along with comparison to MODPSO and DFMO: test problems NHV mean values and (in parenthesis) standard deviation.

MODHA Parameters	Computational Budget Coefficient γ 125	250	500	1000	2000	Average
$\alpha = 1.0, \omega = 1$	0.9458 (9.0934×10^{-2})	0.9641 (7.7112×10^{-2})	0.9725 (7.3252×10^{-2})	0.9751 (7.0314×10^{-2})	0.9770 (6.8226×10^{-2})	0.9669 (7.5968×10^{-2})
$\alpha = 1.0, \omega = 5$	0.9464 (9.2159×10^{-2})	0.9651 (7.7439×10^{-2})	0.9745 (7.1709×10^{-2})	0.9788 (6.8668×10^{-2})	0.9808 (6.6934×10^{-2})	0.9691 (7.5382×10^{-2})
$\alpha = 1.0, \omega = 10$	0.9462 (9.4873×10^{-2})	0.9643 (7.6855×10^{-2})	0.9743 (7.1750×10^{-2})	0.9797 (6.7123×10^{-2})	0.9817 (6.5631×10^{-2})	0.9692 (7.5246×10^{-2})
$\alpha = 1.1, \omega = 1$	0.9634 (5.8652×10^{-2})	0.9704 (5.0271×10^{-2})	0.9854 (3.1711×10^{-2})	0.9896 (2.3128×10^{-2})	0.9915 (1.5723×10^{-2})	0.9800 (3.5897×10^{-2})
$\alpha = 1.1, \omega = 5$	0.9580 (7.0017×10^{-2})	0.9682 (6.6010×10^{-2})	0.9742 (5.6875×10^{-2})	0.9891 (2.4307×10^{-2})	0.9941 (1.4171×10^{-2})	0.9767 (4.6276×10^{-2})
$\alpha = 1.1, \omega = 10$	0.9583 (7.1280×10^{-2})	0.9617 (6.9339×10^{-2})	0.9815 (3.8687×10^{-2})	0.9885 (2.3486×10^{-2})	0.9933 (1.7157×10^{-2})	0.9767 (4.3989×10^{-2})
$\alpha = 1.2, \omega = 1$	0.9675 (5.5413×10^{-2})	0.9741 (4.6050×10^{-2})	0.9859 (3.1321×10^{-2})	0.9893 (2.6102×10^{-2})	0.9910 (1.9876×10^{-2})	0.9815 (3.5798×10^{-2})
$\alpha = 1.2, \omega = 5$	0.9621 (6.6240×10^{-2})	0.9710 (6.0512×10^{-2})	0.9765 (5.1914×10^{-2})	0.9910 (2.3339×10^{-2})	0.9939 (1.5871×10^{-2})	0.9789 (4.3575×10^{-2})
$\alpha = 1.2, \omega = 10$	0.9623 (6.4324×10^{-2})	0.9664 (6.1208×10^{-2})	0.9785 (4.3171×10^{-2})	0.9858 (2.9605×10^{-2})	0.9924 (2.2869×10^{-2})	0.9771 (4.4235×10^{-2})
MODSPO	0.9493 (8.6609×10^{-2})	0.9647 (7.6035×10^{-2})	0.9702 (7.4521×10^{-2})	0.9715 (7.4430×10^{-2})	0.9725 (7.3635×10^{-2})	0.9659 (7.6978×10^{-2})
DFMO	0.9530 (9.0301×10^{-2})	0.9594 (8.6761×10^{-2})	0.9653 (8.2138×10^{-2})	0.9684 (8.0448×10^{-2})	0.9708 (7.9084×10^{-2})	0.9641 (8.3362×10^{-2})

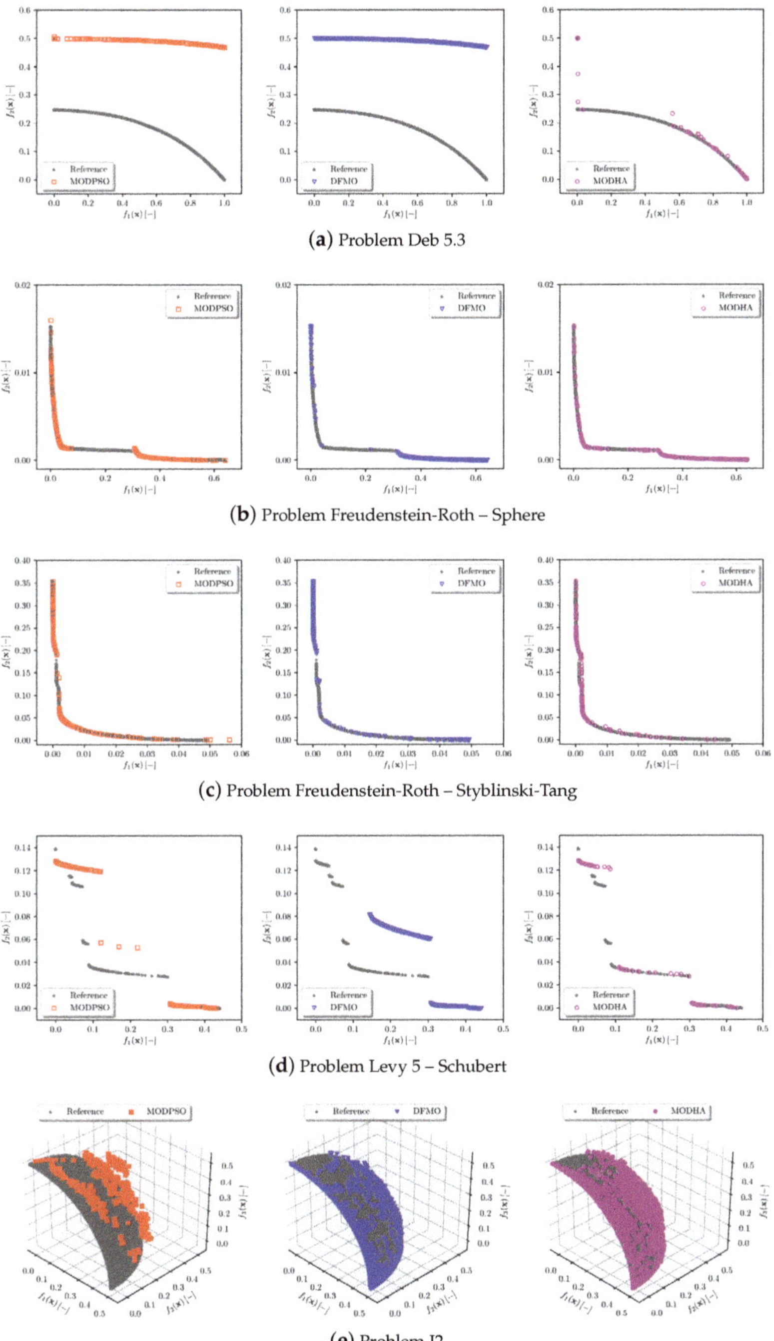

(**a**) Problem Deb 5.3

(**b**) Problem Freudenstein-Roth – Sphere

(**c**) Problem Freudenstein-Roth – Styblinski-Tang

(**d**) Problem Levy 5 – Schubert

(**e**) Problem I2

Figure 9. Analytical test problems: example comparison of the (normalized) non-dominated solution sets obtained by MODPSO, DFMO, and MODHA with the most promising parameters setup.

6.2. Simulation-Based Design Optimization Problems

Figure 10 shows the non-dominated solution sets obtained for the catamaran and the SWATH problems by MODPSO, DFMO, and their hybridization MODHA with the most promising parameters setup ($\alpha = 1.2$, $\omega = 1$). Considering the catamaran problem (see Figure 10a), the hybrid algorithm can cover effectively the reference solution, outperforming its global and local counterparts. It can be noted that MODHA is able to accurately identify the upper right region of the reference solution set $\mathcal{R}$. Considering the SWATH problem, MODHA clearly improves the quality and accuracy of the non-dominated solution set obtained by MODPSO, whereas it is comparable with DFMO. SBDO problems results are summarized in Table 3, showing the NHV metric along with the number of the non-dominated solutions achieved. MODHA achieves the best performance for the catamaran problem, whereas MODHA and its local counterpart DFMO are comparable for the SWATH problem.

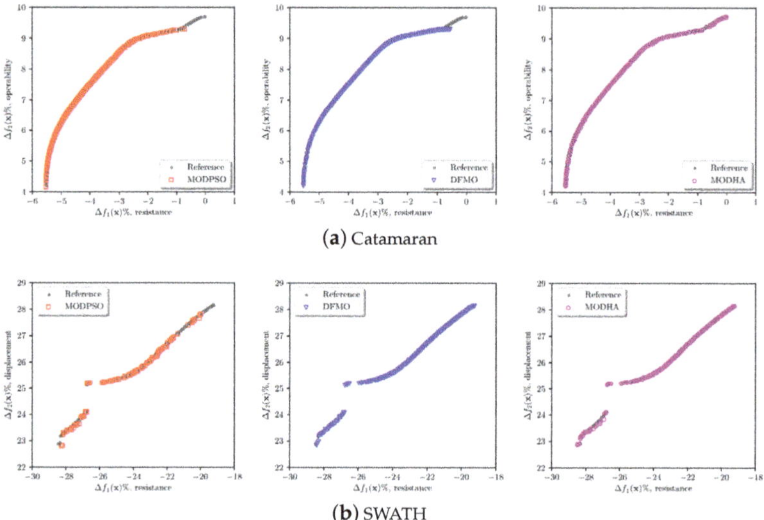

Figure 10. SBDO problems: Comparison of the non-dominated solution sets obtained (from left to right) by MODPSO, DFMO, and MODHA with the most promising parameters setup.

Table 3. SBDO problems, summary of the optimization results in terms of NHV *.

Problem	N	M	MODPSO	DFMO	MODHA
Catamaran	4	2	0.9871	0.9895	0.9935
			(915)	(998)	(1774)
SWATH	4	2	0.9732	0.9947	0.9900
			(210)	(367)	(341)

* In parenthesis is the size of the non-dominated solution set.

7. Conclusions and Future Work

A global/local hybridization of the multi-objective deterministic particle swarm formulation with a local derivative-free multi-objective line-search algorithm is presented, for the effective and efficient solution of multi-objective SBDO for hull-form design. Specifically, the problems of interest are characterized by a number of variables of the order of 10 and up to three objectives. A metric-based memetic scheme is proposed. The hypervolume metric is selected for the current formulation, since it provides information on both convergence and diversity of the non-dominated solution set. The hybridization scheme is "dual" since both global and local algorithms are mutually enriched.

Specifically, MODHA starts with the global exploration of the design variables space; if the HV associated to the current solution set does not increase sufficiently (by a factor α) with respect to the previous iteration, then MODHA activates the local exploitation starting the local search from each point of the current non-dominated solution set and performing a prescribed number of problem evaluations (proportional to the parameter ω). This process iterates until the budget of problem evaluations is consumed.

With the aim of identifying reasonable values for the activation (α) and deepness (ω) parameters, a full factorial combination of these was evaluated and the performance compared. The comparison was performed on a test set of 40 analytical test problems, with 2–3 objectives, 1–12 variables, and several features of variables and function spaces. The test functions were selected as representative of the specific SBDO problem of interest. The results are presented and discussed based on the normalized HV, assessing both the convergence and the diversity of the non-dominated solution set.

The performance of the hybridization scheme depends significantly on the local search activation parameter α. The parameters setup formed by $\alpha = 1.2$ and $\omega = 1$ is the most effective overall. This was compared to its global and local counterpart, whose setting parameters were taken from earlier studies [6,29]. The proposed memetic algorithm showed the best performance on average. MODHA algorithm with the most promising parameters setup was finally applied to the two simulation-based design optimization problems of a high-speed catamaran (aimed at reducing the resistance and increasing the operability in realistic ocean conditions) and a SWATH model (aimed at reducing the calm water resistance and increasing the hull displacement), showing better results than global and local algorithms.

The current achievements motivate further studies and analysis of the proposed metrics-based formulation, focusing on the criterion for the selection of the local-search starting points, such as subsampling of the non-dominated solution set or the use of crowding-distance metrics. Furthermore, the analysis of the effects of the local search maximum and minimum step size (on the overall performance) will be addressed. Future work includes also the investigation of alternative multi-objective PSO formulations, such as the crowding-distance-based multi-objective PSO [58].

Author Contributions: Conceptualization, G.L., S.L. and M.D.; Formal analysis, R.P. and A.S.; Funding acquisition, M.D.; Investigation, R.P.; Methodology, R.P., A.S., G.L., F.R., S.L. and M.D.; Software, R.P.; Supervision, M.D.; Visualization, A.S.; Writing—original draft, R.P. and A.S.; and Writing—review and editing, R.P., A.S., G.L., F.R., S.L. and M.D. All authors have read and agreed to the published version of the manuscript.

Funding: CNR-INM is grateful to Woei-Min Lin, Elena McCarthy, and Salahuddin Ahmed of the Office of Naval Research and Office of Naval Research Global, for their support through NICOP grant N62909-18-1-2033. R.P. is partially supported through CNR-INM project OPTIMAE.

Conflicts of Interest: The authors declare no conflict of interest. The funders had no role in the design of the study; in the collection, analyses, or interpretation of data; in the writing of the manuscript, or in the decision to publish the results.

References

1. Diez, M.; Campana, E.F.; Stern, F. Stochastic optimization methods for ship resistance and operational efficiency via CFD. *Struct. Multidiscip. Optim.* **2018**, *57*, 735–758. [CrossRef]
2. Martins, J.R.R.A.; Lambe, A.B. Multidisciplinary design optimization: A survey of architectures. *AIAA J.* **2013**, *51*, 2049–2075. [CrossRef]
3. Ehrgott, M. *Multicriteria Optimization*; Springer Science & Business Media: Berlin/Heidelberg, Germany, 2005; Volume 491.
4. Zhang, J.; Xing, L. A Survey of Multiobjective Evolutionary Algorithms. In Proceedings of the 2017 IEEE International Conference on Computational Science and Engineering (CSE) and IEEE International Conference on Embedded and Ubiquitous Computing (EUC), Guangzhou, China, 21–24 July 2017; Volume 1, pp. 93–100.
5. Evtushenko, Y.G.; Posypkin, M. A deterministic algorithm for global multi-objective optimization. *Optim. Methods Softw.* **2014**, *29*, 1005–1019. [CrossRef]

6. Pellegrini, R.; Serani, A.; Leotardi, C.; Iemma, U.; Campana, E.F.; Diez, M. Formulation and parameter selection of multi-objective deterministic particle swarm for simulation-based optimization. *Appl. Soft Comput.* **2017**, *58*, 714–731. [CrossRef]
7. Campana, E.F.; Diez, M.; Liuzzi, G.; Lucidi, S.; Pellegrini, R.; Piccialli, V.; Rinaldi, F.; Serani, A. A multi-objective DIRECT algorithm for ship hull optimization. *Comput. Optim. Appl.* **2018**, *71*, 53–72. [CrossRef]
8. Larson, J.; Menickelly, M.; Wild, S.M. Derivative-free optimization methods. *Acta Numer.* **2019**, *28*, 287–404. [CrossRef]
9. Serani, A.; Fasano, G.; Liuzzi, G.; Lucidi, S.; Iemma, U.; Campana, E.F.; Stern, F.; Diez, M. Ship hydrodynamic optimization by local hybridization of deterministic derivative-free global algorithms. *Appl. Ocean Res.* **2016**, *59*, 115 – 128. [CrossRef]
10. Mirjalili, S. Dragonfly algorithm: A new meta-heuristic optimization technique for solving single-objective, discrete, and multi-objective problems. *Neural Comput. Appl.* **2016**, *27*, 1053–1073. [CrossRef]
11. Mirjalili, S.; Gandomi, A.H.; Mirjalili, S.Z.; Saremi, S.; Faris, H.; Mirjalili, S.M. Salp Swarm Algorithm: A bio-inspired optimizer for engineering design problems. *Adv. Eng. Softw.* **2017**, *114*, 163–191. [CrossRef]
12. Połap, D.; others. Polar bear optimization algorithm: Meta-heuristic with fast population movement and dynamic birth and death mechanism. *Symmetry* **2017**, *9*, 203. [CrossRef]
13. Serani, A.; Diez, M. Dolphin Pod Optimization: A Nature-Inspired Deterministic Algorithm for Simulation-Based Design. In *Machine Learning, Optimization, and Big Data: Second International Workshop, MOD 2017, Volterra, Italy, 14–17 September 2017*; Lecture Notes in Computer Science; Nicosia, G., Pardalos, P., Giuffrida, G., Umeton, R., Eds.; Springer International Publishing: Cham, Switzerland, 2018; Volume 10710.
14. Kennedy, J.; Eberhart, R.C. Particle swarm optimization. In Proceedings of the Fourth IEEE Conference on Neural Networks, Perth, Australia, 27 November–1 December 1995; pp. 1942–1948.
15. Serani, A.; Leotardi, C.; Iemma, U.; Campana, E.F.; Fasano, G.; Diez, M. Parameter selection in synchronous and asynchronous deterministic particle swarm optimization for ship hydrodynamics problems. *Appl. Soft Comput.* **2016**, *49*, 313 – 334. [CrossRef]
16. Wang, D.; Tan, D.; Liu, L. Particle swarm optimization algorithm: An overview. *Soft Comput.* **2018**, *22*, 387–408. [CrossRef]
17. Hart, W.E.; Krasnogor, N.; Smith, J.E. Memetic evolutionary algorithms. In *Recent Advances in Memetic Algorithms*; Springer: Berlin/Heidelberg, Germany, 2005; pp. 3–27.
18. Serani, A.; Diez, M.; Campana, E.F.; Fasano, G.; Peri, D.; Iemma, U. Globally Convergent Hybridization of Particle Swarm Optimization Using Line Search-Based Derivative-Free Techniques. In *Recent Advances in Swarm Intelligence and Evolutionary Computation*; Yang, X.S., Ed.; Studies in Computational Intelligence; Springer International Publishing: Cham, Switzerland, 2015; Volume 585, pp. 25–47.
19. Santana-Quintero, L.V.; Ramírez, N.; Coello, C.C. A multi-objective particle swarm optimizer hybridized with scatter search. In Proceedings of the Mexican International Conference on Artificial Intelligence, Apizaco, Mexico, 13–17 November 2006; pp. 294–304.
20. Liu, D.; Tan, K.C.; Goh, C.K.; Ho, W.K. A multiobjective memetic algorithm based on particle swarm optimization. *IEEE Trans. Syst. Man Cybern. Part B (Cybern.)* **2007**, *37*, 42–50. [CrossRef]
21. Izui, K.; Nishiwaki, S.; Yoshimura, M.; Nakamura, M.; Renaud, J.E. Enhanced multiobjective particle swarm optimization in combination with adaptive weighted gradient-based searching. *Eng. Optim.* **2008**, *40*, 789–804. [CrossRef]
22. Kaveh, A.; Laknejadi, K. A novel hybrid charge system search and particle swarm optimization method for multi-objective optimization. *Expert Syst. Appl.* **2011**, *38*, 15475–15488. [CrossRef]
23. Mousa, A.; El-Shorbagy, M.; Abd-El-Wahed, W. Local search based hybrid particle swarm optimization algorithm for multiobjective optimization. *Swarm Evol. Comput.* **2012**, *3*, 1–14. [CrossRef]
24. Xu, G.; Yang, Y.Q.; Liu, B.B.; Xu, Y.H.; Wu, A.J. An efficient hybrid multi-objective particle swarm optimization with a multi-objective dichotomy line search. *J. Comput. Appl. Math.* **2015**, *280*, 310–326. [CrossRef]
25. Cheng, S.; Zhan, H.; Shu, Z. An innovative hybrid multi-objective particle swarm optimization with or without constraints handling. *Appl. Soft Comput.* **2016**, *47*, 370–388. [CrossRef]
26. Qian, C.; Yu, Y.; Zhou, Z.H. Pareto Ensemble Pruning. In Proceedings of the Twenty-Ninth AAAI Conference on Artificial Intelligence, AAAI'15, Austin, TX, USA, 25–30 January 2015; pp. 2935–2941.

27. Qian, C.; Tang, K.; Zhou, Z.H. Selection Hyper-heuristics Can Provably Be Helpful in Evolutionary Multi-objective Optimization. In Proceedings of the International Conference on Parallel Problem Solving from Nature, Edinburgh, UK, 17–21 September 2016; pp. 835–846.
28. Pellegrini, R.; Serani, A.; Liuzzi, G.; Rinaldi, F.; Lucidi, S.; Campana, E.F.; Iemma, U.; Diez, M. Hybrid global/local derivative-free multi-objective optimization via deterministic particle swarm with local linesearch. In *Machine Learning, Optimization, and Big Data: Second International Workshop, MOD 2017, Volterra, Italy, 14–17 September 2017*; Nicosia, G., Pardalos, P., Giuffrida, G., Umeton, R., Eds.; Lecture Notes in Computer Science; Springer International Publishing: Cham, Switzerland, 2018; Volume 10710, pp. 198–209.
29. Liuzzi, G.; Lucidi, S.; Rinaldi, F. A Derivative-Free Approach to Constrained Multiobjective Nonsmooth Optimization. *SIAM J. Optim.* **2016**, *26*, 2744–2774. [CrossRef]
30. Zitzler, E.; Thiele, L. Multiobjective optimization using evolutionary algorithms - a comparative case study. In Proceedings of the 5th Parallel Problem Solving from Nature, Amsterdam, The Netherlands, 27–30 September 1998; pp. 292–301.
31. Audet, C.; Bigeon, J.; Cartier, D.; Le Digabel, S.; Salomon, L. Performance indicators in multiobjective optimization. *Optim. Online* **2018**. Available online: http://www.optimization-online.org/DB_FILE/2018/10/6887.pdf (accessed on 7 April 2020).
32. Haftka, R.T. Requirements for papers focusing on new or improved global optimization algorithms. *Struct. Multidiscip. Optim.* **2016**, *54*. [CrossRef]
33. Jiang, S.; Ong, Y.S.; Zhang, J.; Feng, L. Consistencies and contradictions of performance metrics in multiobjective optimization. *Cybern. IEEE Trans.* **2014**, *44*, 2391–2404. [CrossRef] [PubMed]
34. Campana, E.F.; Diez, M.; Iemma, U.; Liuzzi, G.; Lucidi, S.; Rinaldi, F.; Serani, A. Derivative-free global ship design optimization using global/local hybridization of the DIRECT algorithm. *Optim. Eng.* **2015**, *17*, 127–156. [CrossRef]
35. Pinto, A.; Peri, D.; Campana, E.F. Global optimization algorithms in naval hydrodynamics. *Ship Technol. Res.* **2004**, *51*, 123–133. [CrossRef]
36. Serani, A.; Diez, M. Are Random Coefficients Needed in Particle Swarm Optimization for Simulation-Based Ship Design? In Proceedings of the 7th International Conference on Computational Methods in Marine Engineering (Marine 2017), Nantes, France, 15–17 May 2017.
37. Pinto, A.; Peri, D.; Campana, E.F. Multiobjective optimization of a containership using deterministic particle swarm optimization. *J. Ship Res.* **2007**, *51*, 217–228.
38. Kolda, T.G.; Lewis, R.M.; Torczon, V. Optimization by direct search: New perspectives on some classical and modern methods. *SIAM Rev.* **2003**, *45*, 385–482. [CrossRef]
39. Custódio, A.L.; Madeira, J.F.A.; Vaz, A.I.F.; Vicente, L.N. Direct Multisearch for Multiobjective Optimization. *SIAM J. Optim.* **2011**, *21*, 1109–1140. [CrossRef]
40. Huband, S.; Hingston, P.; Barone, L.; While, L. A review of multiobjective test problems and a scalable test problem toolkit. *IEEE Trans. Evol. Comput.* **2006**, *10*, 477–506. [CrossRef]
41. Fonseca, C.M.; Paquete, L.; Lòpez-Ibà nez, M. An improved dimension-sweep algorithm for the hypervolume indicator. In Proceedings of the IEEE Congress on Evolutionary Computation (CEC'06), Vancouver, BC, Canada, 16–21 July 2006; pp. 1157–1163.
42. Deb, K. Multi-objective Genetic Algorithms: Problem Difficulties and Construction of Test Problems. *Evol. Comput.* **1999**, *7*, 205–230. [CrossRef]
43. Deb, K.; Thiele, L.; Laumanns, M.; Zitzler, E. Scalable multi-objective optimization test problems. In Proceedings of the 2002 Congress on Evolutionary Computation, CEC'02, Honolulu, HI, USA, 12–17 May 2002; Volume 1, pp. 825–830.
44. Jin, Y.; Olhofer, M.; Sendhoff, B. Dynamic weighted aggregation for evolutionary multi-objective optimization: Why does it work and how? In Proceedings of the 3rd Annual Conference on Genetic and Evolutionary Computation, San Francisco, CA, USA, 7–11 July 2001; pp. 1042–1049.
45. Lovison, A. A synthetic approach to multiobjective optimization. *arXiv* **2010**, arXiv:1002.0093.
46. Zitzler, E.; Deb, K.; Thiele, L. Comparison of Multiobjective Evolutionary Algorithms: Empirical Results. *Evol. Comput.* **2000**, *8*, 173–195. [CrossRef]
47. Volpi, S.; Diez, M.; Gaul, N.; Song, H.; Iemma, U.; Choi, K.K.; Campana, E.F.; Stern, F. Development and validation of a dynamic metamodel based on stochastic radial basis functions and uncertainty quantification. *Struct. Multidiscip. Optim.* **2015**, *51*, 347–368. [CrossRef]

48. Huang, J.; Carrica, P.M.; Stern, F. Semi-coupled air/water immersed boundary approach for curvilinear dynamic overset grids with application to ship hydrodynamics. *Int. J. Numer. Methods Fluids* **2008**, *58*, 591–624. [CrossRef]
49. Diez, M.; Campana, E.F.; Stern, F. Design-space dimensionality reduction in shape optimization by Karhunen–Loève expansion. *Comput. Methods Appl. Mech. Eng.* **2015**, *283*, 1525–1544. [CrossRef]
50. He, W.; Diez, M.; Zou, Z.; Campana, E.F.; Stern, F. URANS study of Delft catamaran total/added resistance, motions and slamming loads in head sea including irregular wave and uncertainty quantification for variable regular wave and geometry. *Ocean Eng.* **2013**, *74*, 189 – 217. [CrossRef]
51. Pellegrini, R.; Serani, A.; Broglia, R.; Diez, M.; Harries, S. Resistance and Payload Optimization of a Sea Vehicle by Adaptive Multi-Fidelity Metamodeling. In Proceedings of the 56th AIAA Aerospace Sciences Meeting, SciTech 2018, Kissimmee, FL, USA, 8–12 January 2018.
52. Serani, A.; Pellegrini, R.; Wackers, J.; Jeanson, C.E.; Queutey, P.; Visonneau, M.; Diez, M. Adaptive multi-fidelity sampling for CFD-based optimisation via radial basis function metamodels. *Int. J. Comput. Fluid Dyn.* **2019**, *33*, 237–255. [CrossRef]
53. Broglia, R.; Durante, D. Accurate prediction of complex free surface flow around a high speed craft using a single-phase level set method. *Comput. Mech.* **2018**, *62*, 421–437. [CrossRef]
54. Bassanini, P.; Bulgarelli, U.; Campana, E.F.; Lalli, F. The wave resistance problem in a boundary integral formulation. *Surv. Math. Ind.* **1994**, *4*, 151–194.
55. Pellegrini, R.; Serani, A.; Liuzzi, G.; Lucidi, S.; Rinaldi, F.; Campana, E.F.; Diez, M. Hull-form optimization via hybrid global/local multi-objective derivative-free algorithms. In Proceedings of the 21st Conference of the International Federation of Operational Research Societies, Québec City, QC, Canada, 17–21 July 2017.
56. Wong, T.T.; Luk, W.S.; Heng, P.A. Sampling with Hammersley and Halton Points. *J. Graph. Tools* **1997**, *2*, 9–24. [CrossRef]
57. Clerc, M. *Stagnation Analysis in Particle Swarm Optimization or What Happens When Nothing Happens*; Technical Report; hal-00122031; 2006. Available online: https://www.researchgate.net/publication/247636881_Stagnation_Analysis_in_Particle_Swarm_Optimisation_or_What_Happens_When_Nothing_Happens (accessed on 7 April 2020).
58. Raquel, C.R.; Naval, P.C., Jr. An effective use of crowding distance in multiobjective particle swarm optimization. In Proceedings of the 7th Annual Conference on Genetic and Evolutionary Computation, Washington, DC, USA, 25–29 June 2005; pp. 257–264.

Article

Social Network Optimization for WSN Routing: Analysis on Problem Codification Techniques

Alessandro Niccolai *, Francesco Grimaccia, Marco Mussetta, Alessandro Gandelli and Riccardo Zich

Dipartimento di Energia, Politecnico di Milano,Via Lambruschini 4, 20156 Milano, Italy; francesco.grimaccia@polimi.it (F.G.); marco.mussetta@polimi.it (M.M.); alessandro.gandelli@polimi.it (A.G.); riccardo.zich@polimi.it (R.Z.)
* Correspondence: alessandro.niccolai@polimi.it

Received: 25 February 2020; Accepted: 9 April 2020; Published: 14 April 2020

Abstract: The correct design of a Wireless Sensor Network (WSN) is a very important task because it can highly influence its installation and operational costs. An important aspect that should be addressed with WSN is the routing definition in multi-hop networks. This problem is faced with different methods in the literature, and here it is managed with a recently developed swarm intelligence algorithm called Social Network Optimization (SNO). In this paper, the routing definition in WSN is approached with two different problem codifications and solved with SNO and Particle Swarm Optimization. The first codification allows the optimization algorithm more degrees of freedom, resulting in a slower and in many cases sub-optimal solution. The second codification reduces the degrees of freedom, speeding significantly the optimization process and blocking in some cases the convergence toward the real best network configuration.

Keywords: wireless sensor networks; routing; Swarm Intelligence; Particle Swarm Optimization; Social Network Optimization

---

1. Introduction

The Internet of Things paradigm is increasing the importance of Wireless Sensor Networks (WSN) in which a set of small simple sensors are interconnected for creating a very complex structure. The information sensed by all the nodes of the network should be sent to the cluster head that processes them and exploits the information [1].

The design of a WSN arises some issues in terms of deployment and operational costs. In this framework, Evolutionary Optimization Algorithms (EAs) can be very important tools for the system design because of their flexibility and ease of use.

The research on EAs has two main preferential directions: on the one hand, more complex operators are introduced and analyzed in the algorithms for improving their performance [2]; on the other hand, their field of applicability is analyzed and enlarged [3].

The most used EAs are Genetic Algorithms (GA) and Particle Swarm Optimization (PSO) [4]. PSO and Ant Colony Optimization (ACO) [5] are the most common algorithms that belong to Swarm Intelligence.

Among EAs, Differential Evolution (DE) algorithm has been widely applied and studied [6,7]. This algorithm is not-biologically inspired; while it was originally designed for real-value problems [8], it has been implemented also for discrete problems [9]. This algorithm has been applied to a wide range of multimodal problems, such as neural network training [10]. Biogeography-Based Optimization (BBO) is a more recently developed EA based on the survival mechanism of species in different environments [11].

The design of a WSN is a very challenging task: in fact, several problems can be faced and should be solved for reducing the operational cost of them [12]. The first problem that can be faced in the design of a WSN is the deployment of sensors in the sensing field: the typical objectives are the coverage maximization. Another important problem in WSN is clustering the sensors for reducing the number of cluster heads, for reducing the information delay, and for reducing the energy consumption. This problem has been widely addressed with Evolutionary Algorithms [13] and Swarm Intelligence Algorithms [14].

With recent advancements in massive parallel computing technologies, the problem of scheduling resources and tasks in WSN is becoming more and more important. The topic of task allocation is faced with evolutionary algorithms in [15], where the task scheduling in heterogeneous distributed systems is evaluated comparing a multi-objective evolutionary algorithm with a hybrid GA, and, in [16], where GA is applied for a similar purpose, performing tests on a Java scheduler with a homogeneous set of processors. More recently, in [17], a Logic Based evolutionary algorithm compared to a binary PSO is applied to task allocation in wireless sensor networks. In [18], Social Network Optimization (SNO) is used for solving the task allocation problem.

Among the different techniques that can be used in WSN design and operation optimization, Swarm Intelligence algorithms have been extensively adopted. In [19], an algorithm based on ACO is exploited to find the optimal information path from a source node to a sink node in a multi-hop WSN. ACO is also applied in [20] for the path design of a mobile data collection node.

Lifetime maximization is a critical issue in WSN, especially when the sensors are deployed in a field in which maintenance is very difficult. In particular, this problem can be managed in different ways.

In [21], the authors proposed an enhanced Hybrid Genetic Algorithm for maximizing the WSN lifetime by means of an optimal scheduling of disjoint set of sensors.

The lifetime maximization is achieved in [22] by selecting the proper routing in the WSN by means of Genetical Swarm Optimization. The same optimization algorithm is implemented in [23], in which a real encoding of the chromosome is adopted.

In [24], Particle Swarm Optimization algorithm is used for a multi-objective problem in which the network lifetime maximization achieved by a proper selection of the routing is combined with the quality of service maximization.

From the analysis of all the presented papers, it is clear that the problem codification, i.e., how the network configurations are codified in the optimization variables, is a key aspect for the applicability of EAs in the IoT framework: in fact, the possible network configurations significantly increases with the number of sensors involved. A proper selection of the codification method can reduce the computational time required for the optimization: in particular, it can reduce the size of the search space and the number of unfeasible solutions. On the other hand, the reduction of the search space can be detrimental because it can eliminate the optimal solution.

In this paper, two problem codifications are analyzed: in the first one, a basic heuristic logic is used for reducing the search space size without losing degrees of freedom. In the second codification, a more complex heuristic is used for completely avoiding unfeasible solutions, but the degrees of freedom are highly reduced.

The problem of routing was solved with both the codifications on 27 different networks. Two swarm intelligence optimization algorithms were tested: SNO and PSO. The objective of this design problem is the maximization of the network lifetime, in which the transmitting and receiving energy requirements are considered.

The paper is structured as follows. Section 2 provides a brief description of PSO and SNO. Section 3 contains the description of the adopted WSN model, the two problem codification techniques, and the definition of the problem environment. In Section 4, the obtained results are presented and discussed. Finally, in Section 5, some conclusions are drawn.

2. Swarm Intelligence

Swarm Intelligence Algorithms are an important class of Evolutionary Optimization Algorithms. In this paper, Particle Swarm Optimization and Social Network Optimization are used. In the following, a brief description of both these algorithms is presented.

2.1. Particle Swarm Optimization

Particle Swarm Optimization (PSO) is a milestone in Swarm Intelligence algorithms [25]. Its operators are derived from the concept of collective intelligence, which can be summarized in the following three sentences [26]:

- Inertia: 'I continue on my way'.
- Influence of the past: 'I've always done in this way'.
- Influence by society: 'If it worked for them, it would also work for me'.

Each individual of the population represents a particle, which is characterized by a position and velocity.

At each iteration of the algorithm, the personal best position found by each particle (PB) and the global best (GB) found by the entire swarm are computed. These points are the attractors of each individual. In particular, the velocity is updated in the following way:

$$\mathbf{v}_i(t+1) = \omega \mathbf{v}_i(t) + c_1(\mathbf{PB}_i - \mathbf{p}_i(t)) + c_2(\mathbf{GB} - \mathbf{p}_i(t)) \tag{1}$$

where $\mathbf{v}_i$ is the velocity of the ith particle and $\mathbf{p}_i$ its position in the search space. $\mathbf{PB}_i$ is the personal best while **GB** is the global best. Finally, ω, c_1, and c_2 are three user defined parameters.

At each iteration, the position is updated in the following way:

$$\mathbf{p}_i(t+1) = \mathbf{p}_i(t) + \mathbf{v}_i(t+1) \tag{2}$$

PSO performance are greatly influenced by the selection of the algorithm working parameters. The implemented PSO version is characterized by the following parameters: the inertia weight (ω), the personal learning coefficient (c_1), the global learning coefficient (c_2), and the velocity clamping (V_{rsm}) [27,28]. Moreover, the algorithm performance depends on the population size (n_{pop}) that is related to the number of iterations (n_{iter}) and the number of objective function calls (n_{call}) by the following equation:

$$n_{pop} \cdot n_{iter} = n_{call} \tag{3}$$

2.2. Social Network Optimization

Social Network Optimization is a recently developed population-based, Swarm Intelligence algorithm that takes its inspiration from the information sharing process in Online Social Network [29].

The population of the algorithm is composed by *users* of the social network who interact online by publishing some *posts*. In fact, at each iteration, the users express their opinions by means of the *status* contained in the post. In addition to this information, the post contains the name of the user that have posted it, the time at which it is posted, and a visibility value. The process of passing from *opinions* to a post *status* is called *linguistic transposition*.

The status corresponds to the candidate solution of the optimization problem, while the visibility value is created by means of a proper mapping of the cost function associated to the specific candidate solution, as shown in Figure 1.

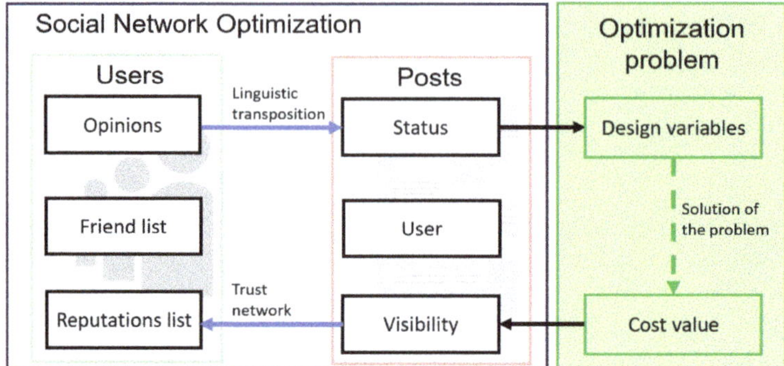

Figure 1. Composition and interaction of data structures in SNO.

The visibility values of the entire population are used to update the reputations of all the users of the social network. This operation transfers global information among the entire population. The reputations are used for creating one of the two interaction structures of SNO: the trust network.

The second interaction network is the friend network. These two networks are very different one from each other: the friend network leads to more consistent interactions, its variability is lower, and its modifications are related to out-of-the-social elements. The trust network creates weaker connections, it varies quickly, and it is modified according to the visibility values, as shown in Figure 2.

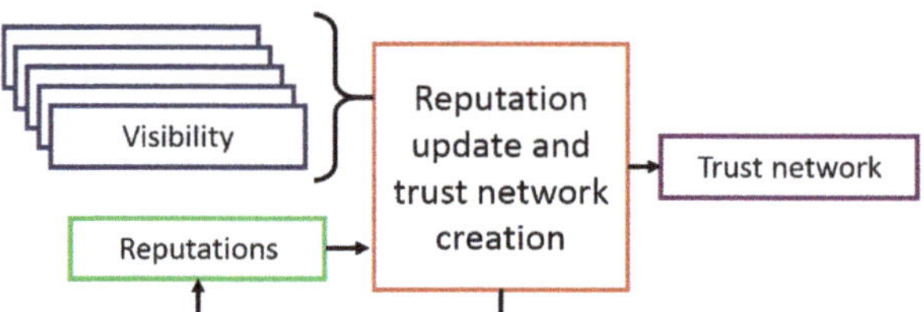

Figure 2. Schematization of the process of reputation update and creation of trust network.

From the interaction networks, each user extracts some ideas that compose the attraction point that is used for creating the new opinions. The implemented operator emulates the assumption of a complex contagion, which guarantees a better trade-off between exploration of the domain and the exploitation of the acquired knowledge:

$$\mathbf{o}_u(t+1) = \mathbf{o}_u(t) + \alpha[\mathbf{o}_u(t) - \mathbf{o}_u(t-1)] + \beta[\mathbf{a}_u(t) - \mathbf{o}_u(t)] \tag{4}$$

All these operators make SNO a quite complex algorithm, but they give to it the possibility to work very well in different kinds of problems and they reduce the risk of stagnation in local minima, which is a well-known issue of PSO.

The most important algorithm parameters are α and β because they tune the effective behavior of the population. Two different analyses have already been performed for properly selecting these parameters: The first one is an analytical analysis of a simplified version of the complex contagion. The second one is a numerical parametric analysis performed on different benchmarks. More details on these analyses can be found in [30].

The parametric analysis on the parameters α and β has been performed on two standard benchmarks for EAs: the Ackley and Schwefel functions. The termination criterion is set to 5000 objective function call and 50 independent trials have been done on 20D functions. Figure 3 shows the results of the parametric analysis: the color is proportional to the cost value, where red is high cost and blue low cost. The results show that the high quality solutions area is different, but it is possible to find a good set of parameters for both the functions. In particular, the selected set of parameters is $\alpha = 0.1$ and $\beta = 0.45$.

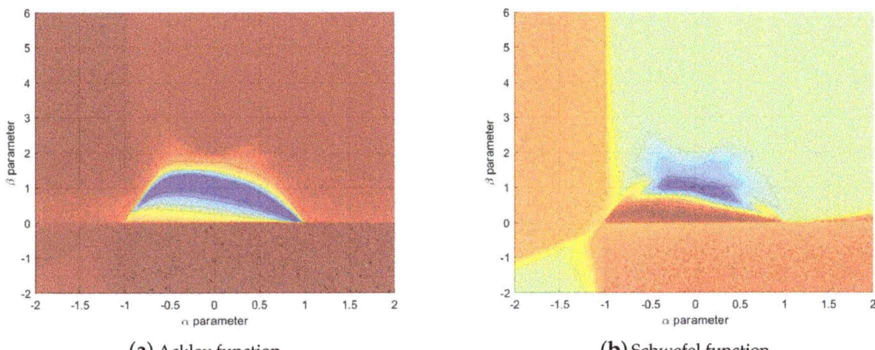

(a) Ackley function. **(b)** Schwefel function.

Figure 3. Parametric analysis on α and β on two different benchmarks.

In [31], it is possible to find a comparison between SNO and other EAs, in which the performance of SNO is assessed on standard benchmarks.

3. Wireless Sensor Network

A Wireless Sensor Network is composed of a set of sensors, deployed in a field, that can communicate by means of a multi-hop protocol to a central node, called the cluster head.

In the following, the adopted network model is described and then the two proposed problem codification techniques are described. Finally, the optimization environment for this problem is analyzed.

3.1. Network Energy Model

The analyzed WSN is composed by a set of equal sensor nodes deployed randomly in the space. Each of these nodes has a maximum communication distance that is imposed by the acceptable signal-to-noise ratio.

The communication between the nodes and the central node, the cluster head, happens by means of a multi-hop protocol: in this way, even nodes quite far from the cluster head can send to it the sensed information.

Each sensor but the cluster head is fed by batteries with a total capacity of 9kJ, thus the network lifetime is limited by the energy consumption of the most stressed sensor.

The energy consumption of each node can be computed as function of the transmitted and received bits: in fact, it is assumed that the sensing energy is negligible with respect to the communication energy [20]. Thus, the total energy consumption is composed by two terms:

$$E_i = E_{\mathrm{trx},i} + E_{\mathrm{amp},i} \tag{5}$$

where $E_{\mathrm{trx},i}$ is the total energy required for transmission and reception of information:

$$E_{\mathrm{trx},i} = b_{\mathrm{trx},i} \cdot e_{\mathrm{trx}} \tag{6}$$

where e_{trx} is the energy required to keep on the communication equipment and $b_{\mathrm{trx},i}$ is the total number of bit transmitted and received.

The second term of the total energy consumption is the amplification energy required for the transmission of the information. It depends on the number of transmitted bits (b_{ij}) and on the transmission distance (d_{ij}) between the two nodes i and j:

$$E_{\text{amp},i,j} = E_{\text{amp}} \cdot b_{ij} \cdot d_{ij} \tag{7}$$

Both the sensing energy and the energy consumed in idle time are considered negligible with respect to the other terms.

The transmitted and received bits are functions of the network topology, i.e., of the interconnections between sensors and can be easily calculated.

The network is composed by a set of randomly deployed sensors: as introduced above, the feasible connections between sensors is determined by the maximum communication distance between nodes.

The feasible connections are bidirectional if they involve only sensor nodes, while are considered monodirectional if one of the two nodes is the cluster head, as shown in Figure 4a, where the directed connections are indicated with the arrow.

The feasible connections are more than the required ones, thus it is possible to select the effective final network topology by identifying the *active* connections among the feasible ones, as shown in Figure 4b.

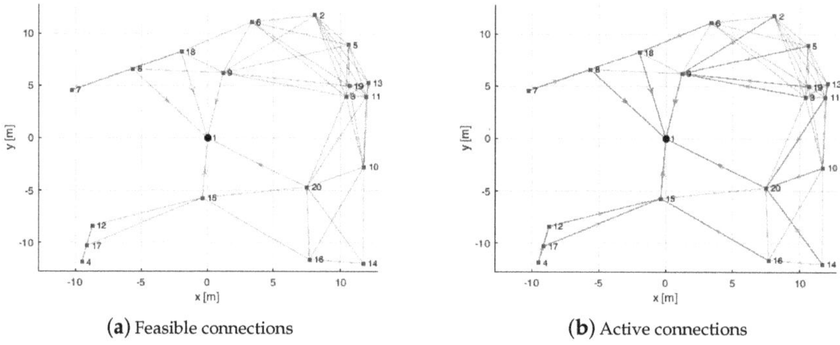

(**a**) Feasible connections (**b**) Active connections

Figure 4. Example of a 20-node network. The feasible connections are shown in grey (**a**), while the active connections are in green in (**b**). Node 1 is the cluster head.

The selection of the active connection should be performed also selecting the direction of the communication. This is an important aspect for the problem codification, as analyzed below.

Thus, when the active connections are selected, the network can be represented by a directed graph; by means of a graph search starting from the nodes with higher depth, it is possible to determine the flow of information in the network, and thus the number of received and transmitted bits.

This network model was implemented in Matlab, adopting the parameters shown in Table 1.

Table 1. List of network parameters.

Parameter	Symbol	Value	Parameter	Symbol	Value
ine Communication package size	b_{pack}	100 bit	Communication energy	E_{TRX}	5 pJ/b
Stored energy	E_s	9 kJ	Amplification energy	E_{amp}	0.01 pJ/b

The objective of this optimization problem is the maximization of the network lifetime, i.e., the number of working cycles in which all the sensors have a residual stored energy.

The design variables of this problem are the active connections in the WSN; in fact, by changing the active connections, the number of bits received and transmitted by the sensors is modified and, thus, the sensor lifetime is changed. It is possible to codify these design variables in different ways in the optimization framework, changing the search space size and the number of unfeasible solutions. A deeper analysis of the possible codification methods is provided in the following.

To the best of the authors' knowledge, there are some papers that have proven that the problem of the lifetime maximization by properly selecting the routing paths in a WSN is a NP-hard problem [32,33].

3.2. Problem Codification

The presented model of WSN can be codified in the optimization environment in different methods that affect the number of optimization variables, their type, the possibility to find unfeasible solutions, and the search space size.

The most basic codification of this problem can be performed using binary design variables, each one representing if a feasible connection is also active. This formulation is very simple and generic, but it leads to a high number of non-connected networks, i.e., networks in which at least one node cannot communicate with the cluster head. Moreover, many solutions are surely suboptimal, such as ones with loops or with a node that transmits its information to more than one single sensors. This possibility, which can be important for reliability analysis of the communication, is not optimal in this context because the transmission and receiving energy of some nodes is doubled.

In this strategy, the number of design variables is equal to the number of feasible connections in the network, and thus generally grows more than linearly with the number of nodes.

It is possible to design more complex problem codifications by analyzing some features of optimal solutions. First, each sensor should be able to transmit its information to another one.

In this framework, it is possible to associate to each node its adjacency list (as shown in Figure 5) and select one target node from this list. With this codification, the problem is characterized by $N-1$ integer design variables, where N is the number of nodes.

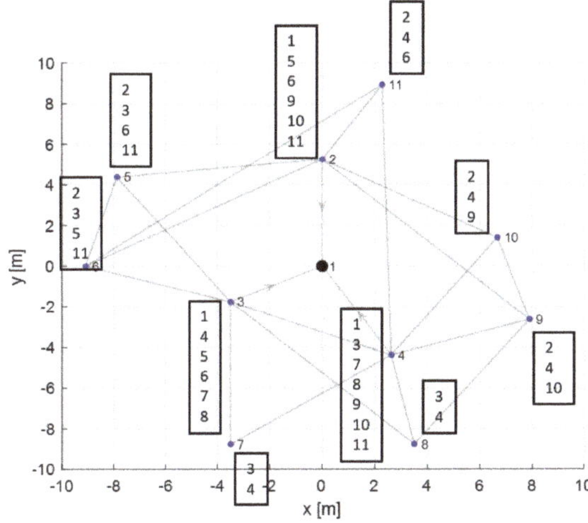

Figure 5. WSN with 11 nodes in which, for each node, its adjacency list is defined.

The drawback of this codification (called in the following adjacency-based codification) concerns the possibility of creating non-connected networks during the optimization process or looped graphs.

To avoid these drawbacks, it is possible to consider the node depth, that is the number of hops needed for communicating from the analyzed sensor to the cluster head. Figure 6a shows the same graph seen before in which the node position is related with node depth. Each circle corresponds to a different depth.

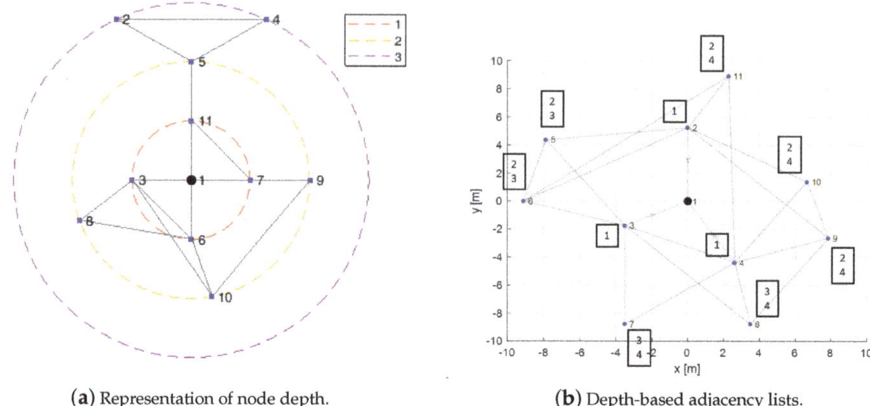

(a) Representation of node depth.

(b) Depth-based adjacency lists.

Figure 6. WSN with 11 nodes: (a) the node depth is underlined with the circles; and (b) to each node its depth-based adjacency list is defined.

For the second codification method, depth-based adjacency lists have been created: in each list, only the node with smaller node depth are included. Figure 6b shows the same network of Figure 5 in which the depth-based lists are shown. In addition, in this case, the design variables are integer numbers.

As it is possible to clearly see, the number of degrees of freedom for the algorithm is significantly reduced; this eliminates completely the possibility to have non-connected nodes in the network, but it eliminates some paths that can be optimal.

These two codification techniques make this problem combinatorial. However, in most real cases, it is impossible to test all of them due to the search space size.

3.3. Performance Calculation

In this section, an example of how the system works is presented: it starts from the optimization variables and shows all the steps required for the calculation of the cost value.

The optimizer works with real variables that are translated into integer values by means of a set of thresholds [23]. Each variable corresponds to a node and the number obtained by the translation of the optimization variables is the index in the adjacency list for the first codification technique or in the depth-based list for the second. In this example, the adjacency-based codification of the network presented in Figure 5 is used, but it can be extended easily to the other method.

Figure 7 shows the decodification of two different sets of optimization variables: in the first one (Figure 7a), the resulting network is connected, while, in the second one (Figure 7b), it is not and, thus, the solution is unfeasible.

If the network is feasible, the calculation of the cost is run. The first activity performed is the estimation of the information packages through each node. This is computationally performed in the graph represented by the active connections by means of a customized search that starts from the deeper nodes and then goes upward. Each node creates one information package and forward the received ones. Figure 8 shows the transmission packages for each edge.

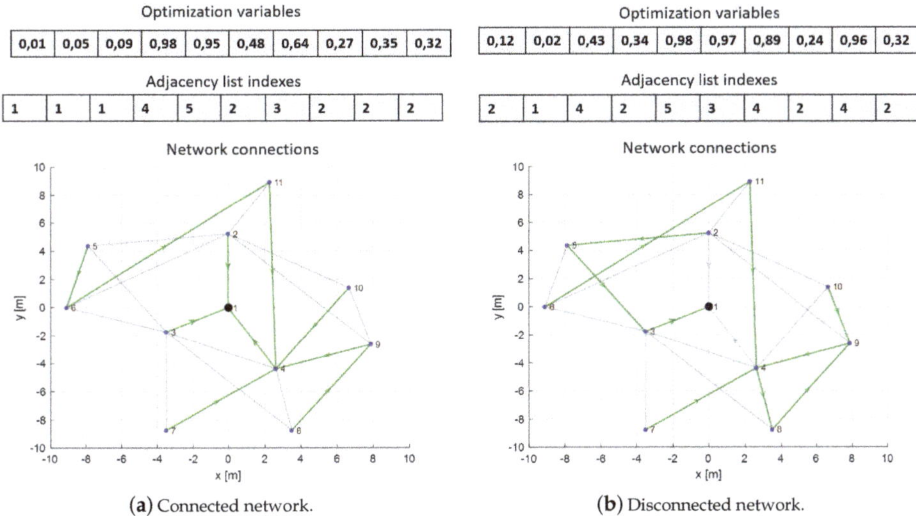

(**a**) Connected network. (**b**) Disconnected network.

Figure 7. Example of the decodification of the optimization variables: (**a**) feasible network; and (**b**) disconnected network.

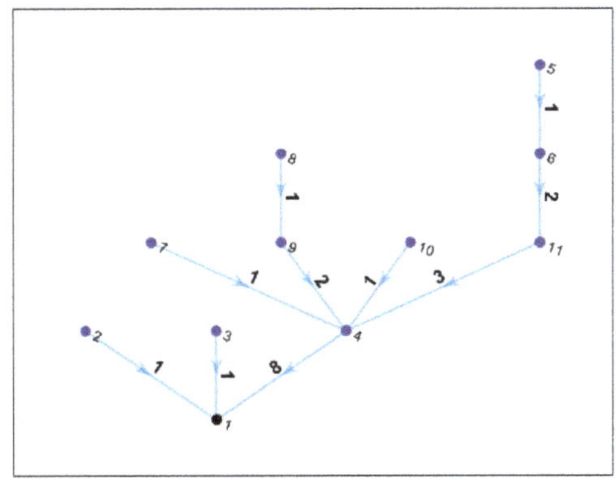

Figure 8. Representation of the active paths in the WSN with the number of transmitted packages for each edge.

Once the number of packages transmitted and received is computed, the energy model shown in Section 3.1 can be easily applied.

3.4. Optimization Environment

After having defined the WSN model and the codification technique, it is possible to design the optimization environment, i.e., all the interactions between the different elements involved in the solution of this problem.

The optimization environment designed for the routing problem is the one shown in Figure 9. Two evolutionary algorithms are used, PSO and SNO. The optimization variables used in these algorithms are decodified to a specific network configuration with the techniques shown in the previous section.

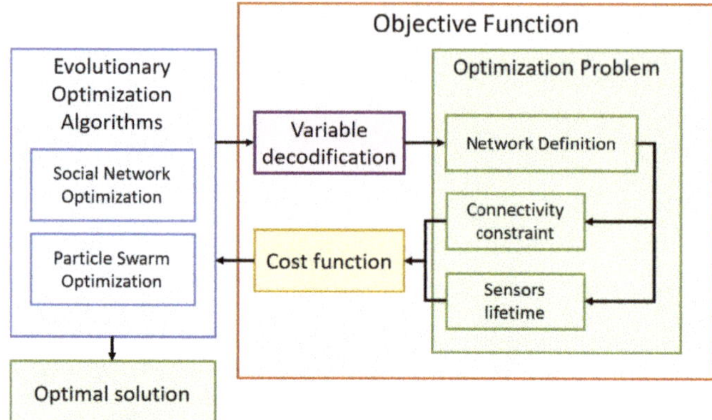

Figure 9. Optimization environment for the analyzed problem.

While a candidate solution of the algorithm is translated in a network configuration, two values are computed: the number of disconnected sensors, which is used as a constraint on the optimization problem, and the energy required by each sensor, which is the major objective of this problem.

The cost function is constructed for imposing the constraint and for helping the convergence of the algorithm.

In particular, it is defined as follows:

$$c = \begin{cases} 70 + 20 \cdot n_{disc} & n_{disc} > 0 \\ 10^5 \cdot \left(\max_i E_i + \dfrac{\sum_i E_i}{\lambda \cdot N} \right) & n_{disc} = 0 \end{cases} \tag{8}$$

The first condition is related to the compliancy with the constraints. The offset value (70) is used to avoid feasible solutions having higher costs than unfeasible ones: this creates a step in the convergence curves that helps the identification of the time in which the algorithm has reached the feasible region of the search space.

The second cost term is the one related to feasible solution. The scale factor 10^5 does not affect the optimization process; the cost term $\max_i E_i$ is the maximum energy consumption of the sensors and it is the real objective of the optimization problem, while the cost term $\sum_i E_i$ is the total energy consumed, and it is added, properly weighted by λ and the number of sensors N, for improving the convergence.

4. Results and Discussion

The optimization environment defined in the previous section has been applied for the routing optimization.

In this section, the results obtained are shown and discussed: in particular, the selected test cases are firstly analyzed; then, the optimization results of PSO and SNO are provided; and, finally, a peculiar case is analyzed in depth.

4.1. Test Cases

Due to the fact that the performance of the algorithms and codification methods can be highly case-dependant, several test cases are here created for the performance analysis.

Each network is characterized by a different number of sensors and a different deployment of them in the space, as shown in Figure 10, where a network with 50 nodes and one with 85 are compared.

The two networks are very different: in particular, the network with 85 nodes has a higher maximum node depth due to the upper area in which several nodes are not connected with the cluster head.

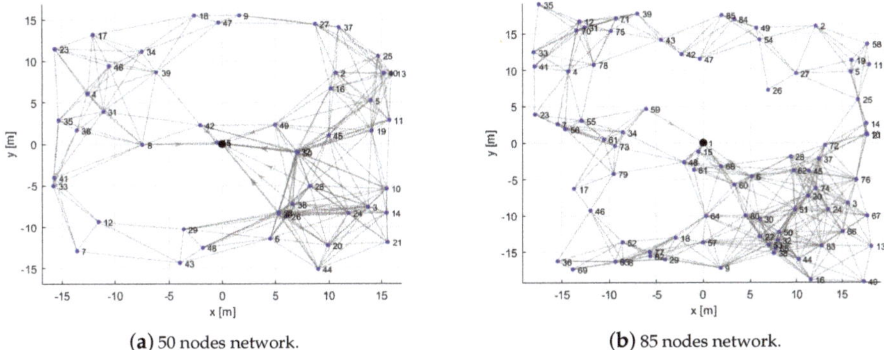

(**a**) 50 nodes network. (**b**) 85 nodes network.

Figure 10. Examples of tested networks: network with: (**a**) 50 nodes; and (**b**) 85 nodes.

For each tested network, it is possible to calculate the maximum node depth, the total number of possible configurations with the adjacency-based codification, and the number of configurations with the depth-based codification.

Table 2 shows the parameters of the tested networks. The characteristics of each network does not depend just on the number of nodes: thus, the selected networks explore different possible combinations between node depth, number of connections, and number of nodes.

Analyzing the table, it is clear that the depth-based codification reduces drastically the size of the problem.

Table 2. Summary of all the tested networks.

Number of nodes	**Maximum** node depth	**Number** of connections	**Number of Adjacency-Based** configurations	**Number of Depth-Based** configurations
ine 20	4	54	1.51×10^{13}	192
25	4	102	1.08×10^{21}	2.4×10^{6}
30	3	157	4.3×10^{28}	1.63×10^{7}
35	4	143	4.36×10^{29}	8.82×10^{9}
40	4	192	4.44×10^{35}	4.12×10^{10}
45	4	226	3.87×10^{42}	2.93×10^{12}
50	5	253	4.64×10^{47}	4.46×10^{15}
55	6	314	4.8×10^{54}	9.63×10^{18}
60	4	341	3×10^{60}	7.1×10^{20}
65	6	250	2.17×10^{53}	1.59×10^{12}
70	7	215	1.6×10^{52}	9.78×10^{13}
75	4	547	7.94×10^{84}	8.8×10^{25}
80	5	549	2×10^{87}	1.16×10^{28}
85	7	410	2.22×10^{79}	1.29×10^{27}
90	5	650	1.95×10^{101}	1.33×10^{38}
95	6	501	2.08×10^{92}	3.58×10^{32}
100	5	792	2.96×10^{116}	2.61×10^{46}
105	4	831	3.51×10^{122}	1.92×10^{46}
110	6	506	2.02×10^{102}	4.79×10^{31}
115	5	987	2.64×10^{137}	1.52×10^{52}
120	8	436	7.5×10^{98}	8.62×10^{28}
125	4	1129	3.7×10^{152}	9.16×10^{63}
130	5	1000	6.63×10^{149}	4.62×10^{62}
135	8	581	1.23×10^{120}	4.45×10^{36}
140	5	1423	1.68×10^{179}	3.93×10^{75}
145	5	1459	3.27×10^{184}	2.83×10^{77}
150	8	784	2.5×10^{147}	5.38×10^{52}

4.2. Results

To compare the results of the two different codification techniques and the two algorithm, the number of objective function calls was selected as termination criterion.

In fact, this parameter is the one that mainly drives the computational time in the optimization due to the fact that the self-time of the optimization algorithm can be usually neglected. For example, in the optimization of an 85-node network with 14,000 objective function calls, the self-time of SNO is 1.3 s for a total optimization time of 26 s, i.e., 5%. PSO is slightly faster and its self-time in the same conditions is 0.9 s, i.e., 3.5%.

Since the problem complexity grows with the number of sensors, the allowed number of objective function calls was also increased with the following empirical rule:

$$n_{call} = 1000 \cdot \left(\frac{N-20}{5} + 1 \right) \tag{9}$$

where N is the number of sensors in the network.

The population size for both these algorithms was selected according to a preliminary parametric analysis on some standard mathematical benchmarks. In particular, for PSO, the optimal population size is 50 individuals, while for SNO it is 100.

Due to the intrinsic stochastic nature of both these algorithms, 50 independent trials were performed for each configuration.

Figure 11 shows the results of the optimization. In particular, the average cost value obtained in the 50 independent trials is reported for each test case with the two codification techniques.

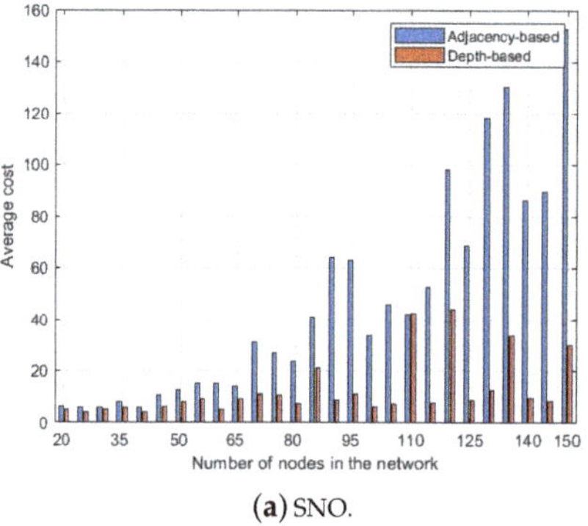

(**a**) SNO.

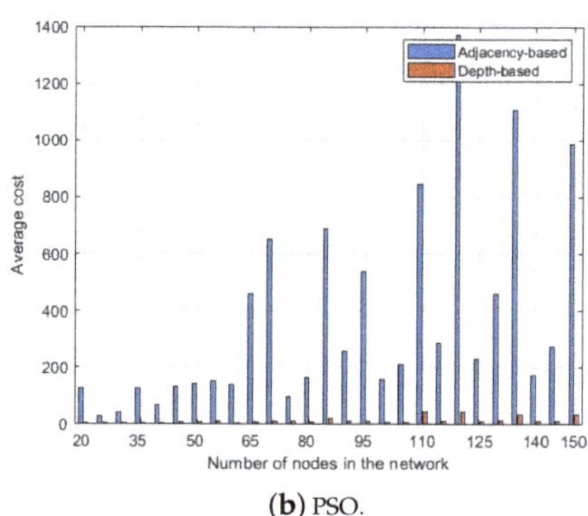

(**b**) PSO.

Figure 11. Average cost value of 50 independent trials as function of the number of nodes in the network: (**a**) results of SNO; and (**b**) results of PSO.

Analyzing these results, it is possible to make some considerations. The effect of the problem codification technique impacts on the performance of the algorithms in different ways: in fact, for PSO, the depth-based codification is always much better than the adjacency-based one, while, for SNO, the superiority of the depth-based codification is less important. This is due to the fact that SNO usually performs better than PSO in handling high-dimensional problems with many local minima.

To introduce a numerical estimation of the difference between these two codification techniques, the gap value was calculated for each case:

$$\text{GAP} = \frac{c_{adj} - c_{depth}}{c_{adj}} \tag{10}$$

This value represents the improvement of the depth-based codification expressed as function of the adjacency-based value.

The average GAP value for the mean value of SNO over all the test cases is 0.58 meaning that the depth-based mean value is, on average, half of the adjacency-based one. For PSO, the average GAP value over the mean value is 0.98.

Figure 12 shows the best results obtained by both SNO and PSO. Analyzing this figure, two additional considerations can be done. The minimum values obtained by PSO are much lower than the average ones, indicating a difficulty of this algorithm in achieving the global minimum of the function. Secondly, the difference between the two codifications is reduced, in particular for the results of SNO: in fact, for this algorithm, in eight cases, the adjacency-based codification is better, and in only eight other cases the depth-based is drastically better. This can also be seen from the average GAP calculated on the minimum value, that is 0.18 for SNO. This means that the optimal solution requires some hops that violate the depth rule used to create the second codification.

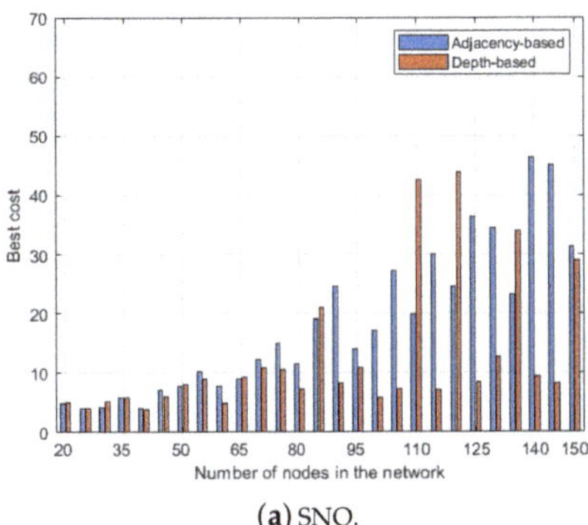

(**a**) SNO.

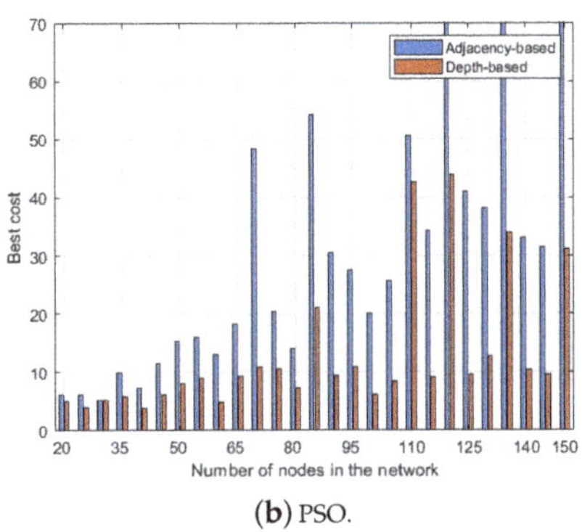

(**b**) PSO.

Figure 12. Best cost value of 50 independent trials as function of the number of nodes in the network: (**a**) results of SNO; and (**b**) results of PSO.

For comparing the results of the two algorithms, in particular, the best value obtained in the 50 independent trials, the results are plotted in Figure 13 underlining the comparison between the two algorithms.

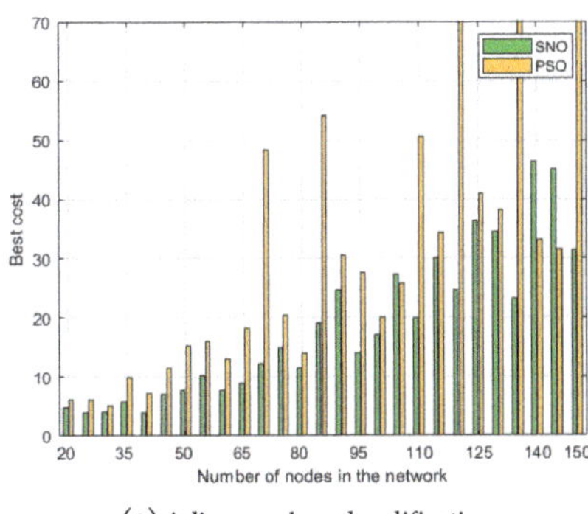

(**a**) Adjacency-based codification.

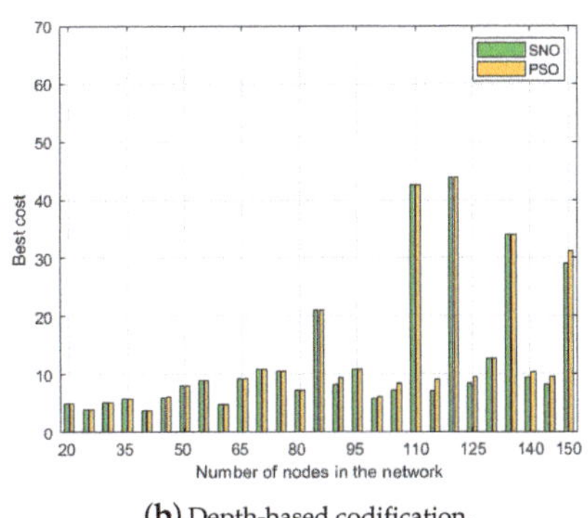

(**b**) Depth-based codification.

Figure 13. Best cost value of 50 independent trials as function of the number of nodes in the network: (**a**) adjacency-based codification; and (**b**) depth-based codification.

Figure 13a shows that the performance of SNO with the adjacency-based codification is drastically better with respect to the ones of PSO, especially in nine cases in which the gap between the two algorithms overcomes 1; in only one case, PSO is better than SNO (gap -0.05). The average gap between the two algorithms is 2.95, highly biased by three cases in which it is over 10.

On the other hand, the results of Figure 13b show that with the depth-based codification the performance of the two algorithm becomes more similar. In fact, the average gap is 0.04 and in 18 cases the solutions obtained by the two algorithms have the same cost value.

To highlight some differences between the optimizers in this case, in which the results seem the same, the termination criterion has been modified in the simulations.

In single-objective optimization, the choice of the termination criterion is not as critical as in multi-objective optimization [34], however different possibilities are analyzed in the literature.

The most common criterion is the number of objective function calls: in this case, the computational time required by the optimization is limited a priori. A second possibility is related to the population diversity: when it drops, the algorithm has reached a minimum. The effectiveness of this criterion depends highly on the mutation operators of the algorithm and it is hardly applicable for SNO. Finally, another common possibility is to fix a maximum number of iterations in which the best solution is not improved [35]. In this last analysis, the non-improving iterations criterion was adopted.

Figure 14 shows the results when the termination criterion was set to 10 non-improving iterations. In addition, in this case, 50 independent trials were performed for both algorithms.

The number of objective function calls required before the termination criterion was applied are shown in Figure 14a. It clearly shows that the effective number of objective function calls is much smaller to the time given in the standard optimization. It is interesting to see that PSO seems slightly faster than SNO.

The time performance should also be combined with the final cost values, as shown in Figure 14b, in which the final cost and the delta with respect to the previously found optimum are shown. These data show that the final cost of PSO is almost always greater than the one of SNO: this means that SNO performs more exploration than PSO. Finally, from the delta values, it is possible to notice that this second termination criterion often is not able to provide the algorithm enough time to find the optimum of the function.

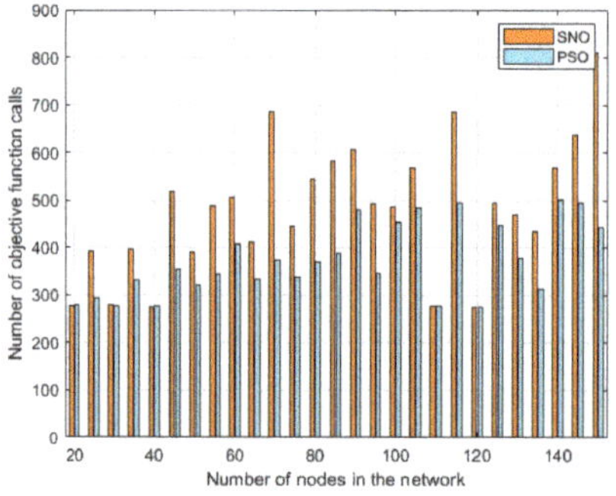

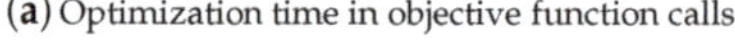

(**a**) Optimization time in objective function calls.

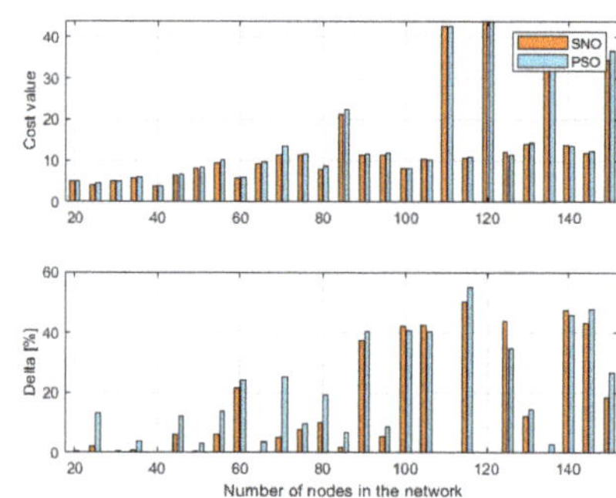

(**b**) Final cost and delta with respect to the standard optimization.

Figure 14. Optimization with 10 non-improvement iterations termination criterion: optimization time (**a**); and optimization cost (**b**).

To analyze the effect of the termination criterion, the number of non-improving iterations was increased to 50. The optimization time (Figure 15a) is more than doubled for both the algorithms. In addition, in these tests, in many cases, SNO requires more iterations than PSO. In Figure 15b, it is clear that, in almost all cases, SNO is able to achieve the optimal solutions, while the errors of PSO are still high.

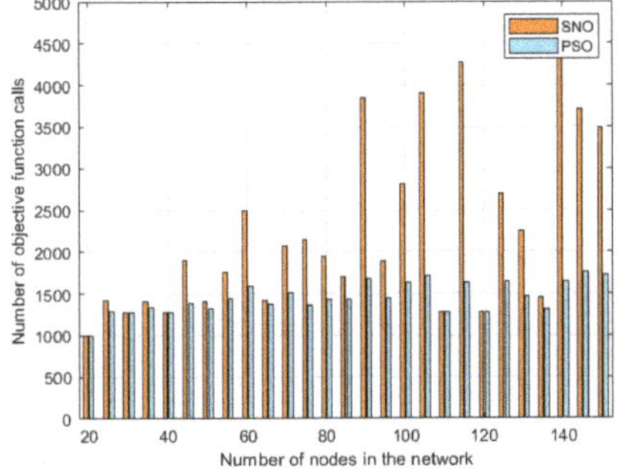

(a) Optimization time in objective function calls.

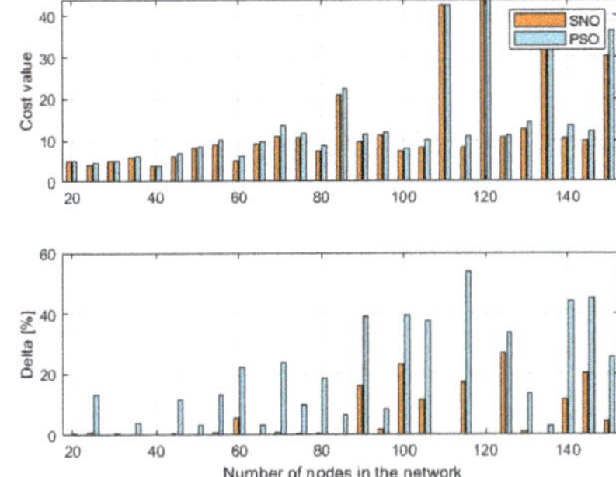

(b) Final cost and delta with respect to the standard optimization.

Figure 15. Optimization with 50 non-improvement iterations termination criterion: optimization time (**a**); and optimization cost (**b**).

In the following, a peculiar network is deeply analyzed: its peculiarity is that the adjacency-based codification gives better results.

4.3. Analysis of a Peculiar Case

The network with 50 nodes is here analyzed as a peculiar case, representative of all the networks in which the adjacency-based codification gives better results with respect to the depth-based.

Here, the results of the four optimizations on this network (two codifications and two algorithms) are analyzed. All these results were obtained performing 50 independent trials and using as termination criterion 7000 objective function calls.

Figure 16a shows the convergence curves obtained with SNO. Each grey line is a single trial, while the blue line is the average convergence. The figure also provides a zoom of the last 6000 calls. The convergence curves show some jumps in the first iterations (before 1500 objective function calls), corresponding to the achievement of feasible solutions.

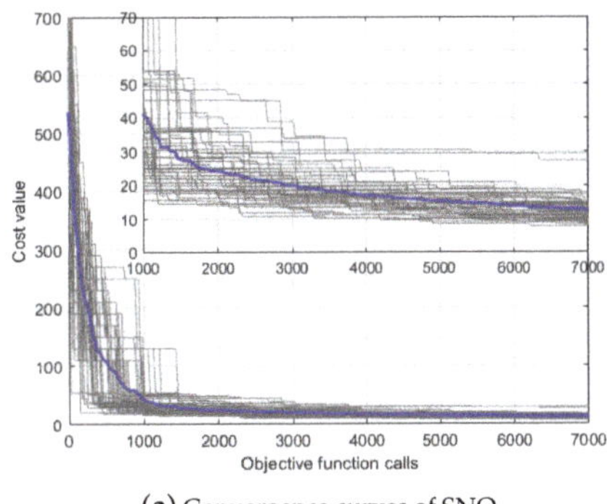

(a) Convergence curves of SNO

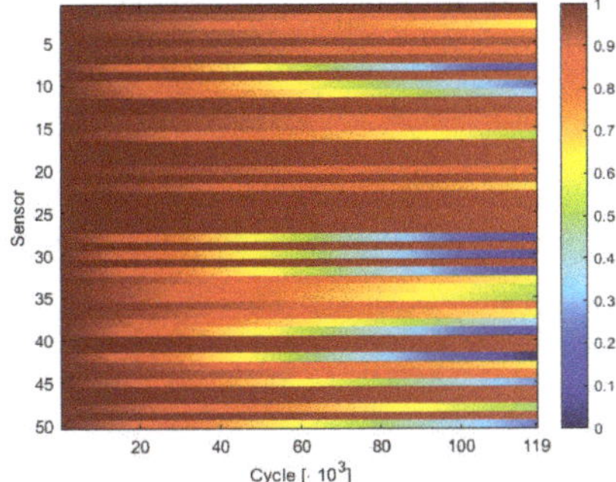

(b) Residual battery capacity for all the sensors

Figure 16. SNO optimization of the 50-nodes network with the adjacency-based codification: (**a**) convergence curves; and (**b**) residual battery capacity.

The performance of the best network achieved with this optimization is shown in Figure 16b, where the evolution of the sensors' battery capacity is represented. In this figure, the dark red color means full charge and blue empty battery. In this figure, it is possible to notice that the network lifetime is around 119 KCC and that six sensors are more critical than the others.

The active paths of the optimal solution are shown in Figure 17 with the green lines. In this figure, two areas have been highlighted. The red circles underline some sub-optimal configurations of the network: for example, in the left one, node 41 hops on 36 that hops on 33 that send all this information to sensor 35. This is sub-optimal because all these nodes could be connected directly to node 35.

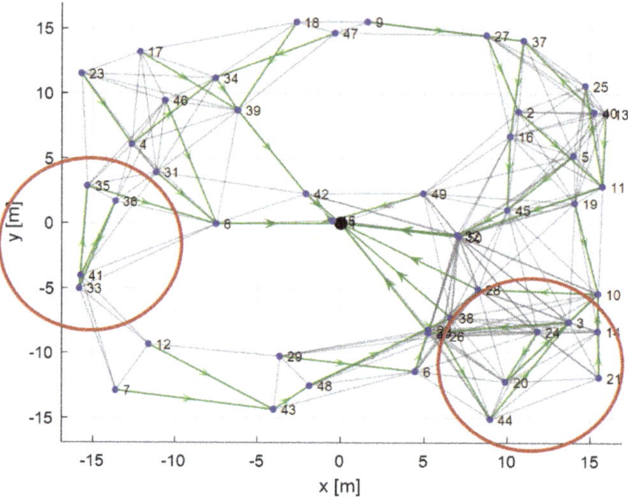

Figure 17. Routing in the optimal network. The green lines are the selected paths and the red circle underline some sub-optimal areas of the network.

The reason this sub-optimality appears is because all the involved nodes are not critical in the definition of the lifetime: in fact, as shown in Figure 16b, all of them achieve a minimum energy greater than the 50% of the initial capacity.

The convergence curves of PSO, represented in Figure 18a, shows that this algorithm could not find feasible solutions in all the independent trials. The stagnation in high-cost local minima makes the final optimal not competitive with the one found by SNO: in fact, as expressed by Figure 18b, the network lifetime is limited to 61 KCC. The reason of this very limited lifetime can be understood analyzing the active paths in the optimal solutions found by PSO shown in Figure 19. From the energy evolution it is possible to notice that the loaded nodes are the 28 and 8. In fact, they aggregates the information of large clusters of nodes: all the nodes highlighted in red transmit their information through node 8, while all the orange nodes hops on node 28.

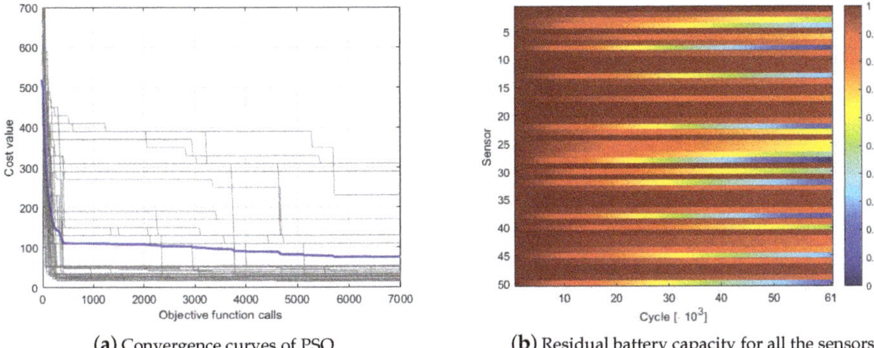

(**a**) Convergence curves of PSO

(**b**) Residual battery capacity for all the sensors

Figure 18. PSO optimization of the 50-nodes network with the adjacency-based codification: (**a**) convergence curves; and (**b**) residual battery capacity.

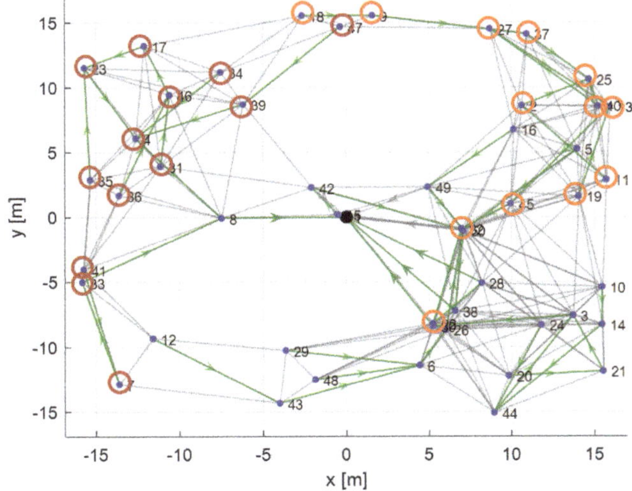

Figure 19. Routing in the optimal network. The green lines are the selected paths. The sensors with red circle all hop on node 8, and the orange circles correspond to nodes that hop on node 28.

The convergence of the 50 independent trials of SNO with the depth-based codification are shown in Figure 20a: the optimization process is very effective because all the trials converge to the same solution before 500 objective function calls out of the 7000 allowed. The fast convergence of all the trials to the same solution suggests that it is the global minimum for this function.

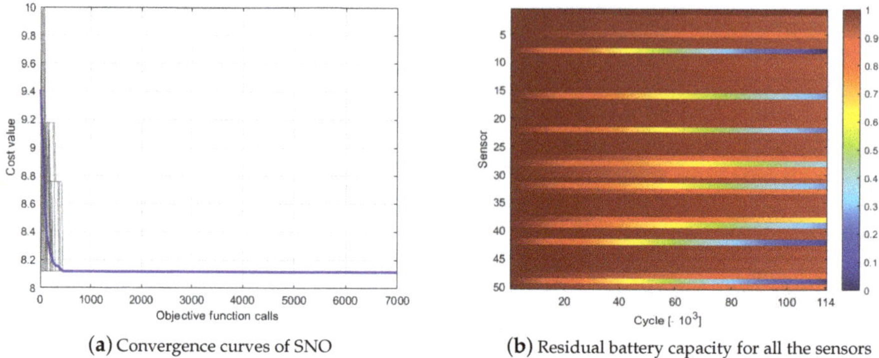

(**a**) Convergence curves of SNO (**b**) Residual battery capacity for all the sensors

Figure 20. SNO optimization of the 50-nodes network with the depth-based codification: (**a**) convergence curves; and (**b**) residual battery capacity.

This assumption can be confirmed analyzing the convergence curves of the 50 trials of PSO (Figure 21a): in fact, most of the solutions converge toward the same minimum value. From these curves, it is possible to highlight that, even if the best solution of PSO is equal to the SNO one, the first algorithm is not able to converge in all the trials.

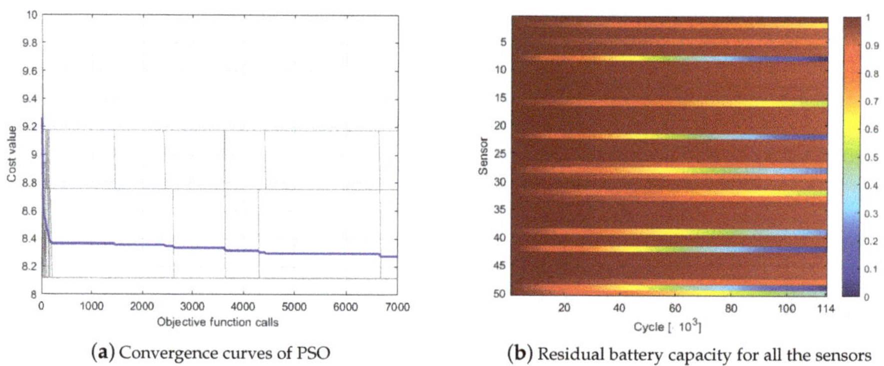

(**a**) Convergence curves of PSO (**b**) Residual battery capacity for all the sensors

Figure 21. PSO optimization of the 50-nodes network with the depth-based codification: (**a**) convergence curves; and (**b**) residual battery capacity.

From the energy evolution of SNO (Figure 20b) and PSO (Figure 21b), it is possible to see that the network lifetime (114 KCC) is lower than the best solution achieved by SNO in the adjacency-based codification (119 KCC). In fact, in these solutions, the transmission load is concentrated mostly just in three nodes: nodes 8, 42, and 49.

The overload of these nodes can bee seen from the optimal solution obtained with the depth-based codification shown in Figure 22. The colored circles underline the cluster of nodes that transmits through the highly loaded nodes. In particular, the red circles belong to the cluster of node 8, the yellow ones to sensor 42, and the orange ones to node 49.

This solution is the global minimum with the depth-based codification because the search space has been drastically reduced and, thus, the algorithm has no degree of freedom to reduce the load of critical nodes.

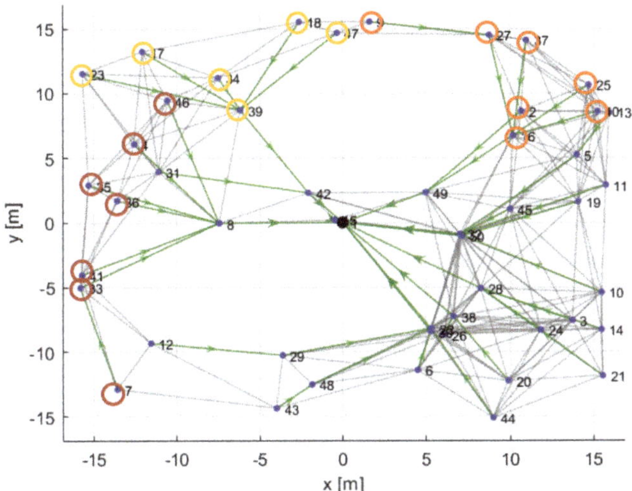

Figure 22. Routing in the optimal network with the depth-based codification. The sensors with red circle all hop on node 8, the orange circles correspond to nodes that hop on node 49, and the yellow ones hop on node 42.

5. Conclusions

In this paper, the maximization of the network lifetime of a Wireless Sensor Network is achieved by properly selecting the routing paths. This problem is faced with two different evolutionary optimization algorithms: the traditional PSO and the more recent SNO.

The objective of this paper is to analyze the difference between two different problem codification methods: with the first one (the adjacency-based codification), the algorithm can choose an "exit way" among all the adjacent nodes, while, in the second one (the depth-based codification), the selection is restricted to sensors with lower depth in the network.

Two important findings were obtained with the test campaign conducted over more than 25 test cases. Firstly, the reduction of the search space size due to the depth-based codification improves drastically the optimization convergence, in terms of both the quality of the final solution and the optimization time. This improvement is more evident in the PSO algorithm: in fact, with the adjacency-based codification, the performance of this algorithm is very poor, while, with the depth-based codification, it is comparable with SNO.

The drawback of the depth-based codification is due to the lower degrees of freedom that are given to the optimizer: in fact, while this reduces the problem complexity, it is likely to eliminate some good network configurations from the search space. This phenomenon was deeply inspected and it was confirmed by the results of the optimization of the 50-nodes network.

Author Contributions: Conceptualization, A.N., F.G., and M.M.; Data curation, A.N. and R.Z.; Formal analysis, A.G. and R.Z.; Funding acquisition, F.G. and A.G.; Investigation, M.M.; Methodology, A.N. and M.M.; Project administration, R.Z.; Resources, R.Z.; Software, A.N. and M.M.; Supervision, F.G. and A.G.; Validation, A.N. and F.G.; Visualization, A.N.; and Writing—original draft, A.N., F.G., M.M., A.G., and R.Z. All authors have read and agreed to the published version of the manuscript.

Funding: This research received no external funding.

Conflicts of Interest: The authors declare no conflict of interest.

Abbreviations

The following abbreviations are used in this manuscript:

WSN	Wireless Sensor Network
KCC	Kilo Clock Cycles
SNO	Social Network Optimization
PSO	Particle Swarm Optimization

References

1. Buratti, C.; Conti, A.; Dardari, D.; Verdone, R. An overview on wireless sensor networks technology and evolution. *Sensors* **2009**, *9*, 6869–6896. [CrossRef] [PubMed]
2. Iacca, G.; Neri, F.; Caraffini, F.; Suganthan, P.N. A differential evolution framework with ensemble of parameters and strategies and pool of local search algorithms. In *European Conference on the Applications of Evolutionary Computation*; Springer: Berlin, Germany, 2014; pp. 615–626.
3. Grimaccia, F.; Gruosso, G.; Mussetta, M.; Niccolai, A.; Zich, R.E. Design of tubular permanent magnet generators for vehicle energy harvesting by means of social network optimization. *IEEE Trans. Ind. Electron.* **2017**, *65*, 1884–1892. [CrossRef]
4. Hassan, R.; Cohanim, B.; De Weck, O.; Venter, G. A comparison of particle swarm optimization and the genetic algorithm. In Proceedings of the 46th AIAA/ASME/ASCE/AHS/ASC Structures, Structural Dynamics and Materials Conference, Austin, TX, USA, 18–21 April 2005; p. 1897.
5. Baioletti, M.; Milani, A.; Poggioni, V.; Rossi, F. An ACO approach to planning. In *European Conference on Evolutionary Computation in Combinatorial Optimization*; Springer: Berlin, Germany, 2009; pp. 73–84.
6. Caraffini, F.; Neri, F.; Poikolainen, I. Micro-differential evolution with extra moves along the axes. In Proceedings of the 2013 IEEE Symposium on Differential Evolution (SDE), Singapore, Singapore, 16–19 April 2013; pp. 46–53.
7. Caraffini, F.; Kononova, A.V.; Corne, D. Infeasibility and structural bias in differential evolution. *Inf. Sci.* **2019**, *496*, 161–179. [CrossRef]
8. Piotrowski, A.P. Review of differential evolution population size. *Swarm Evol. Comput.* **2017**, *32*, 1–24. [CrossRef]
9. Baioletti, M.; Milani, A.; Santucci, V. Automatic algebraic evolutionary algorithms. In *Italian Workshop on Artificial Life and Evolutionary Computation*; Springer: Berlin, Germany, 2017; pp. 271–283.
10. Baioletti, M.; Bari, G.D.; Milani, A.; Poggioni, V. Differential Evolution for Neural Networks Optimization. *Mathematics* **2020**, *8*, 69. [CrossRef]
11. Simon, D. Biogeography-based optimization. *IEEE Trans. Evol. Comput.* **2008**, *12*, 702–713. [CrossRef]
12. Sandeep, D.; Kumar, V. Review on clustering, coverage and connectivity in underwater wireless sensor networks: A communication techniques perspective. *IEEE Access* **2017**, *5*, 11176–11199. [CrossRef]
13. Kuila, P.; Jana, P.K. A novel differential evolution based clustering algorithm for wireless sensor networks. *Appl. Soft Comput.* **2014**, *25*, 414–425. [CrossRef]
14. Zahedi, Z.M.; Akbari, R.; Shokouhifar, M.; Safaei, F.; Jalali, A. Swarm intelligence based fuzzy routing protocol for clustered wireless sensor networks. *Expert Syst. Appl.* **2016**, *55*, 313–328. [CrossRef]
15. Chen, Y.; Li, D.; Ma, P. Implementation of Multi-objective Evolutionary Algorithm for Task Scheduling in Heterogeneous Distributed Systems. *JSW* **2012**, *7*, 1367–1374. [CrossRef]
16. Page, A.J.; Keane, T.M.; Naughton, T.J. Multi-heuristic dynamic task allocation using genetic algorithms in a heterogeneous distributed system. *J. Parallel Distrib. Comput.* **2010**, *70*, 758–766. [CrossRef] [PubMed]
17. Ferjani, A.A.; Liouane, N.; Kacem, I. Task allocation for wireless sensor network using logic gate-based evolutionary algorithm. In Proceedings of the 2016 International Conference on Control, Decision and Information Technologies (CoDIT), St. Julian's, Malta, 6–8 April 2016; pp. 654–658.
18. Niccolai, A.; Grimaccia, F.; Mussetta, M.; Zich, R. Optimal Task Allocation in Wireless Sensor Networks by Means of Social Network Optimization. *Mathematics* **2019**, *7*, 315. [CrossRef]
19. Ho, J.H.; Shih, H.C.; Liao, B.Y.; Chu, S.C. A ladder diffusion algorithm using ant colony optimization for wireless sensor networks. *Inf. Sci.* **2012**, *192*, 204–212. [CrossRef]

20. Zhang, H.; Li, Z.; Shu, W.; Chou, J. Ant colony optimization algorithm based on mobile sink data collection in industrial wireless sensor networks. *EURASIP J. Wirel. Commun. Netw.* **2019**, *2019*, 152. [CrossRef]
21. Hu, X.M.; Zhang, J.; Yu, Y.; Chung, H.S.H.; Li, Y.L.; Shi, Y.H.; Luo, X.N. Hybrid genetic algorithm using a forward encoding scheme for lifetime maximization of wireless sensor networks. *IEEE Trans. Evol. Comput.* **2010**, *14*, 766–781. [CrossRef]
22. Caputo, D.; Grimaccia, F.; Mussetta, M.; Zich, R.E. An enhanced GSO technique for wireless sensor networks optimization. In Proceedings of the IEEE Congress on Evolutionary Computation, Hong Kong, China, 1–6 June 2008; pp. 4074–4079.
23. Caputo, D.; Grimaccia, F.; Mussetta, M.; Zich, R.E. Genetical swarm optimization of multihop routes in wireless sensor networks. *Appl. Comput. Intell. Soft Comput.* **2010**, *2010*, 523943. [CrossRef]
24. Omidvar, A.; Mohammadi, K. Particle swarm optimization in intelligent routing of delay-tolerant network routing. *EURASIP J. Wirel. Commun. Netw.* **2014**, *2014*, 147. [CrossRef]
25. Poli, R.; Kennedy, J.; Blackwell, T. Particle swarm optimization. *Swarm Intell.* **2007**, *1*, 33–57. [CrossRef]
26. Simon, D. *Evolutionary Optimization Algorithms*; John Wiley & Sons: Hoboken, NJ, USA, 2013.
27. Engelbrecht, A. Particle swarm optimization: Velocity initialization. In Proceedings of the 2012 IEEE Congress on Evolutionary Computation, Brisbane, QLD, Australia, 10–15 June 2012; pp. 1–8.
28. Trelea, I.C. The particle swarm optimization algorithm: Convergence analysis and parameter selection. *Inf. Process. Lett.* **2003**, *85*, 317–325. [CrossRef]
29. Niccolai, A.; Grimaccia, F.; Mussetta, M.; Pirinoli, P.; Bui, V.H.; Zich, R.E. Social network optimization for microwave circuits design. *Prog. Electromagn. Res.* **2015**, *58*, 51–60. [CrossRef]
30. Niccolai, A.; Grimaccia, F.; Mussetta, M.; Zich, R. Modelling of interaction in swarm intelligence focused on particle swarm optimization and social networks optimization. *Swarm Intell.* **2018**, pp. 551 - 582.
31. Grimaccia, F.; Mussetta, M.; Niccolai, A.; Zich, R.E. Optimal computational distribution of social network optimization in wireless sensor networks. In Proceedings of the 2018 IEEE Congress on Evolutionary Computation (CEC), Rio de Janeiro, Brazil, 8–13 July 2018; pp. 1–7.
32. Park, J.; Sahni, S. Maximum lifetime routing in wireless sensor networks. *IEEE/ACM Trans. Netw.* **2004**, *12*, 609–619.
33. Dong, Q. Maximizing system lifetime in wireless sensor networks. In Proceedings of the IPSN 2005. Fourth International Symposium on Information Processing in Sensor Networks, Boise, ID, USA, 15 April 2005; pp. 13–19.
34. Wong, J.Y.; Sharma, S.; Rangaiah, G. Design of shell-and-tube heat exchangers for multiple objectives using elitist non-dominated sorting genetic algorithm with termination criteria. *Appl. Therm. Eng.* **2016**, *93*, 888–899. [CrossRef]
35. Eiben, A.E.; Smith, J.E. *Introduction to Evolutionary Computing*; Springer: Berlin, Germany, 2003; Volume 53,.

Article

A GPU-Enabled Compact Genetic Algorithm for Very Large-Scale Optimization Problems

Andrea Ferigo and Giovanni Iacca *

Department of Information Engineering and Computer Science, University of Trento, 38122 Trento, Italy; andrea.ferigo@studenti.unitn.it

* Correspondence: giovanni.iacca@unitn.it

Received: 17 April 2020; Accepted: 6 May 2020; Published: 10 May 2020

Abstract: The ever-increasing complexity of industrial and engineering problems poses nowadays a number of optimization problems characterized by thousands, if not millions, of variables. For instance, very large-scale problems can be found in chemical and material engineering, networked systems, logistics and scheduling. Recently, Deb and Myburgh proposed an evolutionary algorithm capable of handling a scheduling optimization problem with a staggering number of variables: one billion. However, one important limitation of this algorithm is its memory consumption, which is in the order of 120 GB. Here, we follow up on this research by applying to the same problem a GPU-enabled "compact" Genetic Algorithm, i.e., an Estimation of Distribution Algorithm that instead of using an actual population of candidate solutions only requires and adapts a probabilistic model of their distribution in the search space. We also introduce a smart initialization technique and custom operators to guide the search towards feasible solutions. Leveraging the compact optimization concept, we show how such an algorithm can optimize efficiently very large-scale problems with millions of variables, with limited memory and processing power. To complete our analysis, we report the results of the algorithm on very large-scale instances of the OneMax problem.

Keywords: compact optimization; discrete optimization; large-scale optimization; one billion variables; evolutionary algorithms; estimation distribution algorithms

1. Introduction

In recent years, several application domains have shown a constantly growing need for efficient optimization algorithms capable of handling problems with a very large number of decision variables, i.e., problems in the order of thousands, or even millions, of variables. Nowadays, this kind of problems occurs for instance in network optimization, logistics, and scheduling: examples of such problems are resource allocation, vehicle routing, and production scheduling, to name a few.

Of note, most of these industrial problems can be formulated in terms of (Mixed) Integer Linear Programming (MILP), and as such can be solved by popular commercial or open-source solvers such as CPLEX [1], Gurobi [2], or `glpk` [3]. While these solvers are guaranteed to find the optimal solutions, when it comes to solve problems with a very large number of variables, they hit a roadblock. As reported in [4], even on some Linear Programming problems, these solvers are not able to find a feasible solution in feasible time: the so-called "curse of dimensionality".

A valid alternative to overcome this issue is represented by meta-heuristics, namely stochastic search algorithms which trade optimality guarantees off for better scalability and numerical complexity. Among these, Evolutionary Algorithms (EAs) and Swarm Intelligence—such as Particle Swarm Optimization (PSO)—have especially attracted a great research attention, due to their flexibility and attributed general-purposedness. This is demonstrated by a rich literature in the field, not to mention the various benchmarks and competitions dedicated to Large-Scale Global Optimization (LSGO) [5].

However, perhaps also because these benchmarks focus on continuous optimization, most of the attempts in this area are limited to the continuous domain, with few exceptions. The most notable of these is represented by a recent work by Deb and Myburgh [4,6], who reported to have broken for the first time the "billion-variable barrier" on an ILP problem concerning metallurgy casting scheduling. This achievement was obtained with a Genetic Algorithm dubbed PILP (Population-Based Integer Linear Programming), consisting of two populations (parent and offspring) of 60 solutions each. As the authors reported, the flip side of this method is its memory consumption, which with one billion variables is in the order of 120 GB of RAM: despite the low-cost availability of RAM to date, this is a fairly large amount of memory resources.

This leads us to the motivation of our work. The main research question we try to answer here is: *Is it possible to reach at least comparable results to those reported by Deb and Myburgh, with less memory?*. Secondly, we try to assess if there is a trade-off between memory consumption and time to solve the problems. As working case, we consider a scenario where one needs to solve very large-scale problems with the additional constraint to save as much as possible the available RAM, while trying to use only one GPU. This memory-constraint scenario further exacerbates the difficulty of solving very large-scale problems, and as such our research question is quite challenging.

A possible answer to our question comes from a concept known as "compact optimization" [7]. Compact optimization is a way of designing stochastic search algorithms belonging to the class of Estimation of Distribution Algorithms (EDAs) [8]. As EDAs, compact optimization algorithms are essentially meta-heuristics that make use of an explicit probabilistic model of the distribution of the solutions in the search space. During the search, new candidate solutions are sampled from the distribution and evaluated; then the model is updated depending on the feedback from the search process itself, such that the next sampling process will be biased towards the most promising areas of the search space. The distinctive feature of compact optimization algorithms [7] with respect to the more general class of EDAs is that they purposely employ a simplistic probabilistic model that handles each variable separately. This model is usually referred to as "Probability Vector" (PV), and its structure and cardinality depend on the problem dimensionality (n) and variable type (binary, integer, or continuous). For instance, in binary problems, the PV simply consists of an n-dimensional vector of probabilities (i.e., each ith element of PV, $i \in \{1, 2, \ldots, n\}$, represents the probability in $[0, 1)$ that the corresponding ith variable is sampled as 0). In continuous optimization, compact algorithms usually employ n independent truncated Gaussian Probability Distribution Functions (one per variable), and as such the PV consists of two n-dimensional vectors, μ and σ, namely means and standard deviations, one per each truncated Gaussian PDF. In this case, sampling involves calculating the inverse of the corresponding Cumulative Distribution Function (see [7] for details).

In the past decade, the compact optimization paradigm has steadfastly advanced. Originally devised mainly as a tool for binary optimization [9,10], many compact algorithms intelligence have been proposed lately, although most of them are meant for continuous optimization only. These include real-valued compact Genetic Algorithm (cGA) [11] (and the similar "Selfish Gene" Algorithm [12]), compact Differential Evolution (cDE) [13,14] and its many variants [15–21], compact Particle Swarm Optimization (cPSO) [22], compact Bacterial Foraging Optimization (cBFO) [23], and, more recently, compact Teaching Learning Optimization (cTLBO) [24], compact Flower Pollination Algorithm (cFPA) [25], compact Firefly Algorithm (cFA) [26], and compact Artificial Bee Colony (cABC) [27,28]. In terms of applications, these algorithms have been successfully applied to a broad range of memory-limited cases, such as embedded control of robots [29–32], intrinsic evolvable hardware [33], neural network training [34], and topology control in Wireless Sensor Networks [35].

Obviously, the use of a limited-memory probabilistic model that completely replaces a population comes at a cost. The first drawback concerns the lack of a population, which in turn poses a number of problems in terms of exploration efficiency and loss of diversity: these issues are in fact implicitly addressed by having a sufficiently large number of—possibly diverse—candidate solutions at any given time during the search process, which is the main reason for the success of population-based algorithms [36]. This is by definition impossible in compact algorithms, even though various mechanisms such as re-initialization of the PV or partial restarts [37,38] have been proposed to partially alleviate this problem and prevent premature convergence.

The second drawback comes from the fact that compact algorithms treat all variables as independent. This is not the case of all problems though: most optimization problems are indeed non-separable, i.e., they are characterized by mutual dependencies between variables. These can be captured only by multivariate distributions, similar to the Covariance Matrix used in CMA-ES for continuous optimization [39]. However, due to quadratic space and time complexity, using such distributions in very large-scale problems is not be feasible, unless iterative transformation matrices are used [40]. In discrete optimization, a possible solution is to use approximations of the objective function [41], although the research in this field has been so far rather limited. Notwithstanding these limitations on separability, compact algorithms have proven successfully at solving even non-separable problems, especially at large dimensionalities [7]. One possible explanation for this—somehow surprising—result was given by Caraffini *et al.* [42], who collected evidence on the fact that the correlation between pairs of variables appears, from the perspective of a stochastic search algorithm, to consistently decrease when the problem dimensionality increases. In other words, non-separable problems in high dimensionalities can be tackled as if they are separable. In [42], the authors conjectured that this effect is due to fact that in high dimensionalities only a very restricted portion of the decision space can be explored: because of this "localness" of the search, the best strategy to make use of the available budget is to exploit any improvement along each variable, which is in accordance with the most popular and successful methods for large-scale optimization.

In this paper, we build on this conjecture and question if, in practice, GPU-enabled cGAs are a viable solution for solving even very large-scale optimization problems. In particular, we focus on the casting scheduling ILP problem tackled by Deb and Myburgh in [4,6] and evaluate the efficacy of a GPU-enabled compact Genetic Algorithm on it. To complete our analysis, we also assess the performance of the cGA on the OneMax benchmark problem [43], both on the original binary formulation, as well as a continuous and an integer formulation with 16 discrete values (reported in Appendix A), on which we evaluate the time consumption of each element of the algorithm. It is worth noting that, apart from [4,6], only few works have tried to apply Evolutionary Algorithms to very large-scale problems: in [44], a one-billion variable noisy binary OneMax problem is tackled with a parallel implementation of cGA on a cluster of 256 CPU cores, with a very weak "relaxed convergence" criterion requiring each variable to reach a fixed-proportion correct value (probability 0.501). In [45], random embeddings are used as a form of dimensionality reduction applied to Bayesian optimization, and tested on a two-variable real-parameter problem embedded into a one-billion variable vector (in which all but two variables have no effect on the fitness). As for a GPU-enabled cGA, thus far the only attempt was made by Iturriaga and Nesmachnow [46], who tested the algorithm on the binary OneMax problem, with and without noise. However, even though their experiments were scaled up to one billion variables, they did not consider the presence of constraints, nor they tackled ILP problems. Therefore, here we advance with respect to the previous literature by: (1) extending the analysis on the OneMax performed in [46], also in the light of the new hardware available (more specifically, Iturriaga and Nesmachnow [46] used 8 CPU cores (Intel Xeon @ 2.33 GHz with 8 GB RAM) in multi-threading, with 4 Tesla C1060 GPUs (each with 4GB GDDR3, 240 cores @ 610 MHz). Here, we use one CPU core (Intel i9-7940x @ 3.10 GHz with 64 GB RAM), with one Titan XP GPU (with 12 GB GDDR5X, 3840 cores @ 1405 Mhz)); and (2) applying, for the first time, a cGA to the very large-scale industrial ILP problem taken from [4,6]. Overall, the main contributions of this work can be summarized as follows:

- We adapt the binary compact Genetic Algorithm in order to handle integer variables, represented in binary form.
- We present a fully GPU-enabled implementation of the compact Genetic Algorithm, where all the algorithmic operations (sampling, model update, solution evaluation, and comparison) can be optionally executed on the GPU.

- We apply the proposed GPU-enabled compact Genetic Algorithm to the OneMax benchmark problem and the casting scheduling ILP problem described in [4,6], in various dimensionalities.
- For the ILP problem, we include in the compact Genetic Algorithm custom (problem-specific) operators and a smart initialization technique inspired by Deb and Myburgh [4] that, by acting on the PV, guide the sampling process towards feasible solutions.
- We show that on the OneMax problem our proposed GPU-enabled compact Genetic algorithm obtains partial improvements with respect to the results reported in [46]; on the casting scheduling problem, we obtain at least comparable–and in some cases better—results with respect to those in [4,6], despite a much more limited usage of computational resources.

The remaining of this paper is organized as follows. In the next section, we describe the formulation of the OneMax and the casting scheduling problems. In Section 3, we introduce the details of the proposed GPU-enabled compact Genetic Algorithm and its various versions. Then, in Section 4, we present the numerical results on the two problems. Finally, in Section 5, we provide the conclusions and highlight possible future research directions.

2. Problem Formulations

Let us introduce the two problems we use in our experimentation. It should be noted that both problems are scalable, i.e., they can be instantiated in any number of variables. In our experiments, we scaled both problems at various dimensionalities, up to one billion variables.

2.1. OneMax

The first problem we used in our experiments is the binary OneMax benchmark function, also called BitCounting [43]. This problem consists in finding a Boolean vector $\mathbf{x}$ such that:

$$\text{Maximize } f(\mathbf{x}) = \sum_{i=1}^{n} x_i \tag{1}$$

where x_i is the ith element of $\mathbf{x}$, $i = \{1, 2, \ldots, n\}$, and $n = |\mathbf{x}|$. The optimum of this function corresponds to a vector $\mathbf{x}^*$ whose all bits are set to one, $f(\mathbf{x}^*) = n$.

In our experiments, we further extended the original binary formulation of the OneMax problem to account for discrete values ($x_i \in \{0, 1, \ldots, 15\} \forall i$) and for continuous variables ($x_i \in [0, 1] \forall i$). In the first case, the optimum is a vector where all variables are set to 15, and its fitness is $15 \times n$. In the second case, the optimum is the same as in the original binary OneMax formulation. We report these additional results in Appendix A.

2.2. Casting Scheduling Problem

The second problem we consider is the casting scheduling ILP problem defined in [4,6], formulated as follows:

$$\text{Maximize } f(\mathbf{x}) = \frac{1}{H} \sum_{i=1}^{H} \frac{1}{W_i} \sum_{j=1}^{N} w_j x_{i,j} \tag{2}$$

Subject to:

$$\sum_{j=1}^{N} w_j x_{i,j} \leq W_i \qquad \text{for } i \in \{1, 2, \ldots, H\} \tag{3}$$

$$\sum_{i=1}^{H} x_{i,j} = r_j \qquad \text{for } j \in \{1, 2, \ldots, N\} \tag{4}$$

where each variable $x_{i,j}$ is in $\{0, 1, \ldots, 15\}$. The goal is to find the optimal scheduling to cast batches of N distinct objects in H heats. The production of each ith heat is bound by the corresponding crucible size W_i. Each jth object has a weight, w_j, and must be produced in a certain number of copies, r_j. However, the total amount of metal required for one batch of copies of the same object does not necessarily result equal to the size of the crucible in the ith heat, W_i. For this reason, a desired efficiency η is set. This is done by finding the minimum number of heats, H, such that $\sum_{i=1}^{H} \eta W_i \geq M$, where M is the total amount of metal required for all batches, $M = \sum_{j=1}^{N} r_j w_j$. Note that the problem must consider both the fact that the metal molten in one heat cannot be greater than the crucible capacity and that an exact number of objects is required. Therefore, the problem has $N \times H$ variables, H inequality constraints, see Equation (3), and N equality constraints, see Equation (4).

To allow the cGA to handle the constraints, we follow the penalty-based approach described in [4,6], by subtracting to the fitness function described in Equation (2) the following quadratic penalty factor, calculated from the constraint violations:

$$P(\mathbf{x}) = \left[\sum_{j=1}^{N} \left(\sum_{i=1}^{H} x_{ij} - r_j \right)^2 + \sum_{i=1}^{H} \left\langle \frac{1}{W_i} \sum_{j=1}^{N} w_j x_{ij} - 1 \right\rangle^2 \right] \tag{5}$$

where $\langle \cdot \rangle$ indicates that the penalty is summed only if the corresponding inequality constraint is violated. Consequently, the cGA should be applied to maximize $F(\mathbf{x}) = f(\mathbf{x}) - R \times P(\mathbf{x})$, where R is a penalty multiplier. However, as suggested in [4,6] and further verified by us, optimal solutions in terms of fitness $F(\mathbf{x})$ can be obtained simply by minimizing the penalty factor, i.e., $argmax(F(\mathbf{x})) = argmin(P(\mathbf{x}))$. Note that, while the range of F(x) is $(-\inf, \eta]$, the function $P(\mathbf{x})$ takes values in $[0, +\inf)$. For this reason, we minimize Equation (5) until it reaches the value 0, which corresponds to $f(\mathbf{x}) = \eta$ in Equation (2).

3. Proposed Algorithms

To better highlight the changes required to the original cGA [9] to make it run on a GPU and handle integer values, we present here separately two versions of the GPU-enabled cGA, namely a binary cGA and a discrete cGA, the latter being an evolution of the first. Their general structure is similar, and is shown in Algorithm 1. Note that the only parameter of both algorithms is `virtualPopulation`, which affects the dynamics of the algorithm [7]. In a nutshell, the differences between the two algorithms are: (1) the different type of variables (binary vs integer); and (2) the introduction of problem-specific operators in the discrete cGA. Furthermore, for the binary cGA, we present three different versions, i.e., synchronous and asynchronous with sub-problem size 1 and 100 (where "sub-problem" indicates a set of variables handled by one GPU thread). In the following, we describe the details of the GPU implementation of each function used in Algorithm 1 for the various versions of the cGA.

Algorithm 1 Binary cGA: general structure.

```
procedure CGA(problemSize, virtualPopulation)
    PV ← vector of size problemSize, with values 0.5
    elite ← generateTrial(PV)
    fitnessElite ← evaluate(elite)
    while ! stopCriteriaMet() do
        trial ← generateTrial(PV)
        fitnessTrial ← evaluate(trial)
        winner ← compete(fitnessTrial, fitnessElite)
        PV ← updatePV(winner, PV, trial, elite, virtualPopulation)
        if winner == 1 then                                   ▷ trial is better
            elite ← trial
            fitnessElite ← fitnessTrial
        end if
    end while
    return elite
```

3.1. Binary cGA

We developed in total three versions of the binary cGA, which we refer to as "cGA-Base", "cGA-A100", and "cGA-A1", respectively. Each version follows the structure of Algorithm 1. A summary of the variables and functions used in these binary cGAs is reported in Table 1. The differences among the three versions are in the parallelization process, in particular in the evaluate() and updatePV() functions. cGA-Base operates synchronously, evaluating solutions atomically and, as result, performing updatePV() considering all variables. The two other versions are based on the asynchronous cGA presented in [46], in which the problem is divided into sub-problems of the same size, which are processed asynchronously. In cGA-A100 and cGA-A1, we consider sub-problems sizes of 100 and 1, respectively. Overall, the three versions of the binary cGA differ not only in terms of performance (as we show in the next section), but also in terms of the resources used, which are summarized in Table 2. Let us now analyze in detail the implementation of the three binary cGA versions. For each function used in the cGAs, we specify if it is specific for the OneMax problem or not. If another problem were to be solved, those functions would need to be changed.

Table 1. Variables and functions used in the binary cGAs.

Name	Description
`problemSize`	Problem dimensionality (scalar)
`virtualPopulation`	Virtual population (scalar)
`PV`	Probability Vector (vector)
`trial`	Solution sampled from `PV` (vector)
`elite`	Best solution found so far (vector)
`fitnessTrial`	Fitness of `trial` (scalar or vector)
`fitnessElite`	Fitness of `elite` (scalar or vector)
`winner`	Flag(s) indicating if `trial` (1) or `elite` (0) is better (scalar or vector)
generateTrial()	Sample `trial` from `PV` (return a solution)
evaluate()	Evaluate a solution (return its fitness)
compete()	Compare two solutions (return `winner`)
updatePV()	Update `PV` (return `PV`)

Table 2. Data type and size for the binary cGAs.

	cGA-Base		cGA-A100		cGA-A1	
Data	**Type**	**Size**	**Type**	**Size**	**Type**	**Size**
`PV`	Float32	`problemSize`	Float32	`problemSize`	Float32	`problemSize`
`trial`	Bool	`problemSize`	Bool	`problemSize`	Bool	`problemSize`
`elite`	Bool	`problemSize`	Bool	`problemSize`	Bool	`problemSize`
`fitnessTrial`	Int32	1	Int8	`problemSize/100`	Int32	1
`fitnessElite`	Int32	1	Int8	`problemSize/100`	Int32	1
`winner`	Int32	1	Bool	`problemSize/100`	-	-

3.1.1. Binary cGA-Base

In this version of the binary cGA, the parallelization occurs at variable level, i.e., each ith variable is handled by a separate GPU thread. The functions in Table 1 are implemented as follows.

- generateTrial(): This function samples, for each variable, a random number in $[0, 1)$, and compares it to the corresponding element of `PV`. If the random number is greater than the element of `PV`, then the relative variable of `trial` is set to 1; otherwise, it is set to 0. Each variable is handled by a separate GPU thread.
- compete(): This function compares `fitnessTrial` and `fitnessElite`, and sets `winner` to 1 (0) if `trial` is better (worse) than `elite`. This operation is performed without GPU-parallelization.
- evaluate(): This function loops through all the variables of the argument and sums 1 if the ith variable is set to 1. This operation is performed without GPU-parallelization, since its output (the

fitness of the argument) depends on all variables.
Note: This function is specific for the OneMax problem.

- updatePV(): This function biases `PV` towards the winner between `trial` and `elite`, i.e., it makes more probable that the next `trial` will be more similar to the winner, increasing or decreasing each element of `PV` accordingly (see Algorithm 2). Each variable is handled by a separate GPU thread.

Algorithm 2 Binary cGA-Base: updatePV().

```
procedure UPDATEPV(winner, PV, trial, elite, virtualPopulation)
    for i in 1...problemSize do                        ▷ each iteration in a separate GPU thread
        if winner == 1 then                                               ▷ trial is better
            if trial[i] == 1 then
                if trial[i] ≠ elite[i] then
                    PV[i] ← PV[i] - 1/virtualPopulation
                end if
            else
                if trial[i] ≠ elite[i] then
                    PV[i] ← PV[i] + 1/virtualPopulation
                end if
            end if
        else                                                              ▷ elite is better
            if elite[i] == 1 then
                if trial[i] ≠ elite[i] then
                    PV[i] ← PV[i] - 1/virtualPopulation
                end if
            else
                if trial[i] ≠ elite[i] then
                    PV[i] ← PV[i] + 1/virtualPopulation
                end if
            end if
        end if
    end for
    return PV
```

3.1.2. Binary cGA-A100

In contrast with the cGA-Base, in this case, there is not a one-to-one relation between variables and GPU threads. Instead, each GPU thread handles a single sub-problem, in this case of 100 variables, in particular in the evaluate(), compete() and updatePV() functions. On the contrary, generateTrial() is implemented as in the cGA-Base (i.e., with parallelization at variable level). Furthermore, as it is necessary to save the partial fitness and winner for each sub-problem, in this case `fitnessTrial`, `fitnessElite`, and `winner` are vectors of size `problemSize/100` (as shown in Table 2), to allow asynchronous updates. The evaluate(), compete(), and updatePV() functions are implemented as follows.

- evaluate(): This function evaluates the argument by processing each sub-problem independently on the GPU threads. Each thread takes a portion of the argument and calculates its partial fitness. The operations executed by each thread, illustrated in the outer for loop in Algorithm 3, can be parallelized since each sub-problem operates over disjoint portions of the argument. The function getSubProblem() returns the *i*th sub-problem, i.e., a vector of 100 variables. Note that in this case evaluate() returns a vector, `fitness`, rather than a scalar. The partial result of each sub-problem is stored by each GPU thread in the corresponding position of `fitness`.
Note: This function is specific for the OneMax problem.
- compete(): This function operates over the vectors `fitnessTrial` and `fitnessElite` returned by the evaluate() shown in Algorithm 3. As shown in Algorithm 4, each separate GPU thread

simply checks the winner on its sub-problem, i.e., it operates on a single index of `fitnessTrial`, `fitnessElite`, and `winner`, which are all vector of size `problemSize`/100.
Note: This function is specific for the OneMax problem.

- updatePV(): This function is implemented similarly to the evaluate(). As shown in Algorithm 5, each GPU thread loads its portion of data, composed by both corresponding sub-problem of `elite` and `trial` and the relative portion of `winner` and `PV`. After that, the procedure is similar to the one described in the cGA-Base (see Algorithm 2), with the only difference being that in this case the **for** loop in Algorithm 2 is performed within the same GPU thread and not in parallel. The ancillary function setPartialPV() updates `PV` with the result of the partial `pPV`.
Note: This function is specific for the OneMax problem.

Algorithm 3 Binary cGA-A100: evaluate().

```
procedure EVALUATE(solution)
    fitness ← vector of size problemSize/100, with values 0
    for i in 1 ... problemSize/100 do            ▷ each iteration in a separate GPU thread
        partialSolution ← solution.getSubProblem(i)
        for j in 1 ... 100 do
            if partialSolution[j] == 1 then
                fitness[i] ← fitness[i] + 1
            end if
        end for
    end for
    return fitness
```

Algorithm 4 Binary cGA-A100: compete().

```
procedure COMPETE(fitnessTrial, fitnessElite)
    winner ← vector of size problemSize/100
    for i in 1 ... problemSize/100 do            ▷ each iteration in a separate GPU thread
        if fitnessTrial[i] > fitnessElite[i] then
            winner[i] ← 1
        else
            winner[i] ← 0
        end if
    end for
    return winner
```

Algorithm 5 Binary cGA-A100: updatePV().

```
procedure UPDATEPV(winner, PV, trial, elite, virtualPopulation)
    for i in 1 ... problemSize/100 do            ▷ each iteration in a separate GPU thread
        pTrial ← trial.getSubProblem(i)
        pElite ← elite.getSubProblem(i)
        pPV ← PV.getSubProblem(i)
        pPV ← updatePV(winner[i], pPV, pTrial, pElite, virtualPopulation)    ▷ Algorithm 2
        PV[i].setPartialPV(i,pPV)
    end for
    return PV
```

3.1.3. Binary cGA-A1

In principle, this version can be obtained simply by setting the sub-problem size equal to 1 in cGA-A100. However, in this way, the size of `winner`, `fitnessTrial`, and `fitnessElite` would match the problem size, thus increasing the memory usage. For this reason, we decided to embed the operations of evaluate() and compete() into the updatePV() function, thus removing the need for the `winner` vector and reducing the `fitnessTrial` and `fitnessElite` vectors to a scalar. As for the parallelization process, in this case, each GPU thread handles a single variable, as it happens in the cGA-Base. The generateTrial() function is implemented as in the cGA-Base. The updatePV() function is implemented as follows, and replaces Lines 7–13 in Algorithm 1.

- updatePV(): This function includes, in this case, also the operations performed in evaluate() and compete() and results quite simpler than the previous cases. The process is shown in Algorithm 6. Since `PV` is updated only when the variables of `trial` and `elite` are different, each GPU thread checks if one of the two variables is set to 1 and the other to 0. Consequently, `PV` is only reduced, which increases the probability of sampling 1. Finally, if the relative element of `trial` is set to 1, each GPU thread adds 1 to the `fitnessTrial` variable to calculate the total fitness. Note that in this case there is no need for using `winner`, which reduces the memory consumption.
  Note: This function is specific for the OneMax problem.

Algorithm 6 Binary cGA-A1: updatePV().

```
procedure UPDATEPV(PV, trial, elite, virtualPopulation, fitnessElite)
    for i in 1 ... problemSize do                    ▷ each iteration in a separate GPU thread
        if trial[i] == 1 and elite[i] == 0 then
            PV[i] ← PV[i] - 1/virtualPopulation
        end if
        if trial[i] == 0 and elite[i] == 1 then
            PV[i] ← PV[i] - 1/virtualPopulation
        end if
        if trial[i] == 1 then
            fitnessTrial ← fitnessTrial + 1                              ▷ thread-safe sum
        end if
    end for
    if fitnessTrial > fitnessElite then                                  ▷ trial is better
        elite = trial
        fitnessElite = fitnessTrial
    end if
    return PV, elite, fitnessElite
```

3.2. Discrete cGA

The discrete version of the cGA can be seen as an evolution of the binary cGA, specifically adapted to handle integers and include problem-specific mechanisms to solve the casting scheduling problem taken from [4,6]. As for the integer handling, the idea is to represent, quite straightforwardly, integer variables in binary format, and then use a binary cGA to evolve the resulting bit-string. For the casting scheduling problem, we consider integer variables in the interval $\{0, 1, \ldots, 15\}$, and represent each variable with 4 bits. In addition to that, we implement on the GPU the problem-specific smart initialization, crossover and mutation operators described in [4]. We further improve the initialization mechanism in order to adapt it to the cGA paradigm. A summary of the variables and functions used in the discrete cGA is reported in Table 3. The overall structure of the algorithm is shown in Algorithm 7. In the following, we describe the main implementation details of the discrete cGA. Note that, unless indicated differently, all operations are fully parallelized on the GPU, maintaining a one-to-one mapping between GPU threads and variables as described in the cGA-Base.

Table 3. Variables and functions used in the discrete cGA.

Name	Description
N	Number of objects (scalar)
`W`	Crucible sizes (vector)
η	Desired efficiency (scalar)
H	Number of heats to achieve (η) (scalar)
`copies`	Copies to be cast for each object (vector)
`weights`	Weights (vector)
`virtualPopulation`	Virtual population (scalar)
`PV`	Probability Vector (vector)
`trial`	Solution sampled from `PV` (vector)
`elite`	Best solution found so far (vector)
`copiesTrial`	Copies cast by `trial` (vector)
`heatsTrial`	Available space in the heats of `trial` (vector)
`fitnessTrial`	Fitness of `trial` (scalar)
`fitnessElite`	Fitness of `elite` (scalar)
`winner`	Flag indicating if `trial` (1) or `elite` (0) is better (scalar)
estimateH()	Estimate the value of H based on η, `copies`, `weights` and `W` (return H)
smartInitialization()	Initialize `elite` and `PV` (return `elite` and `PV`)
initializeElite()	Initialize `elite` (return `elite`)
inhibitor()	Blocks the unusable elements of `PV` (return `PV`)
initializePV()	Initialize `PV` (return `PV`)
generateTrial()	Sample `trial` from `PV` and apply to it mutations and crossover
newTrial()	Sample `trial` from `PV` (return `trial` and `heatsTrial`)
mutationOne()	Repair `trial` with respect to the equality constraints
mutationTwo()	Repair `trial` with respect to the inequality constraints
crossover()	Operate a heat-wise crossover between `trial` and `elite`
evaluate()	Evaluate a solution (return its fitness)
compete()	Compare two solutions (return `winner`)
updatePV()	Update `PV` (return `PV`)

Algorithm 7 Discrete cGA: specific structure for the cast scheduling problem.

```
procedure CGA(N, copies, weights, W, η, virtualPopulation)
    PV, elite, H ← smartInitialization(copies, weights, W, η)
                                                            ▷ calls estimateH(),
                  ▷ initializeElite(), mutationOne(), mutationTwo(), evaluate(),
                                                  ▷ inhibitor() and initializePV()
    while ! stopCriteriaMet() do
        trial, heatsTrial, copiesTrial ← generateTrial(PV)
                         ▷ calls newTrial(), crossover(), mutationOne() and mutationTwo()
        fitnessTrial ← evaluate(heatsTrial, copiesTrial)
        winner ← compete(trial, elite)
        PV ← updatePV(winner, PV, trial, elite, virtualPopulation)
        if winner == 1 then
            elite ← trial
            fitnessElite ← fitnessTrial
            heatsElite ← heatsTrial
        end if
    end while
    return elite
```

- smartInitialization(): This function performs the initialization process, divided into three steps. Firstly, the function estimateH() estimates the value H to get an efficiency equal to η, as described in Section 2.2. This value is also used to define the size of `elite` and `PV`. Secondly, an `elite` of size $N \times H$ is created in initializeElite(), repaired by the two mutations, and evaluated. Finally, `PV` is created, with a size of $4 \times N \times H$ and processed as described in initializePV() below.

- o initializeElite(): This function initializes the `elite` as described in [4], i.e., creates the `elite` by creating N vectors of size H, which are initialized respecting as much as possible the number of copies required for each object. This operation is GPU-parallelized through two steps: firstly, a vector of random numbers is created, and a correction factor is calculated; and, secondly, the correction is applied to the vector. After this, to repair any possible violated constraint, the `elite` is passed to the two mutation operators mutationOne() and mutationTwo(), and finally evaluated by evaluate(). The implementation of these functions is shown below.
- o initializePV(): This function initializes the `PV` such that it generates solutions that satisfy the inequality constraints, and which are biased towards the `elite`. This initialization is performed in two functions. The first function, shown in Algorithm 8, blocks the `PV` values for the variables which, if set to 1, would cause the violation of an inequality constraint. For example, if the crucible size is 600 kg and an object weights 200 kg, obviously no more than three copies can be cast for that object without a penalty: in this case, the inhibitor blocks (i.e., sets to NaN) the third and fourth bit of the variable, preventing it from assuming values greater than 3 for that object. This operation is GPU-parallelized, with each thread checking a single element of `PV`. The second function aims at modifying the `PV` values to produce solutions closer to the `elite`. This operation is done by setting each `PV` element to 0.25 if the bit of the corresponding element in the `elite` is 1, and 0.75 otherwise, unless that element is blocked by inhibitor(), as shown in Algorithm 9.

- generateTrial(): This function is divided into four steps: generation of the trial, crossover and two mutation operators to repair the trial. The problem-specific crossover and mutation operators were implemented as described in [6]. In addition, we operated some modifications in order to parallelize the operators on the GPU and save the information necessary to simplify the successive calculation of the fitness. This information consists of two vectors, one of size H, called `heatsTrial`, and one of size N called `copiesTrial`. The first one saves the available space in the crucible for each heat, and it is created during newTrial() in two steps: firstly, the vector is initialized with the crucible size; then, while `trial` is created, its values are decreased, as shown in Algorithm 10. Moreover, `heatsTrial` is updated also during crossover and mutations, as shown below. The second vector, `copiesTrial`, stores instead the total number of objects cast by the trial, and is calculated during mutationOne() (see Algorithm 12). The other details of the four steps are presented below.

- o newTrial(): This function, as shown in Algorithm 10, is implemented such that each GPU thread handles one variable of `trial` and its corresponding four values of `PV`. More specifically, each thread samples 4 bits based on the corresponding probabilities of `PV`, converts them to a value in $\{0, 1, \ldots, 15\}$, and assigns this value to the corresponding variable in `trial`. Finally, the element of `heatsTrial` relative to this variable is updated accordingly to the value just assigned. Note that this last operation is implemented as thread-safe, as the same heat can be updated asynchronously by different threads.
- o crossover(): The function, as shown in Algorithm 11, is implemented such that each GPU thread handles one variable of `trial`. The crossover operator compares each heat of `trial` with the corresponding one in `elite`. If `elite` has a better heat (i.e., has a higher (lower) value in case both `trial` and `elite` have negative (positive) heats), then all the `elite` variables referred to that heat are assigned to `trial`, updating `heatsTrial` accordingly.

- o mutationOne(): The first mutation is meant to repair `trial` with respect to the equality constraint (see Equation (4)). The procedure is divided into two steps, as shown in Algorithm 12: firstly, the total number of copies cast by `trial` for each object, `copiesTrial`, is calculated, with each GPU thread handling one variable of `trial` and performing a thread-safe sum over `copiesTrial`. Next, for each object, if the copies required (stored in `copies`) are less than the ones that are cast (stored in `copiesTrial`), this means that some copies in excess should be removed from `trial`: to do that, first the heat with the greatest inequality violation (i.e., the lowest value of `heatsTrial`, returned by argmin()) is found, then the corresponding variable in `trial` is decreased by one, and `heatsTrial` and `copiesTrial` are updated accordingly. On the other hand, if the copies required are more than the ones that are cast, this means that some more copies should be added to `trial`: to do that, first the heat with the lowest crucible utilization (i.e., the highest value of available space, stored in `heatsTrial`, returned by argmax()) is found, then the corresponding variable in `trial` is increased by one, and `heatsTrial` and `copiesTrial` are updated accordingly. This operation is repeated until the equality constraint is satisfied. Note that the functions used to find the two heats, i.e., argmin() and argmax(), are GPU-parallelized. Moreover, during the repair process `heatsTrial` and `copiesTrial` vectors are updated in order to be reused later to calculate the fitness.
- o mutationTwo(): The second mutation tries to reduce the heats which use more space than the one available in the crucible, as described in [4]. The procedure consists in finding two heats: the one with the greatest inequality violation, and the one with the greatest available space. After that, a random object is selected and removed from the first heat and added to the second one, in order to preserve the total number of copies. These operations are repeated the number of times indicated by the parameter `iterationLimit`. In the original version in [4], this value was fixed during all the computation (set to 30), while, in our implementation, due to the compact nature of the algorithm, we needed to double this value at each iteration, starting from 30. This process leads to an exponential increase of the time of mutationTwo(), as shown in the next section. As for the parallelization process, similar to mutationOne(), it is obtained by loading `trial` into the GPU and then performing the argmin() and argmax() operations parallelized with a one-to-one mapping between GPU threads and variables.

- evaluate(): This function evaluates the solution in a time-efficient way, using the information already calculated during generateTrial(). The two penalty factors related to the inequality and equality constraints (see Equations (3) and (4)) are determined, respectively, through the `heatsTrial` and `copiesTrial` vectors, and added to `fitnessTrial`, as shown in Algorithm 13. The total time required is $O(H+N)$, and it is further decreased due to the parallelization over H heats. Note that the second loop over N objects is not parallelized as N is typically much smaller than H.
- updatePV(): This function is parallelized on the size of `PV` (one bit per GPU thread) and operates similarly to Algorithm 2, with the only difference that each thread has to extract the relative bit before performing the update.

Algorithm 8 Discrete cGA: inhibitor().

```
procedure INHIBITOR(PV, W, weights, H)
    for i in 1 ... length(PV) do                          ▷ each iteration in a separate GPU thread
        bitIndex ← i % 4                                  ▷ index of the bit
        objectIndex ← (i/4) / H                           ▷ index of the object
        crucibleIndex ← getCrucible(W,i,H)                ▷ index of the crucible
        if 2^bitIndex > W[crucibleIndex]/weights[objectIndex] then
            PV[i] ← NaN
        end if
    end for
    return PV
```

Algorithm 9 Discrete cGA: initializePV().

```
procedure INITIALIZEPV(PV, elite)
    for i in 1 ... length(PV) do                          ▷ each iteration in a separate GPU thread
        index ← i / 4                                     ▷ index of variable
        bitIndex ← i % 4                                  ▷ index of the bit
        bitElite ← getBin(elite[index], bitIndex)
        if bitElite == 1 and PV[i] ≠ NaN then             ▷ bit is not blocked by inhibitor()
            PV[i] ← 0.25
        else
            PV[i] ← 0.75
        end if
    end for
    return PV
```

Algorithm 10 Discrete cGA: newTrial().

```
procedure NEWTRIAL( PV, trial, weights)
    for i in 1 ... length(trial) do                       ▷ each iteration in a separate GPU thread
        heatsTrial ← vector of size H initialized with the relative crucible size
        num ← 0
        for j in 1 ... 4 do
            if PV[4 × i + j] ≠ NaN then                   ▷ bit is not blocked by inhibitor()
                rnd ← random(0,1)
                if rnd ≥ PV[4 × i + j] then
                    num ← num + j
                end if
            end if
        end for
        trial[i] ← num
        heatIndex ← i % length(heatsTrial)
        objectIndex ← i / length(heatsTrial)
        heatsTrial[i] ← heatsTrial[i]-trial[i] × weights[objectIndex]   ▷ thread-safe sum
    end for
    return trial, heatsTrial
```

Algorithm 11 Discrete cGA: crossover().

```
procedure CROSSOVER(trial, elite, heatsTrial, heatsElite, H)
    heatsTrialTmp ← vector of size H
    for i in 1 ... length(trial) do                          ▷ each iteration in a separate GPU thread
        heatIndex ← i % length(heatsTrial)
        if heatsElite[heatIndex] better than heatsTrial[heatIndex] then
            trial[i] ← elite[i]
            heatsTrialTmp[heatIndex] ← heatsElite[heatIndex]
        else:
            heatsTrialTmp[heatIndex] ← heatsTrial[heatIndex]
        end if
    end for
    heatsTrial ← heatsTrialTmp
    return trial, heatsTrial
```

Algorithm 12 Discrete cGA: mutationOne().

```
procedure MUTATIONONE(trial, heatsTrial, copies, weights, N)
    copiesTrial ← vector of size N
    for i in 1 ... length(trial) do                          ▷ each iteration in a separate GPU thread
        objectIndex ← i/H                                                      ▷ index of the object
        copiesTrial[objectIndex] ← copiesTrial[objectIndex] + trial[i]         ▷ thread-safe sum
    end for
    for j in 1 ... length(copies) do
        while copiesTrial[j] ≠ copies[j] do                          ▷ loop until equality is satisfied
            if copies[j] < copiesTrial[j] then
                minHeatIndex ← argmin(heatsTrial)              ▷ each heat in a separate GPU thread
                trial[j × H + minHeatIndex] ← trial[j × H + minHeatIndex] - 1
                heatsTrial[minHeatIndex] ← heatsTrial[minHeatIndex] + weights[j]
                copiesTrial[j] ← copiesTrial[j] - 1
            end if
            if copies[j] > copiesTrial[j] then
                maxHeatIndex ← argmax(heatsTrial)              ▷ each heat in a separate GPU thread
                trial[j × H + maxHeatIndex] ← trial[j × H + maxHeatIndex] + 1
                heatsTrial[maxHeatIndex] ← heatsTrial[maxHeatIndex] - weights[j]
                copiesTrial[j] ← copiesTrial[j] + 1
            end if
        end while
    end for
    return trial, heatsTrial, copiesTrial
```

Algorithm 13 Discrete cGA: evaluate().

```
procedure EVALUATE(copiesTrial, heatsTrial, copies, weights, W, penalty)
    fitnessTrial ← 0
    for i in 1...H do                                   ▷ each iteration in a separate GPU thread
        crucibleIndex ← getCrucible(W,i,H)                                ▷ index of the crucible
        if heatsTrial[i] < 0 then
            fitnessTrial ← fitnessTrial + (heatsTrial[i]/W[crucibleIndex])^2
        end if
    end for
    for j in 1...N do
        if copiesTrial[j] ≠ 0 then
            fitnessTrial ← fitnessTrial + (copiesTrial[j] − copies[j])^2
        end if
    end for
    return fitnessTrial
```

4. Results

We discuss now the performance of the binary and discrete cGA implementations described in the previous section. Firstly, we compare the three binary cGAs (cGA-Base, cGA-A100, and cGA-A1) with the cGA presented in [46] on the OneMax problem. Then, we compare the discrete cGA with the PILP algorithm presented in [4,6]. We provide the code publicly as Colab notebooks (https://drive.google.com/drive/folders/1k6KWtR9ceuW7HneLptK3TlMpfjqaNMBT?usp=sharing).

4.1. OneMax

We performed the experiments on the OneMax problem using the Google®Colab service, which provides a machine powered by an Intel®Xeon^{TM}4 CPU @ 2.20 GHz, 25 GB RAM, with an NVIDIA®P100 GPU. For each algorithm presented, we tested four dimensionalities, varying the size of the problem from one million to one billion variables. For each algorithm and problem size, we performed 10 runs, with a virtual population of 100. Each run ended either when it reached the optimal fitness, or after a predetermined maximum number of iterations (note that in the cGA one iteration corresponds to one solution evaluation). This last parameter was set to 5000 for all the dimensionalities, except for the case with one billion variables, where it was set to 1600 for the cGA-Base and cGA-A100 and 500 for the cGA-A1. A summary of the results, in comparison with the results taken from [46], is shown in Table 4 for the synchronous cGAs, and in Table 5 for the asynchronous cGAs.

Table 4. Synchronous binary cGAs: comparison on the OneMax problem (mean across 10 runs, std. dev. in parentheses). The symbol '-' indicates data not provided in [46].

	cGA-Sync [46]				cGA-Base [Ours]			
	1M	**8M**	**32M**	**1B**	**1M**	**8M**	**32M**	**1B**
Time (s)	600 (-)	10,680 (-)	49,140(-)	348,000(-)	52.022 (1.552)	256.923(2.169)	969.419 (5.971)	9085.140 (1136.925)
Fitness (%)	82.5 (-)	79.1(-)	74.1(-)	62.3 (-)	51.192 (0.0570)	50.750(0.0194)	50.634(0.0053)	50.525 (0.0012)
Iterations	50,000 (-)	50,000 (-)	50,000 (-)	50,000 (-)	5000 (0)	5000 (0)	5000 (0)	1600 (0)

Table 5. Asynchronous binary cGAs: comparison on the OneMax problem (mean across 10 runs, std. dev. in parentheses). The symbol '-' indicates data not provided in [46].

	cGA-Async [46]				cGA-A100 [Ours]				cGA-A1 [ours]			
	1M	**8M**	**32M**	**1B**	**1M**	**8M**	**32M**	**1B**	**1M**	**8M**	**32M**	**1B**
Time (s)	324 (-)	7560 (-)	35,400 (-)	315,060 (-)	58.745 (0.837)	177.476 (2.249)	789.158 (3.314)	6798.995 (1155.062)	9.913 (0.269)	66.898 (1.513)	286.802 (3.452)	2278.10 (18.273)
Fitness (%)	91.0 (-)	82.6 (-)	77.8 (-)	66.8 (-)	99.317 (0.0293)	95.785 (0.174)	92.548 (0.145)	66.968 (0.163)	100 (0.0)	100 (0.0)	100 (0.0)	99.946 (9.946e-5)
Iterations	50,000 (-)	50,000 (-)	50,000 (-)	50,000 (-)	5000 (0)	5000 (0)	5000 (0)	1600 (0)	986.6 (33.242)	1208.5 (19.448)	1357.7 (17.257)	500 (0)

Our cGA-Base shows only marginal improvements with respect to the cGA-sync from [46]. One possible explanation for this is the different value of virtual population used in [46], which however is not reported in the paper. However, as it is quite improbable that a compact genetic algorithm can converge in very large-scale problems [44], our result can be considered satisfactory. In contrast with these results, our asynchronous versions show much better results, in terms of both computing time and fitness. In particular, the cGA-A100 obtains better results in all four dimensionalities, with execution times from 9 to 46 times smaller than the times reported in [46] (also due to more recent GPU hardware). The cGA-A1, as expected, shows even better behavior, solving the problem in roughly 1300 iterations for the first three dimensionalities, and reaching 99.9% in only 500 iterations for the one billion-variable problem. The fitness trend of the three binary cGAs is shown in Figure 1, where it can be seen that, while the synchronous cGA suffers from premature convergence, the asynchronous cGAs do not, and are able to reach high-quality solutions in both versions of the algorithm, although with a very different convergence profile: the cGA-A100 shows an almost-linear behavior, while the cGA-A1 has an exponential convergence.

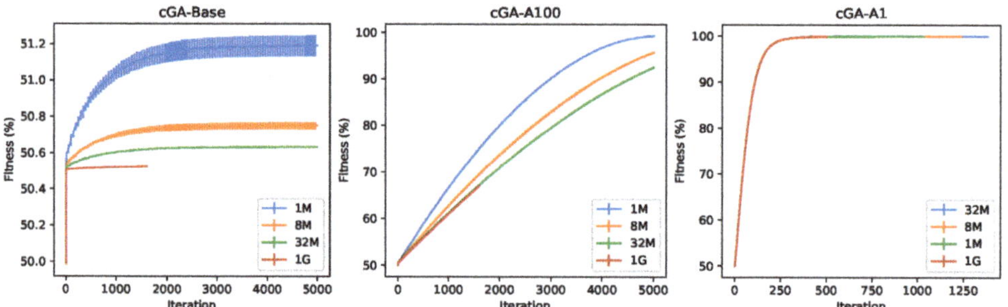

Figure 1. Best fitness value per iteration (mean ± std. dev. across 10 runs per algorithm) for the three binary cGAs on the OneMax problem in four different dimensionalities.

We complete our analysis on the OneMax problem with a profiling of the various functions used in the cGAs. The results for the cGA-A100 are shown in Figure 2, where it can be noted that the time distribution on the four problem dimensionalities is approximately the same, with the updatePV() and generateTrial() functions requiring most of the time, roughly 50% and 30%, respectively. The remaining 20% is divided between evaluate() and compete(). Similar considerations apply to the cGA-Base and cGA-A1, with some caveats: Firstly, as described in the previous section, the compete() function in the asynchronous algorithms is more expensive than in the cGA-Base. Secondly, in the cGA-A1 the operations of compete() and evaluate() are embedded into updatePV(), causing an increase of this function in terms of execution time. Interestingly, these results are quite different from the ones described in [46], where up to 70% of the time is spent in the trial generation phase, while in our setup most of the time is spent to update the probability vector.

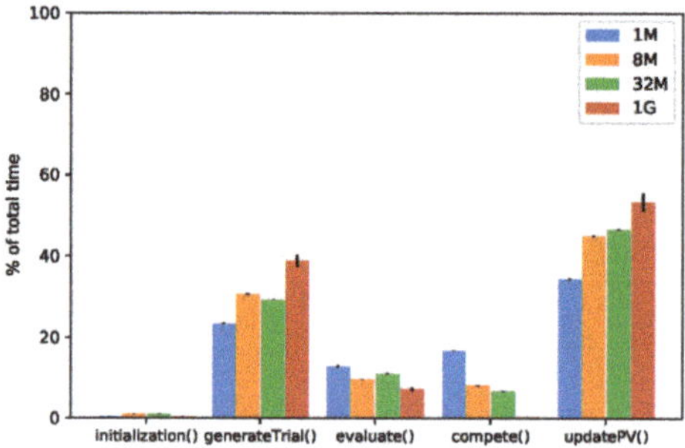

Figure 2. Time profiling (mean ± std. dev. across 10 runs) for the cGA-A100 on the OneMax problem in four different dimensionalities.

4.2. Casting Scheduling Problem

We performed the experiments on the casting scheduling problem on a Linux workstation powered by an Intel®Core™i9-7940x CPU @ 3.10GHz, 64GB RAM, with an NVIDIA®Titan XP GPU. The code was tested on Ubuntu 19.10 (kernel GNU/Linux 5.3.0-40-generic x86_64), CUDA 10.1, and Python3 with the Numba library used to write the GPU kernel for the parallelization process.

Similar to the OneMax case, we ran 10 independent runs of the discrete cGA for three different dimensionalities (100k, 1M, and 10M variables), with $\eta = 0.997$ and the input parameters indicated in Table 6. The `virtualPopulation` and the `iterationLimit` are set to 100 and 30, respectively. Of note, all the runs terminated successfully, finding the optimal solution corresponding to $P(\mathbf{x}^*) = 0$. A summary of the results in comparison with the results taken from [4] is shown in Table 7.

Table 6. Parameters used for the casting scheduling problem. Note that the crucible size, W, and the `weights` are in common for all the problem dimensionalities, therefore the only parameter that changes across dimensionalities is the number of copies per object.

Size	W	{500, 650}									
	Object id	1	2	3	4	5	6	7	8	9	10
	`weights`	79	66	31	26	44	35	88	9	57	22
100K		12560	12,562	12,517	12,567	12,562	12,172	12,076	12,052	12,017	12,012
1M	`copies`	125,600	125,620	125,170	125,670	125,620	121,720	120,760	120,520	120,170	120,120
10M		1255980	1,256,200	1,251,700	1,256,700	1,256,200	1,217,200	1,207,600	1,205,200	1,201,700	1,201,200

Table 7. Casting scheduling problem: number of solution evaluations to reach the optimum, number of heat updates, and execution times (mean across 10 runs, std. dev. in parentheses). The results for the PILP algorithm are taken from Table 6 in [4]; the symbol '-' indicates data not provided.

Algorithm	Size	Solution evals.	Heat Update	Total Time (s)	initialization() (s)	generateTrial() (s)	evaluate() (s)	compete() (s)	update() (s)
discrete cGA [ours]	100K	20.2 (9.249)	431,027.2 (145,724.751)	360.703 (114.444)	5.711 (2.827)	354.707 (114.993)	0.0636 (0.0958)	0.0431 (0.0813)	0.173 (0.266)
	1M	18.1 (7.687)	3,300,602.6 (28,057.973)	2538.747 (405.0527)	38.757 (9.796)	2499.556 (398.775)	0.0444 (0.0396)	0.0230 (0.031)	0.358 (0.315)
	10M	29.3 (14.423)	33,720,256.6 (867,688.014)	29,785.986 (6790.176)	460.459 (136.859)	29,322.144 (6679.297)	0.177 (0.0883)	0.0910 (0.0626)	3.101 (1.367)
PILP [4]	100K	1032 (17)	8,807,564 (167,494)	26 (0.6)	-	-	-	-	
	1M	1080 (35)	91,345,801 (2,048,330)	308 (9)	-	-	-	-	-
	10M	1104 (35)	976,903,439 (19,038,115)	4207 (124)	-	-	-	-	-

It should be noted that our implementation requires only a few tens of solution evaluations to solve the problem (slightly more than 50 in the 10M case), while PILP for the same problem dimensionalities needs above 1000 evaluations. This reflects also in an average number of heat updates (i.e., how many times a variable is modified by the two mutation operators), which is from 10 to 300 times lower than PILP, due to the lower number of trials processed. Additionally, it is important to consider the much smaller amount of memory used with respect to the population-based PILP algorithm: while discrete cGA needs only 18.8 × `problemSize` bytes (More specifically: 4 × `problemSize` float32 (for `PV`), 2 × `problemSize` int8 (for `trial` and `elite`), and 0.2 int32 (for `heats`).) to save `PV`, `trial`, `elite` and its respective `heats` vector, PILP requires 2 × 60 × `problemSize` bytes, where 60 is the population size, excluding some additional memory needed to store the fitness values and other data structures.

As for the time needed to solve the problem, our algorithm results from one to two orders of magnitude slower than PILP. This is mainly due to the lack of a population, which we compensated with an exponential increase of the parameter `iterationLimit` in mutationTwo(). As can be seen in Table 7, the majority of time is spent for the generateTrial() function, particularly for the execution of the mutation operators. As shown in Figure 3, during the first iterations, the execution times for the two mutations is dominated by mutationOne(), but eventually it is mutationTwo() that requires more time. However, it can be seen that the time per iteration of mutationTwo() remains almost constant as the problem dimensionality increases. Therefore, the time increase of mutationTwo() is only caused by the increasing number of iterations required. Overall, it seems then that there is a trade-off between memory consumption (more memory is needed to store more solutions, which allow a better search with fewer iterations needed to repair them) and time to converge (with limited memory, more time is needed to repair the trial generated at each iteration of the cGA).

For completeness, we report the fitness trend of the discrete cGA in Figure 4. The behavior is similar to the one described in [4], with the entire evolution that can be divided into three phases: in the initial galloping phase, the fitness rapidly decreases; then follows a consolidation phase, where new solutions have small improvements; and, lastly, the final solution is created in the culmination phase.

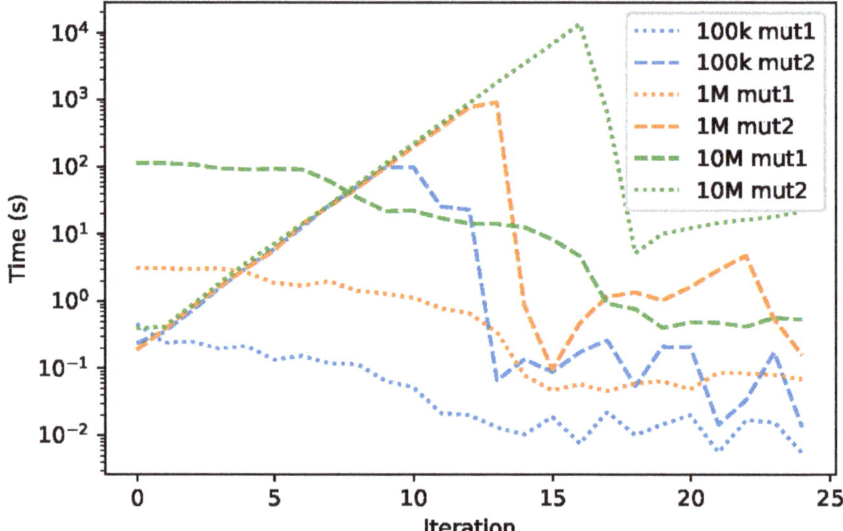

Figure 3. Execution time per iteration (mean across 10 runs) for the two mutation operators on the casting scheduling problem in three different dimensionalities.

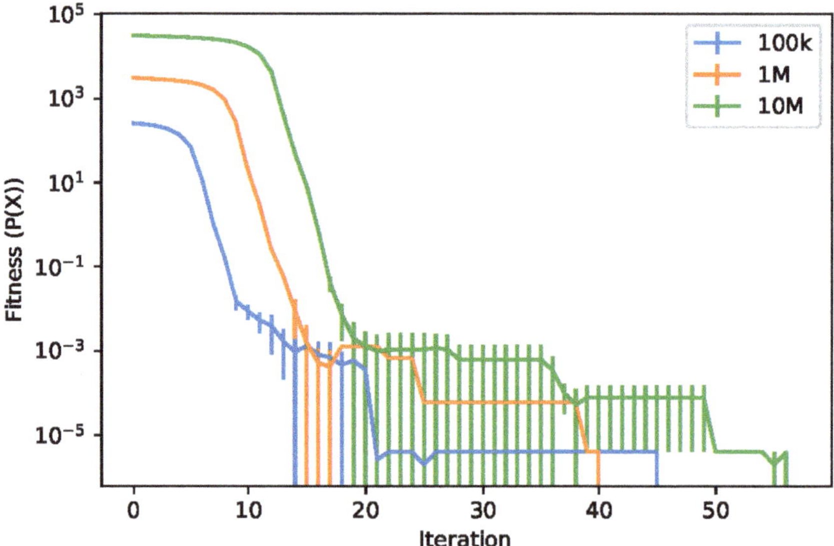

Figure 4. Best fitness value per iteration (mean ± std. dev. across 10 runs) on the casting scheduling problem in three different dimensionalities.

Effect of the Smart PV Initialization

As discussed in the previous section, for the casting scheduling problem, we introduced a smart initialization mechanism aimed at reducing the fitness of the first trial and also avoiding generating unfeasible solutions. To assess the effect of the smart initialization on the algorithmic performance, we compared the effect of inhibitor() and initializePV() with that of a random initialization. As shown in Figure 5, these two functions lead to generate trials that are five times better (in terms of fitness) than those generated by random initialization, indicated as initialization(). This provides the algorithm a "head start" that as seen before allows converging in a very limited number of iterations.

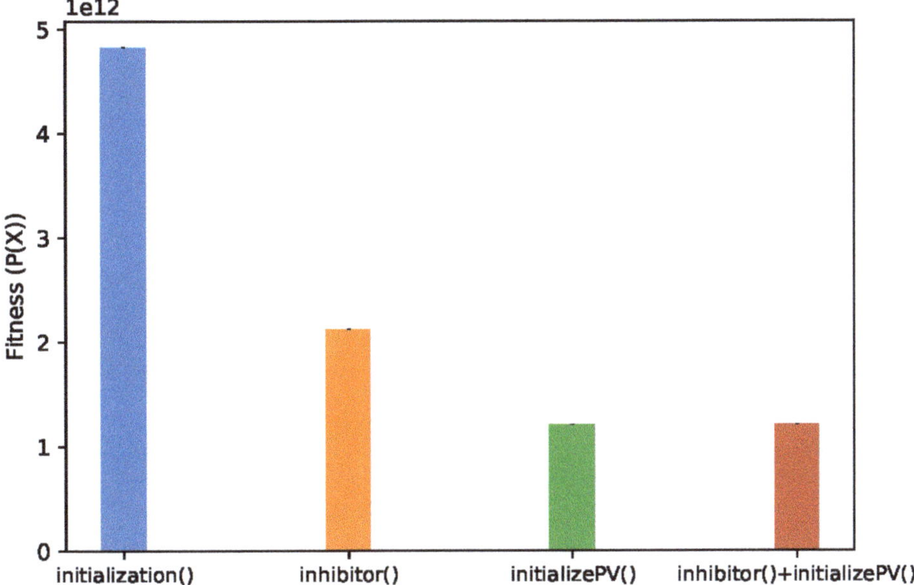

Figure 5. Casting scheduling problem: fitness of trials generated using different PV initialization strategies (mean ± std. dev. across 1000 trials). Here, initialization() indicates a random initialization of PV. Note that in these experiments the mutation operators are not applied, thus any violated constraint is not repaired. While it seems that initializePV() alone is enough to generate good trials at the beginning of the algorithm, inhibitor() blocks the bits also during the evolution.

5. Conclusions

In many application domains, there is a constant demand for ever more efficient optimization techniques. This is especially true for large-scale optimization problems, for which one usually needs large computational resources—in terms of processing power and memory—to obtain a reasonable solution in feasible time. However, in some cases, the available computational resources might be scarce, or should be reserved to other applications. Therefore, it is of great interest to find a trade-off between efficient optimization and resource consumption.

In this study, we tackled very large-scale optimization problems (of up to one billion variables), in both discrete and continuous domains, with special constraints on processing power and memory. In particular, we questioned if it is possible to solve these kinds of problems by fitting efficiently the search algorithm into one GPU. To do that, we considered a compact Genetic Algorithm (cGA) and we adapted it to make it work on the GPU, by splitting the problem and letting multiple GPU cores work in parallel on different sub-problems. We considered two different sub-problem sizes (1 and

100). In addition, we implemented both asynchronous and synchronous schemes, depending on the possibility of updating the best solution as soon as an improvement is found on one sub-problem.

To test the proposed algorithm, we considered binary, integer, and continuous optimization problems. In particular, we first benchmarked the algorithm on the OneMax problem in binary, integer (16 values) and continuous settings. Then, we considered an industrial casting scheduling problem recently presented by Deb and Myburgh [6]. Overall, our numerical results show that: (1) compact optimization techniques are a viable solution for solving very large-scale problems even with limited resources; and (2) they are especially suitable for GPU-parallelization. On the other hand, compact algorithms have some implicit limitations deriving from the fact that they lack a population of candidate solutions, hence being in general less efficient at exploring the search space compared to population-based algorithms. Therefore, to use these algorithms properly in practical applications—without sacrificing too much the optimization performances—it is recommended to couple them with problem knowledge. To show this, here we demonstrated how a base version of the cGA can be customized, for the specific case of the scheduling problem, with a smart initialization of the probability vector (that is, the probabilistic model used in the compact algorithm) aimed at guiding the search towards feasible solutions, as well as problem-specific mutation and crossover operators aimed at repairing constraint violations, adapted from [6]. Furthermore, we adapted the cGA to handle variables of different kinds, as well as equality and inequality constraints.

In future works, we aim to further extend cGAs by hybridizing them with other single-solution optimization algorithms, such as simulated annealing [47], and applying gradient-based methods to perform local search, in a memetic fashion. Furthermore, we will consider the use of decomposition techniques (either problem-aware or problem-agnostic) and restart mechanisms such as the re-sampled inheritance introduced in [37,38]. Another intriguing possibility would be to integrate compact algorithms with a quantum annealer, to obtain a hybrid quantum-classical optimizer. Finally, it will be interesting to apply these algorithms to other domains, e.g. for training deep neural networks, in Wireless Sensor Networks applications [48], or to solve very large-scale instances of TSP and other "NK landscape" problems, as recently discussed in [49,50].

Author Contributions: Conceptualization, G.I.; Data curation, A.F.; Formal analysis, A.F.; Investigation, A.F.; Methodology, G.I.; Software, A.F.; Supervision, G.I.; Validation, A.F.; Visualization, A.F.; Writing—original draft, A.F. and G.I.; Writing—review & editing, G.I. Both authors contributed equally to this work. Both authors have read and agreed to the published version of the manuscript.

Funding: This research received no external funding.

Acknowledgments: We gratefully acknowledge the support of NVIDIA Corporation with the donation of the TITAN Xp GPU used for this research.

Conflicts of Interest: The authors declare no conflict of interest.

Appendix A. Additional Results on OneMax Problem

Appendix A.1. Discrete OneMax

We present here additional results obtained with the discrete cGA on a discrete version of the OneMax problem. For this analysis, we used the general version of the discrete cGA presented in Section 3.2, where we removed all the problem-dependent operations implemented for the casting scheduling problem, namely the smartInitialization(), crossover(), mutationOne() and mutationTwo() functions. The algorithm obtained has a structure similar to Algorithm 1, but it is able to handle integers variables. Similar to the casting scheduling problem, we considered integer variables in the interval $\{0, 1, \ldots, 15\}$.

- Results: We performed 10 runs on four dimensionalities, setting the `virtualPopulation` to 100. The maximum number of iterations was set to 5000 for problem instances up to 32M variables and 1000 for the 1B case. All experiments were executed on the Google®Colab service. The results obtained, reported in Table A1, are similar to the results reported in Section 4.1 for the binary

OneMax problem, although times are (up to) twice as big. The main reason for this time increase is the fact that in this case the algorithm handles 4 bits per variable in the newTrial() and updatePV() functions, instead of just one as in the binary cGA.

Table A1. Results of the cGA on the discrete and continuous OneMax problem.

	Discrete cGA				Continuous cGA			
	1M	**8M**	**32M**	**1B**	**1M**	**8M**	**32M**	**1B**
Time (s)	72.463 (2.038)	395.806 (3.985)	1958.932 (485.963)	13,143.343 (3201.053)	156.647 (1.487)	985.936 (118.716)	3366.755 (20.297)	20,895.186 (4131.433)
Fitness (%)	50.926 (0.0250)	50.658 (0.008)	50.579 (0.002)	50.515 (0.0005)	50.870 (0.039)	50.102 (0.024)	50.034 (0.0072)	50.0055 (0.0003)
Iterations	5000	5000	5000	1000	5000	5000	5000	1000

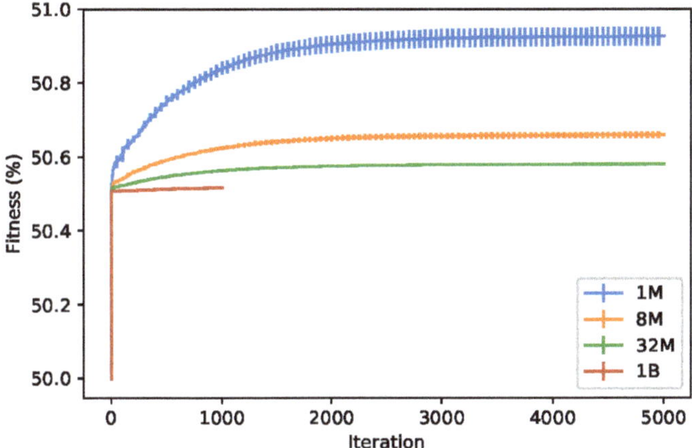

Figure A1. Best fitness value per iteration (mean ± std. dev. across 10 runs per algorithm) for the cGA on the discrete OneMax problem in four different dimensionalities.

Appendix A.2. Continuous OneMax

We also implemented a GPU-enabled continuous cGA, based on the original algorithm proposed in [11], with some modifications. As a benchmark, we considered the OneMax problem where all variables can take real values in $[0, 1]$, instead of binary values. The main difference with respect to the binary and discrete cGAs is the sampling procedure and the update of `PV`. As briefly mentioned in Section 1, the continuous cGA uses for each variable a truncated Gaussian PDF, and to sample new value it calculates the inverse of the corresponding CDF. Therefore, the probability vector in this case consists of 2 vectors which describe the mean μ and standard deviation σ of the Gaussian PDFs (see [7] for details). As for the parallelization process, we implemented a scheme similar to the binary cGA-Base, using a one-to-one mapping between variables and GPU threads, both for the operations in common with the cGA-Base and for the functions described below.

- Algorithm: The main modifications we added to the original cGA scheme illustrated in Algorithm 1 include mutation and crossover operators, inspired by Differential Evolution (DE) [13], and an adaptive restart mechanism. All these mechanisms are problem-independent.

    - Mutation: The value of each variable is obtained by sampling three values from the relative Gaussian PDF, and then combining them as in the rand/1 DE [13]:

$$\texttt{x[i]} = \text{sample}(\mu\texttt{[i]},\sigma\texttt{[i]}) + F \times (\text{sample}(\mu\texttt{[i]},\sigma\texttt{[i]}) - \text{sample}(\mu\texttt{[i]},\sigma\texttt{[i]})) \quad \text{(A1)}$$

    where F is a parameter, and sample() is the procedure to sample from the Gaussian PDF.

  - Crossover: We implemented two different strategies, based on the binomial and exponential crossover used in DE. Both are based on a parameter $CR \in (0,1)$, but their behavior is different. In the binomial crossover, each variable in a trial copies with probability CR the corresponding variable from the elite. In the exponential crossover, starting from a random position, the variables of the trial are copied from the elite, until a random number $p \in [0,1)$ is greater than CR.
  - Restart: The adaptive restart mechanism partially restarts the evolution by resetting the σ values trough two parameters. The first parameter, `invariant`, controls when to apply the restart, which can occur for two reasons: either if for `invariant` consecutive iterations the trial does not improve the elite, or if the trial improves it but the improvement (i.e., the difference between its fitness and that of the elite) is less than 1.0. The second parameter regulates the portion of variables involved, i.e., each variable as a probability `resetPR` to be reset. Finally, the new value of σ after each restart changes dynamically during the evolution process: in particular, it starts from 10, it is doubled every time a restart occurs and it is halved every time there are `invariant` iterations without a restart.

- Results: We tested separately the different behavior of the two crossover strategies, the impact of three different algorithm configurations (base, i.e., without DE-mutation and crossover; with mutation and crossover; and with mutation, crossover and restart), and how the algorithm scales. For this analysis, we used the parameters shown in Table A2. All the experiments were executed on the Google®Colab service.
  - Crossover strategies: We tested the two crossover strategies over 5000 iterations on the continuous OneMax problem with 1M variables. As shown in Figure A2, the two crossover strategies show a similar behavior, although the exponential crossover tends to reach a plateau earlier than the binomial crossover.
  - Algorithm configurations: In this case as well, the different configurations were tested over 5000 iterations on the continuous OneMax problem with 1M variables, using the binomial crossover. As shown in Figure A3, the restart greatly enhances the algorithm's performance, avoiding premature convergence and also increasing the fitness achieved. It is important to consider that the parameters involved in the restart procedure must be chosen carefully in order to avoid making the search ineffective.
  - Scalability: As shown in Figure A4, the results are similar, in terms of behavior, to the binary and discrete cases, shown earlier in Figures 1 and A1. However, the final fitness values are slightly worse. This is mainly due to two reasons: firstly, the continuous domain leads to smaller fitness increases; and, secondly, the chosen restart parameters seem to work well in the 1M case, but not on larger dimensionalities. It is also of note that the total execution time results from two to four times bigger than the equivalent binary and discrete cases.

Table A2. Parameter settings for the cGA applied to the continuous OneMax problem.

Parameter	Value
`virtualPopulation`	100
F	0.7
CR (binomial)	0.5
CR (exponential)	0.9
`invariant`	300
`resetPR`	0.5

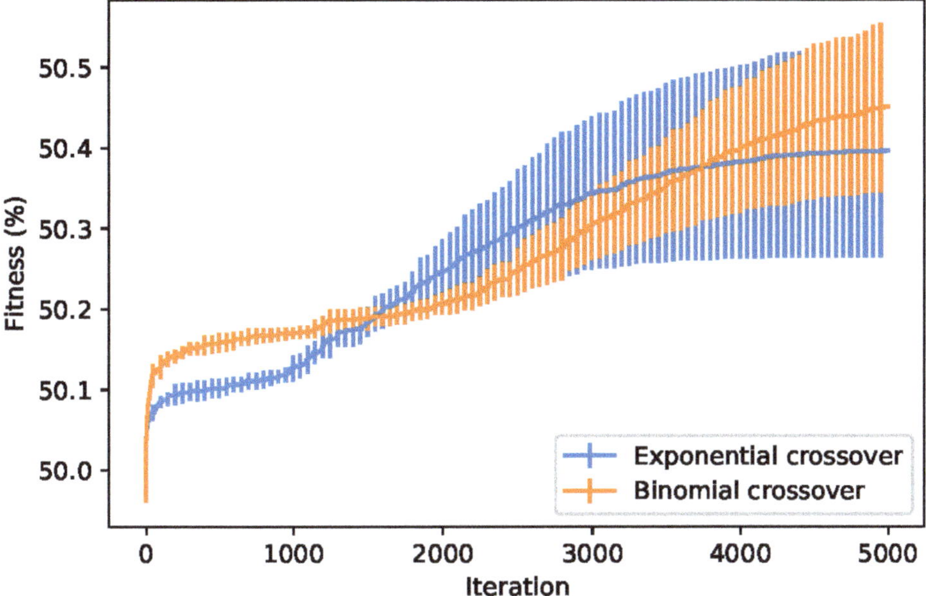

Figure A2. Best fitness value per iteration (mean ± std. dev. across 10 runs) for the cGA with two different crossover operators on the continuous OneMax problem in 1M dimensions.

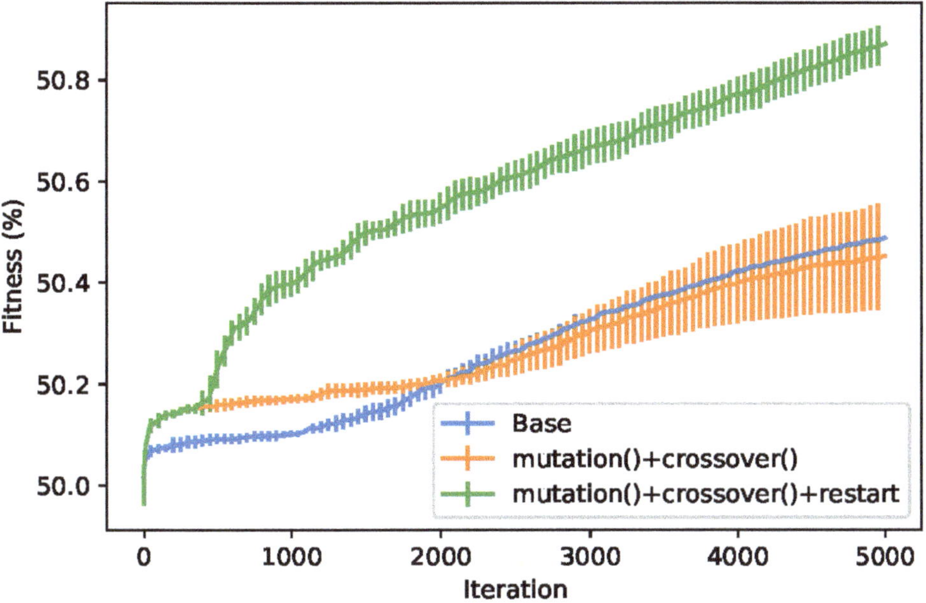

Figure A3. Best fitness value (mean ± std. dev. across 10 runs) for the cGA with different configurations on the continuous OneMax problem in 1M dimensions.

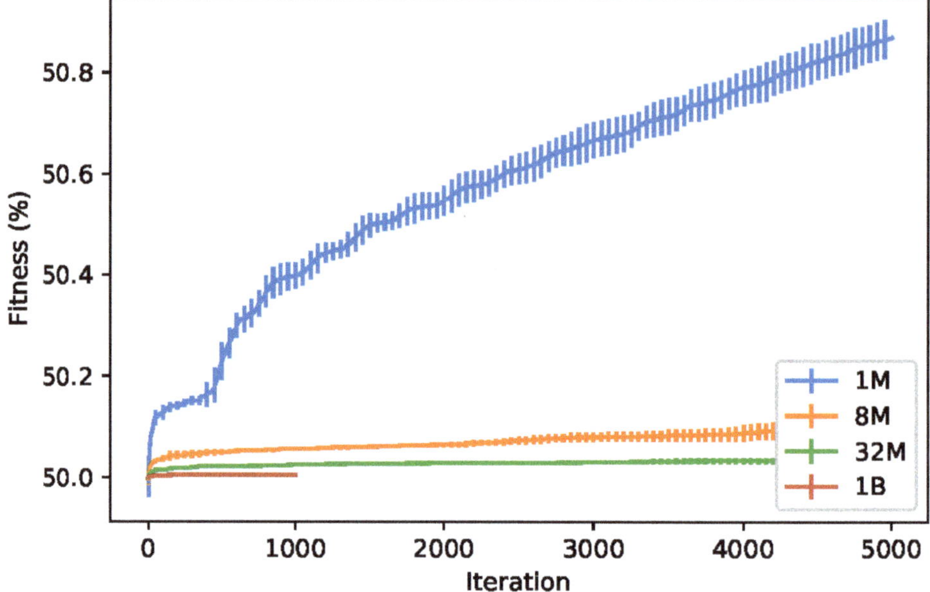

Figure A4. Best fitness value per iteration (mean ± std. dev. across 10 runs per algorithm) for the cGA on the continuous OneMax problem in four different dimensionalities.

References

1. CPLEX IBM ILOG, User's Manual for CPLEX. Available online: https://www.ibm.com/support/knowledgecenter/SSSA5P_12.7.1/ilog.odms.cplex.help/CPLEX/homepages/usrmancplex.html (accessed on 1 April 2020).
2. Gurobi Optimization Inc. *Gurobi Optimizer Reference Manual*. Available online: https://www.gurobi.com/wp-content/plugins/hd_documentations/documentation/9.0/refman.pdf (accessed on 1 April 2020).
3. Makhorin, A. GLPK (GNU Linear Programming Kit). Available online: https://www.gnu.org/software/glpk/ (accessed on 1 April 2020).
4. Deb, K.; Myburgh, C. A population-based fast algorithm for a billion-dimensional resource allocation problem with integer variables. *Eur. J. Oper. Res.* **2017**, *261*, 460–474. [CrossRef]
5. Omidvar, M.N.; Li, X. Evolutionary large-scale global optimization: An introduction. In Proceedings of the Genetic and Evolutionary Computation Conference Companion, Berlin, Germany, 15–19 July 2017; pp. 807–827.
6. Deb, K.; Myburgh, C. Breaking the billion-variable barrier in real-world optimization using a customized evolutionary algorithm. In Proceedings of the Genetic and Evolutionary Computation Conference, Denver, CO, USA, 20–24 July 2016; pp. 653–660.
7. Neri, F.; Iacca, G.; Mininno, E. Compact Optimization. In *Handbook of Optimization: From Classical to Modern Approach*; Springer: New York, NY, USA, 2013; pp. 337–364.
8. Lozano, J.A.; Larra naga, P.; Inza, I.; Bengoetxea, E. *Towards a New Evolutionary Computation: Advances on Estimation of Distribution Algorithms*; Springer: New York, NY, USA, 2006; Volume 192.
9. Harik, G.R.; Lobo, F.G.; Goldberg, D.E. The compact genetic algorithm. *IEEE Trans. Evol. Comput.* **1999**, *3*, 287–297. [CrossRef]
10. Ahn, C.W.; Ramakrishna, R.S. Elitism-based compact genetic algorithms. *IEEE Trans. Evol. Comput.* **2003**, *7*, 367–385.
11. Mininno, E.; Cupertino, F.; Naso, D. Real-Valued Compact Genetic Algorithms for Embedded Microcontroller Optimization. *IEEE Trans. Evol. Comput.* **2008**, *12*, 203–219. [CrossRef]

12. Corno, F.; Reorda, M.S.; Squillero, G. The Selfish Gene Algorithm: A New Evolutionary Optimization Strategy. In *ACM Symposium on Applied Computing*; ACM: New York, NY, USA, 1998; pp. 349–355.
13. Mininno, E.; Neri, F.; Cupertino, F.; Naso, D. Compact differential evolution. *IEEE Trans. Evol. Comput.* **2011**, *15*, 32–54. [CrossRef]
14. Iacca, G.; Caraffini, F.; Neri, F. Compact Differential Evolution Light: High Performance Despite Limited Memory Requirement and Modest Computational Overhead. *J. Comput. Sci. Technol.* **2012**, *27*, 1056–1076. [CrossRef]
15. Mallipeddi, R.; Iacca, G.; Suganthan, P.N.; Neri, F.; Mininno, E. Ensemble strategies in Compact Differential Evolution. In Proceedings of the Congress on Evolutionary Computation, New Orleans, LA, USA, 5–8 June 2011; pp. 1972–1977.
16. Iacca, G.; Mallipeddi, R.; Mininno, E.; Neri, F.; Suganthan, P.N. Super-fit and population size reduction in compact Differential Evolution. In Proceedings of the Workshop on Memetic Computing, Paris, France, 11–15 April 2011; pp. 1–8.
17. Iacca, G.; Mallipeddi, R.; Mininno, E.; Neri, F.; Suganthan, P.N. Global supervision for compact Differential Evolution. In Proceedings of the Symposium on Differential Evolution, Paris, France, 11–15 April 2011; pp. 1–8.
18. Iacca, G.; Neri, F.; Mininno, E. Opposition-Based Learning in Compact Differential Evolution. In Proceedings of the Conference on the Applications of Evolutionary Computation, Dublin, Ireland, 27–29 April 2011; pp. 264–273.
19. Iacca, G.; Mininno, E.; Neri, F. Composed compact differential evolution. *Evol. Intell.* **2011**, *4*, 17–29. [CrossRef]
20. Neri, F.; Iacca, G.; Mininno, E. Disturbed Exploitation compact Differential Evolution for limited memory optimization problems. *Inf. Sci.* **2011**, *181*, 2469–2487. [CrossRef]
21. Iacca, G.; Neri, F.; Mininno, E. Noise analysis compact differential evolution. *Int. J. Syst. Sci.* **2012**, *43*, 1248–1267. [CrossRef]
22. Neri, F.; Mininno, E.; Iacca, G. Compact Particle Swarm Optimization. *Inf. Sci.* **2013**, *239*, 96–121. [CrossRef]
23. Iacca, G.; Neri, F.; Mininno, E. Compact Bacterial Foraging Optimization. In *Swarm and Evolutionary Computation*; Springer: Berlin/Heidelberg, Germany, 2012; pp. 84–92.
24. Yang, Z.; Li, K.; Guo, Y. A new compact teaching-learning-based optimization method. In *International Conference on Intelligent Computing*; Springer: Cham, Switzerland, 2014; pp. 717–726.
25. Dao, T.K.; Pan, T.S.; Nguyen, T.T.; Chu, S.C.; Pan, J.S. A Compact Flower Pollination Algorithm Optimization. In Proceedings of the International Conference on Computing Measurement Control and Sensor Network, Matsue, Japan, 20–22 May 2016; pp. 76–79.
26. Tighzert, L.; Fonlupt, C.; Mendil, B. A set of new compact firefly algorithms. *Swarm Evol. Comput.* **2018**, *40*, 92–115. [CrossRef]
27. Dao, T.K.; Chu, S.C.; Shieh, C.S.; Horng, M.F. Compact artificial bee colony. In Proceedings of the International Conference on Industrial, Engineering and Other Applications of Applied Intelligent Systems, Kaohsiung, Taiwan, 3–6 June 2014; pp. 96–105.
28. Banitalebi, A.; Aziz, M.I.A.; Bahar, A.; Aziz, Z.A. Enhanced compact artificial bee colony. *Inf. Sci.* **2015**, *298*, 491–511. [CrossRef]
29. Jewajinda, Y. Covariance matrix compact differential evolution for embedded intelligence. In Proceedings of the IEEE Region 10 Symposium, Bali, Indonesia, 9–11 May 2016; pp. 349–354.
30. Neri, F.; Mininno, E. Memetic compact differential evolution for Cartesian robot control. *IEEE Comput. Intell. Mag.* **2010**, *5*, 54–65. [CrossRef]
31. Iacca, G.; Caraffini, F.; Neri, F. Memory-saving memetic computing for path-following mobile robots. *Appl. Soft Comput.* **2013**, *13*, 2003–2016. [CrossRef]
32. Iacca, G.; Caraffini, F.; Neri, F.; Mininno, E. Robot Base Disturbance Optimization with Compact Differential Evolution Light. In *Conference on the Applications of Evolutionary Computation*; Springer: Berlin/Heidelberg, Germany, 2012; pp. 285–294.
33. Gallagher, J.C.; Vigraham, S.; Kramer, G. A family of compact genetic algorithms for intrinsic evolvable hardware. *IEEE Trans. Evol. Comput.* **2004**, *8*, 111–126. [CrossRef]
34. Yang, Z.; Li, K.; Guo, Y.; Ma, H.; Zheng, M. Compact real-valued teaching-learning based optimization with the applications to neural network training. *Knowl. -Based Syst.* **2018**, *159*, 51–62. [CrossRef]

35. Dao, T.K.; Pan, T.S.; Nguyen, T.T.; Chu, S.C. A compact artificial bee colony optimization for topology control scheme in wireless sensor networks. *J. Inf. Hiding Multimed. Signal Process.* **2015**, *6*, 297–310.
36. Prugel-Bennett, A. Benefits of a Population: Five Mechanisms That Advantage Population-Based Algorithms. *IEEE Trans. Evol. Comput.* **2010**, *14*, 500–517. [CrossRef]
37. Caraffini, F.; Neri, F.; Passow, B.N.; Iacca, G. Re-sampled inheritance search: High performance despite the simplicity. *Soft Comput.* **2013**, *17*, 2235–2256. [CrossRef]
38. Iacca, G.; Caraffini, F. Compact Optimization Algorithms with Re-Sampled Inheritance. In *Conference on the Applications of Evolutionary Computation*; Springer: Cham, Switzerland, 2019; pp. 523–534.
39. Hansen, N.; Müller, S.D.; Koumoutsakos, P. Reducing the time complexity of the derandomized evolution strategy with covariance matrix adaptation (CMA-ES). *Evol. Comput.* **2003**, *11*, 1–18. [CrossRef]
40. Loshchilov, I.; Glasmachers, T.; Beyer, H.G. Large scale black-box optimization by limited-memory matrix adaptation. *IEEE Trans. Evol. Computat.* **2018**, *23*, 353–358. [CrossRef]
41. Halman, N.; Kellerer, H.; Strusevich, V.A. Approximation schemes for non-separable non-linear boolean programming problems under nested knapsack constraints. *Eur. J. Oper. Res.* **2018**, *270*, 435–447. [CrossRef]
42. Caraffini, F.; Neri, F.; Iacca, G. Large scale problems in practice: the effect of dimensionality on the interaction among variables. In *Conference on the Applications of Evolutionary Computation*; Springer: Cham, Switzerland, 2017; pp. 636–652.
43. Schaffer, J.; Eshelman, L. On crossover as an evolutionarily viable strategy. In Proceedings of the International Conference on Genetic Algorithms, San Diego, CA, USA, 13–16 July 1991; pp. 61–68.
44. Goldberg, D.E.; Sastry, K.; Llorà, X. Toward routine billion-variable optimization using genetic algorithms. *Complexity* **2007**, *12*, 27–29. [CrossRef]
45. Wang, Z.; Hutter, F.; Zoghi, M.; Matheson, D.; de Feitas, N. Bayesian optimization in a billion dimensions via random embeddings. *J. Artif. Intell. Res.* **2016**, *55*, 361–387. [CrossRef]
46. Iturriaga, S.; Nesmachnow, S. Solving very large optimization problems (up to one billion variables) with a parallel evolutionary algorithm in CPU and GPU. In Proceedings of the IEEE International Conference on P2P, Parallel, Grid, Cloud and Internet Computing, Victoria, BC, Canada, 12–14 November 2012; pp. 267–272.
47. Xinchao, Z. Simulated annealing algorithm with adaptive neighborhood. *Appl. Soft Comput.* **2011**, *11*, 1827–1836. [CrossRef]
48. Iacca, G. Distributed optimization in wireless sensor networks: an island-model framework. *Soft Comput.* **2013**, *17*, 2257–2277. [CrossRef]
49. Whitley, D. Next generation genetic algorithms: A user's guide and tutorial. In *Handb. Metaheuristics*; Springer: Cham, Switzerland, 2019; pp. 245–274.
50. Varadarajan, S.; Whitley, D. The massively parallel mixing genetic algorithm for the traveling salesman problem. In Proceedings of the Genetic and Evolutionary Computation Conference, Prague, Czech Republic, 13–17 July 2019; pp. 872–879.

 *mathematics*

Article

The SOS Platform: Designing, Tuning and Statistically Benchmarking Optimisation Algorithms

Fabio Caraffini [1,*] and Giovanni Iacca [2]

[1] Institute of Artificial Intelligence, School of Computer Science and Informatics, De Montfort University, Leicester LE1 9BH, UK

[2] Department of Information Engineering and Computer Science, University of Trento, 38123 Trento, Italy; giovanni.iacca@unitn.it

* Correspondence: fabio.caraffini@dmu.ac.uk

Received: 31 March 2020; Accepted: 5 May 2020; Published: 13 May 2020

Abstract: We present Stochastic Optimisation Software (SOS), a Java platform facilitating the algorithmic design process and the evaluation of metaheuristic optimisation algorithms. SOS reduces the burden of coding miscellaneous methods for dealing with several bothersome and time-demanding tasks such as parameter tuning, implementation of comparison algorithms and testbed problems, collecting and processing data to display results, measuring algorithmic overhead, etc. SOS provides numerous off-the-shelf methods including: (1) customised implementations of statistical tests, such as the Wilcoxon rank-sum test and the Holm–Bonferroni procedure, for comparing the performances of optimisation algorithms and automatically generating result tables in PDF and LaTeX formats; (2) the implementation of an original advanced statistical routine for accurately comparing couples of stochastic optimisation algorithms; (3) the implementation of a novel testbed suite for continuous optimisation, derived from the IEEE CEC 2014 benchmark, allowing for controlled activation of the rotation on each testbed function. Moreover, we briefly comment on the current state of the literature in stochastic optimisation and highlight similarities shared by modern metaheuristics inspired by nature. We argue that the vast majority of these algorithms are simply a reformulation of the same methods and that metaheuristics for optimisation should be simply treated as stochastic processes with less emphasis on the inspiring metaphor behind them.

Keywords: algorithmic design; metaheuristic optimisation; evolutionary computation; swarm intelligence; memetic computing; parameter tuning; fitness trend; Wilcoxon rank-sum; Holm–Bonferroni; benchmark suite

1. Introduction

Experimentalism plays a major role in all fields of metaheuristic optimisation, such as evolutionary computation (EC) and swarm intelligence (SI). It can be a time-demanding, repetitive and tedious process to fine-tune optimisers, compare them or validate new algorithmic strategies on benchmark problems. Due to their stochastic nature, and the lack of theoretical knowledge on the internal dynamics of many of these nature-inspired approaches, their algorithmic design is partially blind and often empirically performed through a series of trial-and-error phases [1]. This involves using several artificially built benchmark problems simulating characteristics present in real-world scenarios, but for which some knowledge is available. The use of such testbed problems is preferable since they usually have a more reasonable execution time than real-world applications, they can be run on any machine, this being e.g., a modest personal computer or a high-performance computing system, and they are not subject to time and other limitations typical of real-world scenarios, unless these are purposely imposed to reproduce a specific situation. Thus, these problems are suitable choices for performing the algorithmic design phase of novel algorithms and evaluating the performances of specific operators and algorithmic variants. By design, these problems display particular characteristics reported in the corresponding technical reports in term of, e.g., ill-conditioning, separability, m-separability, noise, modality (i.e., multimodal or unimodal function), etc., allowing also for studying the algorithmic behaviour of certain operators [2–4].

In this light, the research community benefits from platforms featuring a large number of algorithms and benchmark problems, such as the Decision Tree for Optimisation software [5], the COCO platform [6], IOHprofiler [7], jMetal [8] or PlatEMO [9]. However, the availability of code implementing complete benchmark suites and popular optimisation algorithms is only one amongst the many requirements an algorithm designer has to work efficiently and focus algorithmic issues rather than having to deal with the implementation of ancillary software. The other functionalities that a research-oriented platform for metaheuristic optimisation should have are:

- performance evaluation methods, based on predefined metrics such as algorithmic overhead, scalability, best results, average best result, convergence (e.g., in terms of fitness and population diversity [10,11]), structural bias [1,12], etc.;
- statistical methods for comparing the performances of algorithms [13,14];
- result visualisation methods (in both tabular and graphical formats).

On top of that, such a platform should provide:

- a vast and heterogeneous number of benchmark functions and real-world testbed problems;
- several state-of-the-art algorithms to be used for comparisons with newly-designed algorithms;
- libraries with implementations of the most popular algorithmic operators such as mutation methods, crossover methods, parent and survivor selection mechanisms, heuristics selection methods for memetic computing (MC) and hyperheuristic approaches, etc.;
- utility methods for generating random numbers, new solutions, keeping track of the best solutions and saving them to plot fitness trends and handling infeasible solutions [1,15,16];
- a system to spread the runs over multiple cores and threads, thus speeding up the data generation process, and automatically collecting and processing raw data into meaningful information.

Additionally, the platform should be based on:

- open source software, so that researchers can freely use it and build upon it;
- a simple and flexible structure so that new problems and algorithms can be easily added.

The Stochastic Optimisation Software, henceforth SOS, exhibits the aforementioned key features, thus facilitating the design, the evaluation and the use of metaheuristic optimisation algorithms. Thanks to its simple and extendible structure, SOS can be easily used by researchers to implement new algorithms and compare their performances with those of several other benchmark algorithms. The most recent SOS release, available in the Zenodo repository [17], contains over fifty ready-to-use algorithms amongst single-solution and population-based metaheuristics, as those described in [18], all belonging to established optimisation paradigms such as evolutionary algorithms (EA) [19], swarm intelligence (SI) optimisation [20] and other hybrid schemes. Hence, it provides users with a variety of techniques to explore multiple flavours of metaheuristic/stochastic optimisation. More algorithms will be added in future releases. Researchers can benefit in several ways from having such a large availability of algorithms. Firstly, this means that comparison algorithms do not need to be implemented as they are already present and ready to be executed. Secondly, their source code is accessible and can be modified to create new variants or simply visioned as a source of inspiration for implementing other algorithms or algorithmic operators. Thirdly, they can combine the existing algorithms or embed them into new algorithms, thus creating novel hybrid methods as done, e.g., in memetic computing and hyperheuristics [3,21–24]. It must be highlighted that implementing algorithms in SOS is quite simple since most algorithmic operators are provided by the platform. Moreover, the "template" class `Algorithm`, which is equipped with several ancillary methods, can simply be extended to implement any metaheuristic, regardless of the optimisation family. In this light, SOS also represents an excellent pedagogical tool for teaching purposes and can be used to give guidance to students while implementing optimisation algorithms. Indeed, since 2015 the "student" version of SOS has been used to teach the master module "Computational Intelligence Optimisation" led by Dr Fabio Caraffini at De Montfort University (Leicester, UK), for which it was originally designed before being extended to the current form. Researchers willing to collaborate and add code (algorithms, real-world problems, etc.) to the platform are welcome and invited to get in touch to be added to the SOS GitHub repository (whose link is provided in Table 1). Finally, it is worth mentioning that due to the simplicity in adding problems and algorithms, SOS represents a useful tool for practitioners who might have less interest in the algorithmic design but a high need for off-the-shelf implementations of optimisation algorithms. The remainder of this article is structured as follow:

- Section 2 presents the SOS software platform and provides detailed descriptions of its features;
- Section 3 focuses on benchmarking optimisation algorithms with SOS and lists the available benchmark suites, including established benchmarks and one customised benchmark;
- Section 4 first provides a literature review on statistical methods for comparing stochastic optimisation algorithms, then describes the implementation and working mechanisms of three statistical analyses available in SOS;
- Section 5 reports on the result visualisation capabilities of SOS and provides "how-to" examples based on studies from the current literature in metaheuristic optimisation;
- Section 6 summarises the key points of this article and remarks on the novel aspects of SOS;
- Appendix A gives additional details on the routine used to produce average fitness trend graphs.

Table 1. Relevant links to source code and software documentation.

Description	Content	Link
SOS web page	software documentation	https://tinyurl.com/FabioCaraffini-SOS
GitHub repository	source code	https://github.com/facaraff/SOS
Zenodo repository [17]	SOS releases	https://doi.org/10.5281/ZENODO.3237023

2. The SOS Platform

For portability reasons, SOS is coded in Java, thus being platform-independent. To speed up the execution of extensive experiments, some benchmark suites are also available in the form of C/C++ native libraries (compiled for MS Windows, MAC OS and Linux machines) using the Java Native Interface (JNI). By default, SOS spreads the "runs" (i.e., the optimisation processes in which one algorithm optimises one problem) over the available threads and executes them in parallel. In order to avoid that, e.g., if the experiment is executed on a laptop or personal computer and CPU power must be saved of other applications, it is possible to switch to the single-thread mode by means of the setter method `setMT(boolean MT)` of the class `Experiment` (i.e., its Boolean flag variable must be set to `false`). By looking at the code available in [17], one can notice that in the current release, a strong emphasis is given to real-valued box-constrained single-objective optimisation, also known as "single-objective continuous optimisation" in the EC community. Nevertheless, SOS can be used for addressing other optimisation domains such as discrete or combinatorial ones.

2.1. Functional Overview

SOS is research-oriented and is designed to have an intuitive structure making it straightforward to use for both expert algorithm designers in the research community and, most importantly, for practitioners and students from different subject fields. Indeed, to execute an experiment, it is sufficient to add a class, which extends the abstract `Experiment` class, in the homonym folder (package). Such a class will automatically inherit the `add` methods, which can be used to add algorithms and problems to the experiment, and other methods to set, e.g., the allowed computational budget (if not set, by default, this is assumed to be equal to 5000 × n with n being the dimensionality of the problem) and the number of runs to be performed. Subsequently, a `main` method must be written to launch one or multiple experiments. To do this, the methods of the utility class `RunAndStore` can be imported to, e.g.,:

- execute a list of experiments having different budgets, number of runs, problems and algorithms, or the same experiments repeated for different dimensionalities (if the problems are scalable);
- generate log files describing the performed experiments (i.e., the list of problems and their corresponding details, e.g., dimensionality, as well as the parameter setting, number of runs, etc.);
- store raw data and process them into results tables and plottable formats, as shown in Section 5.

It must be said that the computational budget can be arbitrarily allocated for each experiment using the provided setter methods (described in the SOS documentation). These can be used to either set a multiplicative factor (which multiplies n, thus adjusting the budget of each function in the experiment based on its dimensionality) or set a fixed amount of functional calls. This poses limitations on the maximum number of allowed functional evaluations, but does not prevent the algorithms from stopping the search earlier. Indeed, each algorithm can be equipped with a customised "stop" criterion based, e.g., on the number of improvements on the fitness values, the use of thresholds specifying a desired fitness tolerance, an incremental improvement value, etc.

Figure 1 shows an example of a main method used for executing a list of experiments and displaying the final outputs, i.e., tables and graphs, generated by SOS after processing the results. With reference to the latest release in [17], a large number of experiment files can be found in the `experiments` package, while several examples of main files (showing different configurations and options for storing and processing results in different formats) are available in the `mains` package. Among these examples, the most convenient one to be used as a template is the class `RunExperiments`, shown in Figure 1, which is located in the default package of the SOS platform. More indications on how to write an experiment class are given in Section 2.2 and graphically shown in Figure 2.

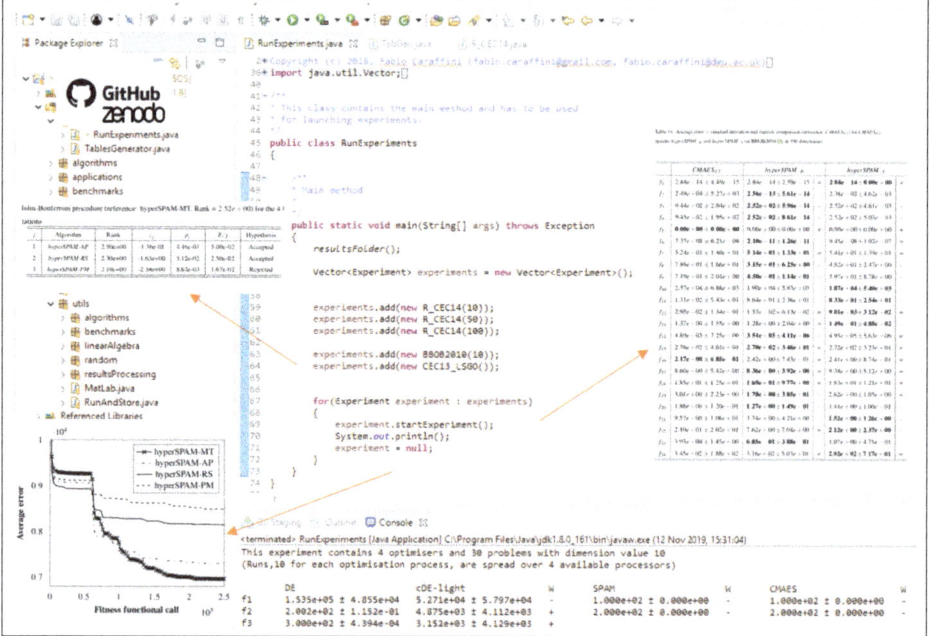

Figure 1. A graphical functional overview of the Stochastic Optimisation Software (SOS) platform.

To minimise the need for referring to a detailed document, SOS is equipped with a high number of self-explanatory example classes. Moreover, the clear nomenclature used for methods, variable types and class names and the user-friendly structure of the proposed platform give guidance to the users, who may not even need further indications. Nonetheless, the software documentation is available online to provide users with further information on SOS packages, classes and methods. Table 1 displays relevant links to SOS web pages, from which SOS documentation can be accessed. These web pages are kept up-to-date, and the documentation itself is constantly updated to reflect any major changes of the platform. This documentation plays a key role for those who intend to modify SOS source code, extend it or customise it to a specific need.

SOS makes it very easy to add new algorithms and new problems that are fully compatible with the platform, so as to exploit its full capabilities. For the sake of good organisation, it is suggested to place their implementation into the three provided packages: `algorithms`, `benchmarks` and `applications`.

The `algorithms` folder currently contains a vast list of optimisation algorithms for continuous optimisation, which, in may cases, can be run under different combinations of variation and/or selection operators, thus providing an even higher number of choices. This results in a good balance between established optimisation methods, such as differential evolution (DE) [1,25], simulated annealing (SA) [26], genetic algorithms (GAs), particle swarm optimisation (PSO) [27] and evolution strategies (ES) [28–30], and more modern optimisers such as single particle optimisation [31], self-adaptive algorithms [32–38] and MC/hyperheuristic approaches [3,23,24,39–41]. For a complete list, one can refer to the `algorithms` folder from [17] or check the online documentation. More information on the SOS algorithms is given in Section 2.3.

The `benchmarks` folder contains the implementation of problems taken from some established benchmark suites for optimisation. SOS implements several testbed problems and complete benchmarks suites that can be added in the experiment files. A complete list of these optimisation problems and their relevant documentation is provided in Section 3. Furthermore, a novel benchmark suite is presented in Section 3.

The `applications` folder is where specific real-world applications should be implemented. Some examples from the literature, e.g., the code for the application described in [23], and from a benchmark suite for real-world problems [42] are available in this folder.

Following the execution of an experiment, the produced raw data can be processed by means of the classes in the package `utils.resultsProcessing`. These provide several functionalities, including:

- the execution of statistical tests for comparing optimisation algorithms, described in Section 4, including the advanced statistical analysis (ASA) presented in Section 4.3;
- the generation of plottable files suitable for Tikz, Matplotlib, Gnuplot or MATLAB scripts for displaying best/median/worse/average fitness trends, as shown in Section 5 and Appendix A;
- the generations of files containing further plottable information, such as histograms relative to the distribution of the final best fitness values across several runs and graphs describing infeasibility and structural bias in optimisation algorithms (see [1,16,43–46] for details);
- the generation of LaTeX tables (both source code and PDF files are produced) showing results in terms the average fitness value (or average error w.r.t. the known optimum, if available) with the corresponding standard deviation and further statistical evidence, as explained in Section 4 and graphically shown in Section 5.

The file `TabGen.java`, located in the `Mains` folder, contains the methods for generating the previously mentioned tables and graphical outputs and can be modified to customise the production of tables with different layouts and informative content, as discussed in Section 5.

It is worth stressing the fact that having a fast way to visualise results in multiple formats is a key element to facilitate research in the area of stochastic optimisation. This feature is indeed very useful since the raw data are automatically processed and the LaTeX source of the comparative tables is generated without any user intervention, together with the corresponding PDF files.

Moreover, this also accelerates the algorithmic design phase, which is often performed empirically through several trial-and-error steps. At each step, different variants of the same algorithm must be compared and tables must be generated. This is quite common in the EC filed where using a different combination of variation operators [19] with different combinations of parents and survivors selection schemes [19] can lead to a variety of possible algorithmic behaviours. The same issue arises when designing MC and hyperheuristic algorithms, for which several coordination schemes must be tested to ensure high performances [3,21,22,24]. This aspect suggests that the empirical design approach can only produce algorithms tailored to the problem(s) considered during the design phase. To some extent, this issue can be considered analogous to the training process in machine learning.

In this light, the algorithmic design process is not so different from the process performed to choose the less disruptive correction method for handling infeasible solutions generated while addressing a problem [1,16,43] or from the fine-tuning process performed to find the most appropriate set of parameters for an algorithm A meant to address specific problem P.

2.2. Parameters' Fine-Tuning

Metaheuristics for optimisation must be tuned on the problem at hand to make them achieve their full potential and return high-quality solutions. Against the original common belief that a "universal" algorithm could have been designed, theoretical achievements such as, first and foremost, the "no free lunch theorems" (NFLTs) for optimisation [47,48] highlighted the need for fine-tuning, self-adaptation and the use of problem-specific operators.

However, finding the optimal parameter configuration is an optimisation problem itself. Despite meta-optimisation strategies having been proposed [49,50], these do not completely solve the problem and are arguable, since they introduce further complexity due to the fact that also the meta-optimiser might need to be fine-tuned. Furthermore, the fact that most real-world optimisation problems are black-box systems, i.e., little or no analytical information on the fitness function is available, increases the difficulty in finding an exact method for finding the optimal configuration of parameters.

Under these circumstances, the parameters' tuning process is necessarily performed empirically, thus resulting in a rather time-consuming and tiresome activity. SOS can be used to accelerate and facilitate this task significantly as: (1) the same algorithm can be included multiple times in the same experiment file with different parameter configurations; (2) runs can be spread over the available processors and threads to speed up the generation of results; (3) raw data are statistically analysed and results automatically displayed in compact, yet highly informative tables and graphs (see the examples provided in Section 5). Hence, the only portion of code that one needs to write to apply and tune an algorithm A on a specific real-world problem P with SOS is actually the code implementing P.

It must be pointed out that also artificially built testbed problems, as those already implemented in SOS (see Section 3), can play a role in the parameter tuning process. Indeed, studying how the algorithmic behaviour of a metaheuristic algorithm changes under different configurations of its parameters on these benchmarks can help shed light on how to perform the tuning process. By using testbed problems, whose mathematical properties (e.g., differentiability, separability, modality, ill-conditioning, etc.) are known, one can understand, e.g., the effect of the parameters in relation to these mathematical features. As a result, specific regions of the parameter space could be detected to obtain a robust general-purpose behaviour rather than a very problem-specific one, or to strengthen exploitation capabilities rather than exploration capabilities, or to minimise the rise of algorithmic structural biases, as done in [1,12,43], the generation of a high number of infeasible solutions, as done in [1,16], or premature convergence [51] and stagnation [52], as done in [11].

Most of the aforementioned studies have been performed with SOS, and their experiment files can be found in [17]. Portions of the code from three experiment files are reported in Figure 2, which helps understand how the parameter space of an algorithm can be explored in SOS.

```java
super(probDim,5000,"ExperimentDE");
setNrRuns(30);

Algorithm a;
Problem p;

a = new DE("ro",'b');
a.setID("DEr1bin");
a.setParameter("p0", (double)probDim);//Population size
a.setParameter("p1", 0.7);//F
a.setParameter("p2", -1.0); //CR
a.setParameter("p3", 0.3);//Alpha
add(a);

a = new DE("ro",'e');
a.setID("DEr1exp");
a.setParameter("p0", (double)probDim);//Population size
a.setParameter("p1", 0.7);//F
a.setParameter("p2", -1.0); //CR
a.setParameter("p3", 0.3);//Alpha
add(a);

a = new DE("ctro");
a.setID("DEctr1");
a.setParameter("p0", (double)probDim);//Population size
a.setParameter("p1", 0.7); //F
a.setParameter("p2", Double.NaN);//CR
a.setParameter("p3", Double.NaN);//Alpha
add(a);

a = new DE("ro",'x');
a.setID("DEr1noxo");
a.setParameter("p0", (double)probDim);//Population size
a.setParameter("p1", 0.7);//F
a.setParameter("p2", Double.NaN); //CR
a.setParameter("p3", Double.NaN);//Alpha
add(a);
```

(a)

```java
a = new DE("ro",'e');
a.setID("rDEr1exp01");
a.setParameter("p0", 10.0); //Population size
a.setParameter("p1", 0.4); //F
a.setParameter("p2", 0.1); //CR
a.setParameter("p3", Double.NaN);//Alpha
add(a);

a = new DE("ro",'e');
a.setID("rDEr1exp05");
a.setParameter("p0", 10.0); //Population size
a.setParameter("p1", 0.4); //F
a.setParameter("p2", 0.5); //CR
a.setParameter("p3", Double.NaN);//Alpha
add(a);

a = new DE("ro",'e');
a.setID("rDEr1exp07");
a.setParameter("p0", 10.0); //Population size
a.setParameter("p1", 0.4);//F
a.setParameter("p2", 0.7);//CR
a.setParameter("p3", Double.NaN);//Alpha
add(a);

a = new DE("ro",'e');
a.setID("rDEr1exp09");
a.setParameter("p0", 10.0); //Population size
a.setParameter("p1", 0.4);//F
a.setParameter("p2", 0.9);//CR
a.setParameter("p3", Double.NaN);//Alpha
add(a);

for(int i = 1; i<=30; i++)
      add(new RCEC2014(probDim, i));
```

(b)

```java
char[] corrections = {'s','t','m','c'};
String[] DEMutations = {"ro","rt","bo","bt","ctbo","rtbt"};
char[] CrossOvers = {'b','e'};
int[] populationSizes = {5,20,100};
double[] FValues = new double[10];
double[] CRValues = new double[5];

for (int i=0; i<5; i++)
    FValues[i] = 0.05+i*(1.95/9.0);
for (int i=0; i<5; i++)
    CRValues[i] = 0.05 +i*((0.94/4.0));
```

→

```java
for (char correction : corrections)
    for (String mutation : DEMutations)
        for(char xover : CrossOvers)
            for(int popSize: populationSizes)
                for (double F : FValues)
                    for (double CR : CRValues)
            {
                a = new DE(mutation,xover);
                a.setCorrection(correction);
                a.setParameter("p0", (double)popSize); //Population size
                a.setParameter("p1", F); //F - scale factor
                a.setParameter("p2", CR); //CR - Crossover Ratio
                a.setParameter("p3", Double.NaN); //Alpha
                test.add(a);
                a = null;
            }
```

(c)

Figure 2. Examples from experiment classes (original files were downloadable from [17]) for parameter tuning and algorithmic behaviour testing, i.e., the same algorithms are included in the same experiment under different parameter configurations. The fragment of code reported in (**a**) refers to an experiment performed for the studies in [4,53] located in `SOS→src→experiments→rotInvStudy→CEC11.java`. The file `RCEC14TuningDEroe.java`, from which the segment of code in (**b**) was taken, can be found in the same folder. In this case, the `DE/rand/1/exp` algorithm executes with three different values for the crossover rate parameter C_r over the 30 rotated versions of the problems of the R-CEC14 (R indicates rotation flag) benchmark suite proposed in this article in Section 3. Finally, the fragment of code in (**c**) shows how to define compactly an experiment running 12 DE variants (each with four correction strategies) with several different parameter configurations, over one problem only, to find the most appropriate triplet ⟨population size, scale factor (F), crossover ratio (C_r)⟩. The complete class file for this example is located in `SOS→src→mains→ExampleTuning.java`.

In Figure 2a, four DE variants are added into an experiment named `ExperimentDE` by using the constructor `super(probDim,5000,‘ExperimentDE’)`. Therefore, all the produced text files containing raw data will be saved in a homonym folder, located inside the SOS default `results` folder. From the constructor method, it can be seen that a maximum computational budget of $5000 \times n$ fitness functional calls, with n being the dimensionality of the problem, is used. Clearly, this experiment file contains scalable testbed problems and not real-world applications since the dimensionality of the problems is not fixed and passed as an argument with the variable `probDim`. The number of performed runs, 30 in this case, is specified with the command `setNrRuns(30)` (SOS performs 100 runs if not specified otherwise). It must be noted that all the other parameters, apart from the population size, are fixed. This means that this experiment will only study the effect of varying population sizes, which are in this case equal to the problem dimension as required for the studies in [4,53], which can be read for further details. To execute this experiment with increasing problem dimensionalities, and therefore, in turn, increasing population sizes, it is sufficient to instantiate objects with increasing `probDim` values by calling the class constructor in the file `RunExperiments.java` (as shown in the example of Figure 1). It is interesting to note that, to avoid confusion, a name is assigned to each DE variant (e.g., `a.setID("DErn1bin")`). The assigned name will show up in the generated tables. If a name is not provided, as in the example of Figure 2c, it will be automatically assigned equal to the class name of the algorithm. Therefore, in this case, the assigned names would start with DE, followed by "-i", where "i" represents a counter incremented every time an algorithm is added to an experiment. In a nutshell, if names were not specified in the example of Figure 2a, tables would display DE-1, DE-2, DE-3 and DE-4, since all the algorithms in this experiment are instances of the DE class. Differently, in Figure 2b, the same DE variant, namely DE/rand/1/exp, is added four times to the experiment, with four different values for the so-called "crossover ratio" parameter C_r. As can be seen at the bottom of the figure, all 30 rotated problems from the benchmark suite proposed in Section 3 are added to this experiment. Hence, its purpose is to understand the impact of the C_r value on DE/rand/1/exp when facing different problems.

Finally, Figure 2c shows a very compact and fast way for configuring a large experiment in SOS, in which 12 DE variants are equipped with four different correction strategies for handling infeasible solutions, for a total of 48 different algorithms to tune. From the left-hand side of the figure, it can be seen that the DE variants are obtained by combining the six mutations with the two crossover operators in the `DEMutations` and `CrossOvers` arrays, respectively. The four correction strategies are in the `corrections` array. For details about the adopted notation for DE variants and correction strategies, one can consult the articles [1,12] and the results in [45]. For each algorithm, the explored parameters space is defined as the Cartesian product of the three sets represented by the `populationSizes` array, containing three population sizes, the `FValues` array, containing 10 equally spaced values for the so-called "scale factor" parameter F in the range $[0.05, 2]$, and the `CRValues` array, containing five equally spaced values for C_r in the range $[0.05, 0.99]$. Hence, each one of the 48 algorithms is tuned over a set of 150 possible combinations of the three parameters. In total, this means that $7200 \times N_r$ optimisation processes are executed, with N_r being the number of runs performed for each problem added to this experiment. The full code for this example, named `ExampleTuning`, is located in the `mains` package. For demonstration purposes, N_r is set equal to 10, and the computation budget is set to $1000 \times n$.

2.3. Adding New Algorithms and Problems

The literature is currently saturated with "novel" nature-inspired optimisation metaphors claiming to mimic collaborative or individual behaviours of animals [54–61], human activities [62–64] or other natural phenomena [65–67]. The contribution made by most of these new optimisation paradigms is arguable, as in general, they are very similar to more established EAs, such as GA, DE or ES, or swarm intelligence algorithms such as PSO. Furthermore, these new metaheuristics are by definition subject to the NFLTs, and as such, it can be shown that, unless they are highly fine-tuned, they cannot outperform classic algorithms over all possible problems. It must be remarked that also amongst the most established optimisation frameworks, some similarities can be detected in support of the thesis that taxonomies based on metaphors inspiring the algorithm design could (and should) be indeed avoided. To provide some examples, it is sufficient to observe that DE mutations, being linear combinations of individuals from the population [1,25], operate in the same way as the arithmetic crossover used in real-valued GAs and other EAs [19]. Similarly, it can be observed that the lack of parent selection in SI frameworks, such as in PSO [27], is simply moved in the perturbation operator, since this requires the definition of the neighbourhood of the candidate solution to be perturbed to find its local best solution.

In this light, a scientific approach to describe a metaheuristic for optimisation is by considering it as a stochastic process returning a near-optimal solution for an optimisation problem. Regardless of the metaphor behind them, all metaheuristic methods are the expression of the same concept and aim at reaching a good balance between exploration of the fitness landscape, looking for promising basins of attraction, and their exploitation, to refine the search and find local optimal solutions (i.e., maxima or minima according to the need). This dynamic process can alternate such phases as appropriate to keep looking for the global optimum, without prematurely converging to a locally optimal solution or stagnating without being able to return any satisfactory near-optimal solution.

Hence, it makes sense to analyse the algorithmic structure of modern nature-inspired optimisation algorithms and extrapolate a common, high level, general skeleton to use as a template for their implementation. This is done in SOS by means of the abstract class `Algorithm`, from which all algorithms inherit ancillary and auxiliary methods (e.g., for algorithm execution, storing the fitness trend, setting or generating the initial guess, etc.) and extend only the parts to be customised to implement the desired optimisation framework (e.g., GA, DE, a hybrid method or a completely new optimisation paradigm). Therefore, any class file extending `Algorithm` is already equipped with all the methods needed to add the algorithm to an experiment file and execute it as described in the previous section, while it must implement one method, named `execute`, which returns an object of the kind `FTrend`. To give further details on this, other than referring to the online software documentation, it is worth briefing about the classes of the package `utils` listed below.

- `RunAndStore`: This class contains the implementation of all methods involved in executing an algorithm in single or multi-thread mode, collecting the results, displaying them on a console (unless differently specified) and saving them into text files.
- `RunAndStore.FTrend`: This class instantiates auxiliary objects storing information collected during the optimisation process and that must be returned by all the algorithms implemented in SOS. As shown in Figure 3b, the primary purpose of these objects is to store the fitness function evaluation counter and the corresponding fitness value in order to be able to retrieve the best (max or min according to the specific optimisation problem) and plot the so-called fitness trend graphs like those in Figure 1 and in Figure 8 of Section 5. The class provides numerous auxiliary methods to return the initial guess, the best fitness value, the median value, and so on. Obviously, the near-optimal solution must be also stored, and if needed, several other extra pieces of information can be added in the form of strings, `integer` or `double` values. As an example, in [11], population and fitness diversity measures were collected during the optimisation process.
- `MatLab`: This class provides several methods to initialise and manipulate matrices quickly. These methods include binary operations on matrices and, most importantly, auxiliary methods for generating matrices of indices, for making copies of other matrices and decomposing them.
- `Counter`: This is an auxiliary class to handle counters when several of them are required to, e.g., count fitness evaluations, count successful and unsuccessfully steps, etc. In [1,16], this class was used to keep track of the number of generated infeasible solutions and of how many times the pseudo-random number generator was called in a single run. It must be remarked that for obtaining this information in a facilitated way, it is necessary to extend the abstract class `algorithmsISB` rather than `algorithms` [17].

Moreover, the packages `utils.algorithms` and `utils.random` provide other useful methods helping users to implement algorithms. In particular, these packages contain:

- utility methods for cloning and generating individuals in the search space for population-based, single-solution and estimation of distribution algorithms (with a specific package for compact optimisation [68–70]);
- utility methods for measuring population and fitness diversity according to multiple metrics [11] and for manipulating populations into covariance matrices [71,72];
- the class `Corrections` with the implementation of several correction strategies to handle infeasible solutions, as those from [1,16], which are selected by simply setting a `char` flag of the class `Algorithm`, as shown in Figure 2c;
- the `random` utilities, which provide methods for generating random numbers from different distributions, initialising random (integer and real-valued) arrays, permuting the components of an array passed as input, changing the pseudo-random number generator, changing seed, etc.;
- a vast number of algorithmic operators (sub-package `operators`) implementing simple local search routines, restart mechanisms and most importantly:
  - the class `GAOp`, which provides numerous mutation, parent selection, survivor selection and crossover operators from the GA literature;
  - the class `DEOp`, which provides all the established DE mutation and crossover strategies, plus several others from the recent literature;
  - the classes in `aos`, which implements adaptive heuristic/operator selection mechanisms from the hyperheuristic field [24].

Given the facility of adding new algorithms and the availability of numerous off-the-shelf operators, SOS is a suitable workbench to design hybrid algorithms or variants tailored to the problem at hand. In particular, the possibility of declaring variables of the kind `Algorithm`, in order to instantiate and execute optimisation processes inside another algorithm, leads to the implementation of rather complex algorithms by writing very neat code. This is evident from Figure 3, where the code of the iterated local search algorithm proposed in [73] is shown. At first glance, one can observe that the source code is clear and resembles pseudocode. Indeed, instead of implementing the sophisticated (1+1)–CMA-ES algorithm in [74], which plays the local searcher role, an `Algorithm` variable is instantiated by using the `CMAES_11` class available from the `algorithms` package and implementing the (1+1)–CMA-ES algorithm. The parameters and computational budget for this algorithm, which can be seen as an operator in this context, are passed as described in the previous sections. It can be noticed that the initial solution for each iterated search is passed to the local searcher before it is executed. This solution is generated by applying the exponential crossover operator [1] to the current best solution and a feasible randomly sampled individual. Obviously, the crossover method, as well as the method for generating an individual in the search space are already present in SOS, and therefore, they can be simply imported without the need for writing further code. To monitor the internal dynamics of the resulting algorithm, a `FTrend` object can be used as shown in Figure 3.

Based on similar considerations, also optimisation problems are added to SOS by following a template defined in the abstract class `Problem`. Indeed, regardless of their nature, i.e., black-box, grey-box, real-world or synthetic testbed, all optimisation problems share similar attributes indicating their dimensionality (i.e., the number of design variables), the boundaries delimiting the search space in which the optimal solution must be found and an optional name to describe the problem. All these fields are accessed and changed with setter and getter methods already implemented in `Problem`. This way, no coding is required to deal with such aspects, thus allowing SOS users to focus on writing the body of only one abstract method, `f`, which obviously represents the fitness function. When the `execute` method of an algorithm is called, a reference to a `Problem` variable `P` is passed to the algorithm, which evaluates the fitness value `y` with the code line `y = P.f(x)`, with `x` being a candidate solution. To ensure that the return value is meaningful, algorithms inherit from the superclass `Algorithm` the function `x = correct(x,bounds)`, which can be used before performing the fitness function evaluation. The latter executes the correction strategy specified when the algorithm object is instantiated (the default strategy is the toroidal correction [1]). The `Problem` class comes with multiple constructors so that problems can be instantiated in different ways. Usually, real-world applications have a fixed number of design variables and fixed boundaries, while benchmark functions are expected to be scalable and with adjustable search space boundaries. Thus, being able to select the most appropriate constructor is very convenient. Finally, it is worth reminding that the methods from the `MatLab` class can aid the implementation of a novel problem and that, if a benchmark function displaying particular features is needed, its implementation might already be available amongst those indicated in Section 3.

```
int problemDimension = problem.getDimension();
double[][] bounds = problem.getBounds();

int i;

double[] best;
double fBest;

FTrend FT = new FTrend();

i = 0;
if (initialSolution != null)
{
    best = initialSolution;
    fBest = initialFitness;
}
else
{
    best = generateRandomSolution(bounds, problemDimension);
    fBest = problem.f(best);
    i++;
}
FT.add(i, fBest);

// INITIALISE MEMES//

double globalCR = getParameter("p0").doubleValue();

CMAES_11 cma11 = new CMAES_11();
cma11.setParameter("p0",getParameter("p1").doubleValue());
cma11.setParameter("p1",getParameter("p2").doubleValue());
cma11.setParameter("p2",getParameter("p3").doubleValue());
cma11.setParameter("p3",getParameter("p4").doubleValue());


double maxB = getParameter("p5").doubleValue();

FTrend ft = null;
```

(**a**) Initialisation phase

```
while (i < maxEvaluations)
{
    xTemp = generateRandomSolution(bounds, problemDimension);
    xTemp = crossOverExp(best, xTemp, globalCR);
    fTemp = problem.f(xTemp);
    i++;
    if(fTemp<fBest)
    {
        fBest = fTemp;
        for(int n=0;n<problemDimension; n++)
            best[n] = xTemp[n];
        FT.add(i, fBest);
    }


        cma11.setInitialSolution(xTemp);
        cma11.setInitialFitness(fTemp);

        int  budget = (int)(min(maxB*maxEvaluations, maxEvaluations-i));

        ft = cma11.execute(problem, budget);
        xTemp = cma11.getFinalBest();
        fTemp = ft.getLastF();

        FT.merge(ft, i);
        i+=budget;

        if(fTemp<fBest)
        {
            fBest = fTemp;
            for(int n=0;n<problemDimension; n++)
                best[n] = xTemp[n];
        }

}

FT.add(i,fBest);
finalBest = best;
return FT;
```

(**b**) Optimisation phase

Figure 3. Algorithmic design and implementation in SOS. This example shows portions of code from the algorithm class `RI1p1CAMES`, located in the `algorithms` package, which implements the algorithm proposed in [73]. On the left-hand side (**a**), the initialisation phase, where the object `cma11`, which is an instance of the class `CMAES_11` (which extends `Algorithms`), is initialised and ready to be executed inside another class extending `Algorithms`. On the right-hand side (**b**), the implementation of an iterated local search method using `cma11` as a local searcher. Note that the `FTrend` variable `ft` returned by `cames11` is automatically appended to `FT` to obtain the overall fitness trend.

3. Benchmarking with SOS

Due to the difficulties in dealing with real-world black-box problems, the metaheuristic optimisation research community started developing artificially built functions [75] to:

- significantly reduce the optimisation time;
- investigate algorithmic behaviours over problems with known or partially known properties;
- be able to study the scalability of optimisation algorithms empirically.

Since the publication of the first testbed problems [75], several and ever more challenging benchmark suites have been released on an annual basis. These usually consist of heterogeneous groups of functions displaying similar features in terms of modality, separability and ill-conditioning. Despite the high number of suites in the literature, their technical reports show similarities. In particular, a set of established functions, such as Michalewicz, Ackley, Rastrigin, De Jong and Schwefel functions, is almost always present. Hence, many of these testbeds are basically equivalent, if it was not for the recent tendency to increase the degree of difficulty by also including in the testbeds hybrid functions, obtained by combining or composing the aforementioned basic functions or by shifting and rotating them. Even though the utility of having (often over-complicated) compositions of functions, which are sometimes as cryptic as black-box real-word problems, might be arguable, these have constantly been employed to propose challenging competitions for stochastic optimisation.

To remove the burden of implementing such a high number of testbed problems, the SOS platform comes with full implementations of:

- several basic testbed problems, taken from [17,75];
- several complete benchmark suites amongst those released over the years for competitions on real-parameter optimisation at the IEEE Congress on Evolutionary Computation (CEC), namely:
  - CEC 2005 [76];
  - CEC 2008 for Large Scale Global Optimisation (LSGO) [77];
  - CEC 2010 for Large Scale Global Optimisation (LSGO) [78];
  - CEC 2013 [79];
  - CEC 2013 for Large Scale Global Optimisation (LSGO) [80];
  - CEC 2014 [2];
  - CEC 2015 [81];
- the fully-scalable test suite for the Special Issue of Soft Computing on scalability of evolutionary algorithms and other metaheuristics for large scale continuous optimisation problems [82];
- the Black-Box Optimisation Benchmarking (BBOB) suite [83], as well as all the BBOB suites included in the 2019 release of the COCO platform [6];
- a novel variant of the CEC 2014 benchmark, named "R-CEC14" (the "R" indicates a rotation flag used to activate or deactivate rotation operators), presented in Section 3.

It is worth indicating that some applications from the CEC 2011 benchmark suite for real-world optimisation [42] are also available, implemented in the `applications` package.

The R-CEC14 Benchmark Suite

The original IEEE CEC 2014 benchmark suite [2] consisted of 30 functions for single-objective real-parameter numerical optimisation. These problems are obtained by shifting, rotating and ill-conditioning well-established functions in the fields of computational optimisation. However, only two of the 30 mathematical functions are not subject to rotation, i.e., Function Number 8 (shifted Rastrigin's function) and Function Number 10 (shifted Schwefel's function), but they do have a rotated counterpart in the benchmark suite.

For this reason, one could argue that only these four functions, i.e., Function Numbers 8, 10 and their rotated counterparts, are insufficient to draw interesting conclusions on:

- the efficacy of the so-called rotation-invariant operators in optimising rotated and unrotated landscapes, as done for DE in [4,53], where it was unveiled no difference in the performances of such operators over the two classes of problems (indeed, from the point of view of the algorithm, these are just different problems) thus arguing the need for rotating benchmark functions, if not for transforming separable functions into non-separable ones;
- separability and degrees of separability, measured, e.g., with the separability index proposed in [3], as well as on the suitability of certain algorithms and algorithmic operators for addressing separable and non-separable problems.

Indeed, rotating a separable function does not alter its modality or ill-conditioning features, but does alter its separability. To further investigate this effect, SOS contains a redesign of the original CEC 2014 benchmark (used, for instance, in [53]), in which the rotation can be activated or deactivated by simply setting a flag (R). If R is equal to one, the rotation is active, and the functions are identical to those of CEC 2014. Otherwise, no rotation takes place. To avoid duplicates, Functions 8 and 10 are removed since they are obtainable from their rotated counterparts, i.e., Functions 9 and 11, by setting the rotation flag equal to zero. Thus, the resulting R-CEC14 suite contains 56 functions:

- the first 28 are the functions listed in Table 2, whose analytical expressions can be found in [2];
- the last 28 functions are their rotated versions.

These problems are optimised within a given search space $\mathcal{D}$ defined as $[-100, 100]^n$, with $n \in [10, 30, 50, 100]$ being the admissible dimensionality values. For each problem, the corresponding $n \times n$ rotation matrices are stored in the `benchmarks.problemsImplementation.CEC2014.files_cec2014` package [17] and loaded when the rotation is active. The minimum fitness function value $f_{min} = f(x_{min})$, with $x_{min} = \text{argmin} f(x)$, $x \in \mathcal{D}$, is shown for each problem in Table 2. More detailed information can be found in [2] or by inspecting the source code [17] and the online documentation.

Table 2. R-CEC14 (R stands for rotation flag) benchmark suite. Each problem can be evaluated with and without the action of the rotation. Further details on these problems and the rotation procedure are available at [2].

f Class	f Number	Function Name and Description	f_{min}
Unimodal	1	High Conditioned Elliptic Function	100
	2	Bent Cigar Function	200
	3	Discus function	300
Multimodal	4	Shifted Ackley's Function	400
	5	Shifted Rosenbrock's	500
	6	Shifted Griewank's Function	600
	7	Shifted Weierstrass Function	700
	8	Shifted Rastrigin's	900
	9	Shifted Schwefel's	1100
	10	Shifted Katsuura	1200
	11	Shifted HappyCat	1300
	12	Shifted HGBat	1400
	13	Shifted Expanded Griewank's plus Rosenbrock's	1500
	14	Shifted and Expanded Scaffer's F6 Function	1600
Hybrid	15	Hybrid Function 1	1700
	16	Hybrid Function 2	1800
	17	Hybrid Function 3	1900
	18	Hybrid Function 4	2000
	19	Hybrid Function 5	2100
	20	Hybrid Function 6	2200
Hybrid	21	Composition Function 1	2300
	22	Composition Function 2	2400
	23	Composition Function 3	2500
	24	Composition Function 4	2600
	25	Composition Function 5	2700
	26	Composition Function 6	2800
	27	Composition Function 7	2900
	28	Composition Function 8	3000

4. Statistical Analysis with SOS

Stochastic algorithms can be evaluated, e.g., by measuring their overhead (in SOS, the class `mains.test.TestOverhead` can be used for this purpose), by calculating their time and memory complexity, in terms of scalability, average performances (i.e., final fitness value returned by the algorithm, averaged over multiple runs) and, qualitatively, by visual inspection of the fitness trend graphs. However, in order to claim that an algorithm is capable of outperforming one or more competing algorithms, when tested on a specific problem, or a set of multiple problems, statistical evidence must be sought.

A review of the statistical tests to be used for analysing and comparing stochastic algorithms can be found in [13]. Amongst the suggested methods, non-parametric tests such as the Holm test [84] and the Wilcoxon signed-rank test [85] are commonly employed since results collected over multiple runs of stochastic algorithms are not necessarily normally distributed. Several recent studies adopted similar variants of these methods, namely the Wilcoxon rank-sum test and the Holm–Bonferroni test, which can now be considered quite established in the field [3,4,24,41,86].

To facilitate the use established and advanced statistical tests for evaluating the performance of stochastic optimisation algorithms, SOS provides implementations of:

- a comparison test based on the Wilcoxon rank-sum test, described in Section 4.1
- a comparison test based on the Holm–Bonferroni test, described in Section 4.2;
- the advanced statistical analysis (ASA) test, described in Section 4.3.

4.1. The Wilcoxon Rank-Sum Test

The Wilcoxon rank-sum test [87], also known as the Mann–Whitney U-test [88], is a non-parametric test used to understand whether two independent samples belong to populations having the same distribution. Unlike similar parametric counterparts, such as the two-sample unpaired t-test [89], it does not operate on data, but scores, referred to as ranks, associated with the actual values and for which assumptions on their distribution can be made. This makes it suitable for analysing the results of stochastic algorithms, whose distributions may significantly vary over different problems and combinations of parameters, while their ranks' distributions can be assumed to follow a normal model.

To describe the procedure implemented in SOS, let us consider two generic algorithms A and B running on a generic problem P, respectively n_A and n_B times (usually, $n_A = n_B$, although having an unbalanced number of runs would not prevent one from using the Wilcoxon rank-sum test). The null-hypothesis is then formulated as $H_0 : A = B$, which means that, regardless of their working logic, the two algorithms are statistically equivalent, i.e., they are instances of the same population as the solutions provided over multiple runs are equally distributed. This is graphically shown in Figure 4. The Wilcoxon rank-sum test can then be used to produce evidence to reject H_0. Failing this task would instead support the so-called alternative hypothesis, which is formulated as $H_1 : A < B$ or $H_1 : A > B$, if a one-sided (also known as one-tailed) test is performed, or $H_1 : A \neq B$, if a two-sided (also known as two-tailed) test is performed.

By default, SOS performs the two-tailed test. However, both the one-sided and the two-sided tests are available in the platform. Indications on how to run these tests can be found in the online documentation. Regardless of the specific variant, i.e., single or two-sided, the null-hypothesis has always the same formulation and can be either accepted or rejected as shown in Figure 4. From the outcome of the test, conclusions can then be drawn on the comparison of the performance of the algorithms A and B. For example, if $H_0 : A = B$ is rejected in a two-sided test, indicators such as the average fitness value or the median fitness value across the performed runs can be used to understand which algorithm outperforms the other on a given problem P. Obviously, this is not necessary if a one-sided variant of the test is employed.

The steps in Sections 4.1.1–4.1.5 describe the decision-making process implemented in the SOS platform in order to apply the Wilcoxon rank-sum test.

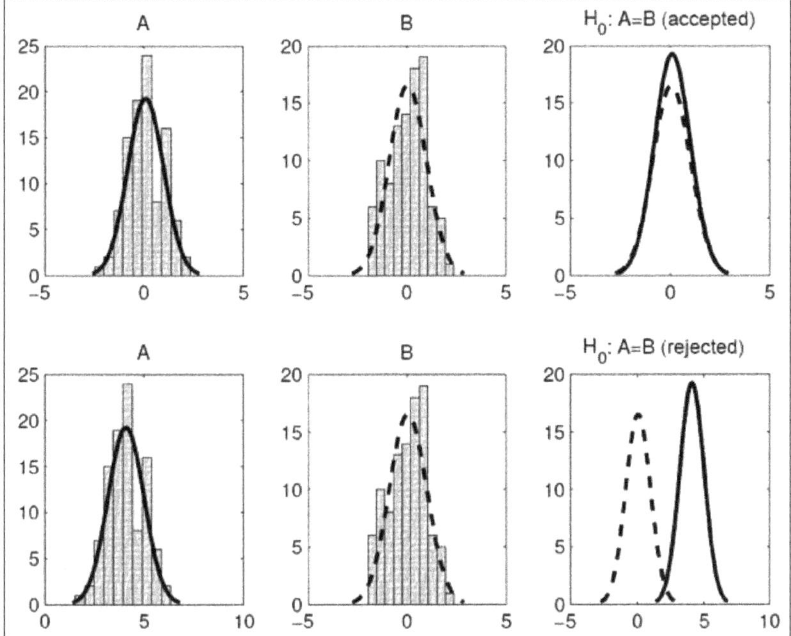

Figure 4. Hypothesis testing. On the top row, the test fails at rejecting the null-hypothesis (i.e., H_0 is accepted) as the algorithms A and B have statistically similar distributions. On the bottom row, H_0 is rejected as A and B are two different stochastic processes with statistically different distributions.

4.1.1. Reference and Comparison Algorithms

To be performed a statistical test, a "reference" is needed. When asked to perform the Wilcoxon rank-sum test on the results generated with an experiment E, the list of algorithms executed in E is shown, and the reference can be indicated. If not specified, SOS assumes that the reference algorithm is the first one added to E. Without loss of generality, let us indicate the reference algorithm with A. The remaining algorithms in E will form a set of "comparison" algorithms. If more than two comparison algorithms are present, SOS will iteratively schedule a Wilcoxon rank-sum test to compare the reference algorithm A with each comparison algorithm B, taken from such a set, for each problem P in E. The order of the comparison algorithms can be specified. This will be the order of appearance of the algorithms on the automatically generated result table, as those shown in the graphical examples included in Section 5. By pressing "c" during the order selection process, a table will be created if the comparison set is not empty. This way, multiple and smaller results tables can be incrementally generated from a single experiment. If the "a" (i.e., all) option is instead used, a single long table for the whole experiment E will be generated. This will contain as many columns as algorithms in E and as many rows as problems in E.

4.1.2. Assigning Ranks

Let us consider the totality of the observations obtained from the $N = n_A + n_B$ runs. Without loss of generality, let us refer to minimisation problems, for which the lower the fitness value, the better the algorithm's performance. Observations are then sorted in ascending order to assign ranks $r_i = i$ (with $i = 1, 2, \ldots, N$) so that the smallest value has Rank 1 (i.e., $r_1 = 1$), the second smallest value has Rank 2 (i.e., $r_2 = 2$), and so on, until the observation with the greatest value, which has rank $r_N = N$.

This process needs to be slightly modified if the dataset presents "ties", i.e., observations with identical numerical value, for which SOS will assign the same rank by computing the average of their position index i in the ordered sequence. This intermediate value will indeed represent better the distribution of the original results. To clarify this case with a numerical example, let us suppose that four observations have ordinal Ranks 4, 5, 6, and 7, but an equal fitness value. To make sure that the rank distribution would be a good representation of the real distribution, these four ties will be assigned with the same rank value $r_j = \frac{4+5+6+7}{4} = 5.5$ ($j \in \{4,5,6,7\}$).

4.1.3. The Rank Distributions

Let us indicate with W_A a random variable associated with A. Its distribution of probability is tabulated only for small sample sizes [90], i.e., $N < 20$, which can be very small for optimisation experiments. To overcome this limitation, SOS implements such a distribution to deal also with larger sample sizes and get more accurate evaluation results. To do this, see [87,90], it is sufficient to define the normal distribution $\mathcal{N}(\mu_A, \sigma_A)$ whose mean value μ_A and standard deviation σ_A are calculated as:

$$\mu_A = \sqrt{n_A n_B \frac{N+1}{12}} \quad \text{and} \quad \sigma_A = n_A \frac{N+1}{2}.$$

4.1.4. Wilcoxon Rank-Sum Statistic and p-Value

Let us fill a set R_A with the n_A ranks r_i associated with the observations from the reference algorithm A. The so-called Wilcoxon rank-sum statistic w_A is calculated by summing all these ranks:

$$w_A = \sum_{r \in R_A} r$$

and is then used to define the p-value. For the default case, i.e., the two-sided test with $H_0 : A = B$ versus $H_1 : A \neq B$, this is formulated as follows:

$$p\text{-value} = \begin{cases} 2 \cdot \text{Prob}\{W_A \geq w_A\} & \text{if } w_A > \mu_A \quad \text{(i.e., } w_A \text{ is in the upper tail)} \\ 2 \cdot \text{Prob}\{W_A \leq w_A\} & \text{otherwise} \quad \text{(i.e., } w_A \text{ is in the lower tail)} \end{cases}$$

where the probability of falling into the tail of the distribution closest to w_A is doubled in order to consider both the cases in which the two algorithms differ because $A > B$ and $A < B$ simultaneously. This is not needed for the one-sided test, for which the null-hypothesis $H_0 : A = B$ is either tested against the alternative hypothesis $H_1 : A < B$, with a corresponding p-value $= \text{Prob}\{W_A \leq w_A\}$, or against the alternative hypothesis $H_1 : A > B$, with a corresponding p-value $= \text{Prob}\{W_A \geq w_A\}$. A graphical explanation of the null-hypothesis testing process is given in Figure 5.

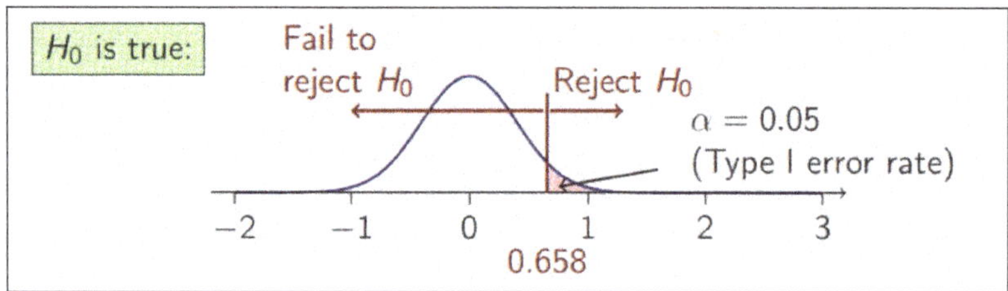

Figure 5. A graphical explanation of the hypothesis testing process for a generic one-sided test with alternative hypothesis $H_1 : A > B$ and $\alpha = 0.05$. The area under the normal distribution, highlighted in red, is equal to α and indicates the rejection zone. Indeed, any $x > 0.658$ would return a p-value (i.e., area under the distribution) lower than α, thus rejecting H_0. Conversely, for all $x < 0.658$, the corresponding p-value would be greater than α, thus failing to reject H_0. In a two-sided test, the red area on the right-hand side should have a symmetric counterpart on the left-hand side. The two resulting tails, each one casting an area of 0.025 (i.e., $\frac{\alpha}{2} = \frac{0.05}{2}$, so that the total significance level is 0.05 and the corresponding confidence is $1 - \alpha = 95\%$) would form the rejection zone.

The p-value provides an indication of the truthfulness of H_0 by calculating the probability of obtaining test results similar to those observed experimentally, assuming that H_0 is correct. This probability can then be used to decide whether the null-hypothesis can be trusted to be true or it must be rejected. This probability can be calculated by numerically integrating the statistic for the specific test, in this case $\mathcal{N}(\mu_A, \sigma_A)$, with μ_A and σ_A calculated as indicated in Section 4.1.3, and then using the appropriate definition of the p-value for the specific test, i.e., one- or two-sided, as described before.

4.1.5. Decision-Making

To test whether or not H_0 is rejected, the calculated p-value is compared to a threshold α, commonly referred to as the significance level. This value represents the probability of making a so-called "Type I" error, which occurs when a true null-hypothesis is incorrectly rejected, as indicated in Table 3.

The α value is arbitrarily chosen. Usually, this is a small number amongst 0.10 (one Type I error chance in 10 decisions is tolerated), 0.05 (one Type I error chance in 20 decisions is tolerated), and 0.01 (one Type I error chance in 100 decisions is tolerated). It is worth mentioning that if a decision is made with a probability α of making a Type I error, its correctness can be trusted with a probability of $1 - \alpha$. This figure is referred to as the confidence level, and it is sometimes provided in the literature instead of α.

Table 3. Table of truth, also known as the confusion matrix.

	H_0 is true	**H_0 is false**
Test rejects H_0	Type I error (False Positive)	Correct inference (True Positive)
Test fails to rejects H_0	Correct inference (False Negative)	Type II error (True Negative)

By default, SOS performs the Wilcoxon rank-sum test with $\alpha = 0.05$, which is a common value for non-parametric tests, unless differently specified before running the test. This results in a confidence level of 95%. To employ a different α value, a setter method is provided (see the online documentation). Other methods are also available to switch between two-sided to one-sided tests and decide whether or not p-values are shown on screen.

To conclude, once the p-value is computed and the significance level α is chosen, a final decision is made by means of the following logic:

- if p-value $\geq \alpha$, the test fails at rejecting $H_0 : A = B$, i.e., the two algorithms are equivalent on P;
- if p-value $< \alpha$, the test rejects $H_0 : A = B$, and one can be $(1 - \alpha)\,\%$ confident that a significant difference does exist between A and B on P.

4.2. The Holm–Bonferroni Test

The Holm test is a non-parametric and sequentially-rejective procedure for multiple-hypothesis testing, which aims at rejecting one hypothesis at a time until no further rejection is possible [84].

In the optimisation field, this test can be used to compare the reference algorithm with more than one comparison algorithm over multiple problems simultaneously. This provides a different and more global view of the comparison with respect to the one provided by the Wilcoxon rank-sum test, which is instead problem-specific. In this light, the two tests complement each other, and it is suggested to always use both (or equivalent tests), to analyse the results obtained statistically with empirical experimentation.

The original Holm test was designed with the intent of guaranteeing a reasonably low family-wise error rate (FWER) [13,84], an error plaguing multiple-hypothesis testing and known to increase when the number of hypotheses increases. To further improve upon this aspect and keep the FWER low even when the number of hypotheses is high, several "corrections" for obtaining more informed p-values have been proposed, such as the well-established Bonferroni's correction [91].

SOS implements a simple Holm–Bonferroni test for comparing stochastic algorithms consisting of the steps described in Sections 4.2.1–4.2.4.

4.2.1. Choosing the Reference Algorithm

Let us consider a generic experiment E containing NA algorithms and NP problems. When the test is performed on E, SOS will display the list of available algorithms and ask the user to select the reference algorithm. Unlike the case described in Section 4.1.1 for the Wilcoxon rank-sum test, this step can change the output of the test. Indeed, in the one-to-one comparison performed with a Wilcoxon rank-sum test, exchanging A with B would not alter the rank distributions obtained on P. Conversely, the $\text{NA} - 1$ statistics computed in the Holm–Bonferroni test do depend on the rank of the reference algorithm. In this light, if the goal is to test the performance of an algorithm A in E against the other algorithms, A should play the role of the reference algorithm. On the contrary, if the goal is to test which algorithm has the best overall performance in E, it could be necessary to run the test twice. The first time, a random reference is chosen. The second time, the algorithm with the highest rank must be found and selected to be the reference for a second round of the test. Once a reference algorithm is selected, SOS proceeds with the test.

4.2.2. Assigning Ranks

First, the average final fitness values must be computed based on the values returned by the NA algorithms after the execution of multiple runs on the *NP* problems. SOS automatically collects the text files containing the information stored in `FTrend` objects and provides the NP final average values for each algorithm. It is worth mentioning that SOS processes this information only if it was not previously requested for another test or graphical procedure (in which case the values are retrieved and immediately returned, thus minimising overheads). Then:

- for each problem in E, a score equal to NA is assigned to the algorithm displaying the best average performance in terms of fitness function value, i.e., the greatest if it is a maximisation problem or the smallest if it is a minimisation one, while a score equal to $\text{NA} - 1$ is assigned to the second-best algorithm, a score equal to $\text{NA} - 2$ to the third-best algorithm, and so on. The algorithm with the worst final average fitness value gets a score equal to one;
- for each algorithm in E, the scores assigned over the NP problems are collected and averaged:
  - the average score of the reference algorithm is referred to as rank R_0;
  - the remaining $\text{NA} - 1$ average scores are used to sort the corresponding algorithms in descending order and constitute their ranks, which are indicated with R_i ($i = 1, 2, \ldots, \text{NA} - 1$). These ranks provide a first indication of the global performance of the algorithms on E, and their order will be automatically displayed in the form of a "league" table.

4.2.3. Z-Statistics and p-Values

For each ith comparison algorithm ($i = 1, 2, \ldots, \text{NA} - 1$), this test requires the use of the so-called "z-value" statistic [13], calculated with the formula:

$$z_i = \frac{R_i - R_0}{\sqrt{\text{NA}\frac{\text{NA}+1}{6\text{NP}}}}$$

This allows for the determination of $\text{NA} - 1$ p-values through the normalised cumulative normal distributions z_i for each comparison algorithm.

4.2.4. Sequential Decision-Making

When all the previous steps have been completed, SOS proceeds with a sequential decision-making process in which A is tested against each comparison algorithm, following the order obtained as described above. A similar procedure to that presented in Section 4.1.5 for the Wilcoxon rank-sum test is therefore iterated NA − 1 times. However, the rejection threshold requires an adjustment due to the multiple-hypothesis settings. In detail, for each ith comparison algorithm ($i = 1, 2, \ldots, \text{NA} - 1$), taken in the order above, the corresponding p_i-value and threshold α/i are compared and a decision on the null-hypothesis is made as follows:

- if p_i-value $\geq \frac{\alpha}{i}$, the test fails at rejecting the null-hypothesis $\text{H}_0 : \text{A} = \text{B}$;
- if p_i-value $< \frac{\alpha}{i}$, the null-hypothesis is rejected, and the rank can be used as an indicator to establish which algorithm displayed a better performance on E.

Finally, a table displaying the outcome of the test is automatically generated. Some examples of Holm–Bonferroni tables are shown in Section 5. It is worth remarking that SOS employs a default $\alpha = 0.05$ also for this test. However, this can be changed as previously pointed out in Section 4.1.

4.3. Advanced Statistical Analysis

The SOS platform provides a novel advanced statistical analysis procedure, referred to as the ASA procedure, for comparing couples of algorithms on a single problem. This procedure makes use of several statistical tests, and it is based on a simple workflow, displayed in Figure 6. The main rationale of ASA is that non-parametric tests are preferable if the assumption of normality cannot be made on the results' distributions, whereas parametric tests should be used if statistical evidence is found in support of such an assumption. Therefore, the ASA procedure first checks the distributions of the results of two selected algorithms, A and B, by looking for such evidence, and then applies the most appropriate tests accordingly.

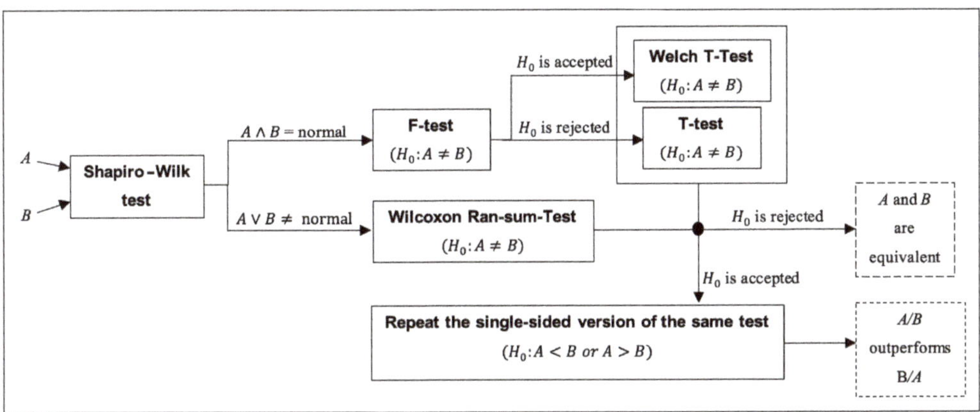

Figure 6. ASA workflow diagram.

To launch the ASA routine, a `TableAvgStdStat` object must be instantiated with the `UseAdvancedStastic` Boolean flag set equal to `true`; see Figure 7. It must be remarked that if such a flag is not activated, the Wilcoxon rank-sum test is performed. It should be noted that the two tests can differ especially when the number of runs is inferior to 100. In such a case, it is strongly recommended to use ASA for a more accurate caparison. Conversely, when a very high number of runs is available, ASA and Wilcoxon rank-sum are comparable. For this reason, the default number of runs performed by SOS is 100. However, this number of runs might be unpractical when facing time-consuming real-world optimisation problems. Hence, as explained in the previous sections (see Figure 2), SOS provides a setter method for specifying the number of runs to be performed in a given experiment. The most common values used in the literature range from 30 to 60 runs. When the ASA procedure is activated, SOS performs the following steps:

- the Shapiro–Wilk test [92] is performed on the reference algorithm A and subsequently on the comparison algorithm B with null-hypothesis H_0: "results are normally distributed";
- if both A and B are normally distributed (i.e., the Shapiro–Wilk test fails at rejecting H_0 on both algorithms), the homoscedasticity of the two distributions (i.e., homogeneity of variances) is tested with the F-test [93], in order to check whether the normal distributions have identical variances and:
    - if variances are equivalent, it is concluded that both A and B have normal distributions, which suggests the use of a two-sided T-test [94,95] with null-hypothesis $H_0 : A \neq B$ to make a decision according to the following logic:
        - if H_0 is rejected, A and B are two equivalent stochastic processes. The test terminates.
        - if the test fails at rejecting if H_0, a one-sided T-test [94,95] is performed to test if A outperforms B, i.e., $H_0 : A < B$ is rejected, or B outperforms A, i.e., $H_0 : A > B$ is rejected. The test terminates.
    - if variances are not equivalent, the Welch T-test [96,97] is used to test the null-hypothesis $H_0 : A \neq B$;
        - if H_0 is rejected, A and B are two equivalent stochastic processes. The test terminates.
        - if the test fails at rejecting if H_0, a one-sided Welch T-test [96,97] is performed to test if A outperforms B, i.e., $H_0 : A < B$ is rejected, or B outperforms A, i.e., $H_0 : A > B$ is rejected. The test terminates.
- if at least one between A and B is not normally distributed (i.e., the Shapiro–Wilk test rejects H_0 on at least one algorithm), a non-parametric test is necessary, and a two-sided Wilcoxon rank-sum test is performed to test $H_0 : A \neq B$ and:
    - if the null-hypothesis is rejected, A and B are two equivalent stochastic processes. The test terminates.
    - if the test fails at rejecting H_0, a one-sided Wilcoxon rank-sum test is performed to test if A outperforms B, i.e., $H_0 : A < B$ is rejected, or B outperforms A, i.e., $H_0 : A > B$ is rejected. The test terminates.

It should be noted that the ASA procedure formulates the null-hypothesis as $H_0 : A \neq B$, while the stand-alone version of the Wilcoxon rank-sum test explained in Section 4.1 is implemented by considering $H_0 : A = B$. This should not generate confusion. The null-hypothesis can be indeed arbitrarily chosen, and it is formulated differently here to facilitate the implementation of the test.

5. Visual Representation of Results

With SOS, results can be displayed in several formats. After performing an experiment, raw data are located in the SOS `results` folder, unless differently indicated, and can be processed straight away or merged with those from other experiments before being processed. In the `results` folder, data are arranged in sub-folders with self-explanatory names to be easily individuated. The main folder comes with the name of the experiment and contains a text file describing it, as explained in Section 2.1, as well as sub-folders whose names refer to the benchmark suite used, the specific function identifier and the dimensionality of the problem. Inside each folder, a fitness trend file is stored for each performed run. On top of the fitness values' trend, these files also report the variables of the best solution found.

From the raw data, SOS can extrapolate information and create graphs and tables thanks to several auxiliary methods available in the platform. The class `Experiments` provides some of these routines for collecting data and transforming them into more useful formats. Some of these can be seen in Figure 7, which shows the main method of the `TablesGenerator` class located in the default package.

```
public class TablesGenerator
{
    public static void main (String args[]) throws Exception
    {
        String workingDir = "";
        if (args.length < 1)
            workingDir ="/home/workingDirectory";
        else
            workingDir = args[0];

        Experiment experiment = new Experiment();
        experiment.setDirectory(workingDir);
        experiment.setTrendsFlag(true, true);

        experiment.importData();
        experiment.describeExperiment();


        experiment.computeAVG();
        experiment.computeSTD();
        experiment.computeMedian();
        experiment.deleteFinalValues();


        TableCEC2013Competition T1 = new TableCEC2013Competition(experiment);
        T1.setErrorFlag(true);
        T1.execute();

        TableBestWorstMedAvgStd T2 = new TableBestWorstMedAvgStd(experiment);
        T2.setErrorFlag(true);
        T2.execute();


        TableHolmBonferroni T3 = new TableHolmBonferroni(experiment);
        T3.setReferenceAlgorithm();
        T3.execute();

        TableStatistics T4 = new TableAvgStdStat(experiment, true, true);
        T4.setErrorFlag(true);
        T4.setReferenceAlgorithm();
        T4.execute();
    }
}
```

Figure 7. An example of the `TablesGenerator` class containing methods for visually displaying results in several different formats.

With reference to Figure 7, let us focus first on the `experiment.setTrendsFlag(true,true)` method. The first Boolean value, in the example set to `true`, indicates that while scanning the raw data for producing tables, SOS will simultaneously save a further text file with the data needed to plot an average fitness trend graph. The second Boolean value, i.e., the error flag, also set to `true` in the example , indicates that the graph will show the fitness trend in terms of average error w.r.t. the known optimum, rather than average fitness. Obviously, this flag is mainly activated when optimising benchmark functions, for which the optimum is usually known. The outputs will look like the examples reported in Figure 8. More details regarding the generation of the fitness trends are given in Appendix A.

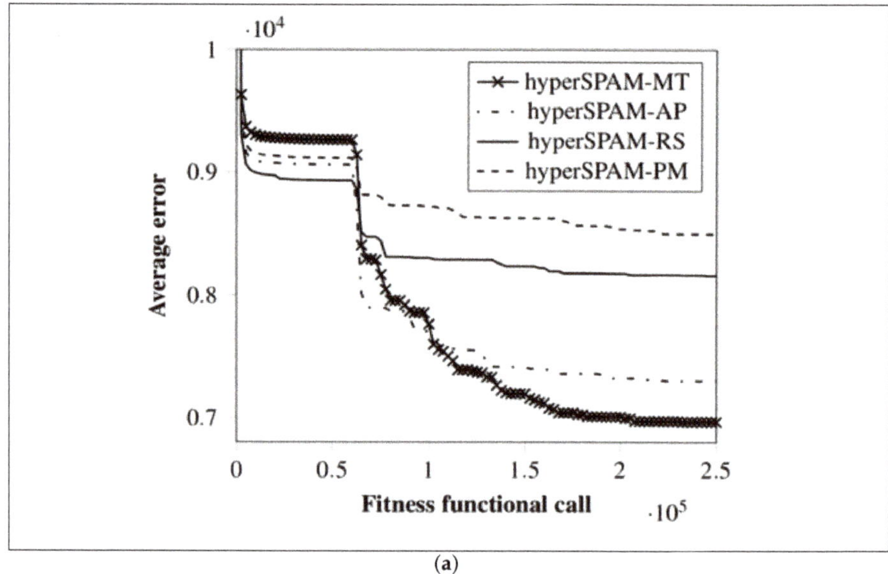

(a)

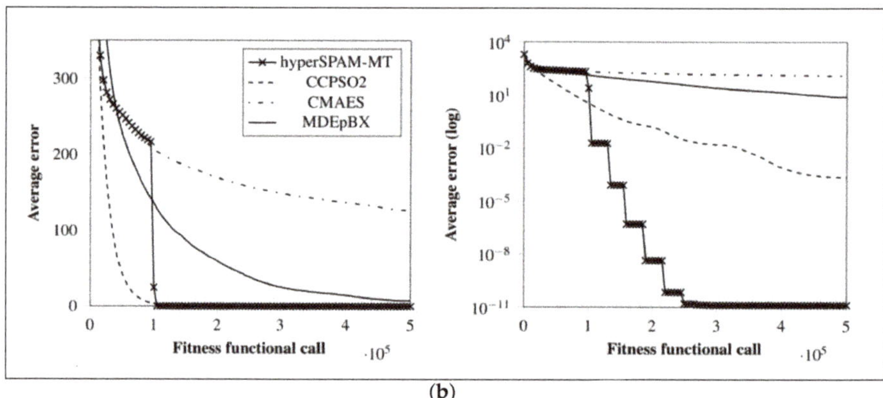

(b)

Figure 8. Example of average error trends produced with SOS for the study in [24]. In (**a**), the full image with the caption. In (**b**), a further example of an average error trend from the same study, plotted in both linear and logarithmic scale. More examples from several other studies were gathered in [98].

The next two methods, `experiment.importData()` and `experiment.describeExperiment()`, start this process and describe it (in the generated log file) iteratively. If the first flag passed to `experiment.setTrendsFlag()` is not active, the `importData()` method will collect data from the corresponding folders and process them, but will store in memory only the information required for the generation of tables, without saving the average trends' information into plottable files. Once the raw data are loaded in memory, SOS can extrapolate the information to be displayed in tables. For example, the commands `experiment.computeAVG()` and `experiment.computeSTD()` will lead to a table where the average fitness value (or the average error value if the error flag is active) $\pm$ the corresponding standard deviation are shown. The command `experiment.computeMedian()` will instead return the median fitness value (or median error depending on the error flag) amongst the available runs. In the example, the three methods are used, but this is not compulsory. It is indeed possible to call only some of or none of them. Moreover, other methods not included in this example are also available, e.g., those for displaying the best or the worst run. If the methods are not called at all, they will be called automatically (if needed) when the tables are generated, as shown in the following part of the example in the figure. Conversely, if they are called, it is suggested to use `experiment.deleteFinalValues()` to free some memory by discarding the final results from memory and keeping only their average values (and the corresponding standard deviations). Before freeing the memory, one may also want to save the final values to plot histograms and distributions, which might be useful in some cases. This can be done in SOS by simply activating some additional flags. For more details on these advanced features, we refer the reader to the online software documentation.

At this point, the statistical tests described in Section 4 can be performed to produce tables in PDF and LaTeX source code formats. To structure compact, but highly informative tables, SOS provides specific classes. Let us keep following the example in Figure 7. One can see that four classes are used to produce tables for the `experiment` object passed as an argument to the constructor. These classes are:

- `TableCEC2013Competition`, which produces tables displaying results according to the guidelines of the CEC 2013 competition [79];
- `TableBestWorstMedAvgStd`, which produces tables displaying the worst, the best, the median, and the average fitness value over the performed runs;
- `TableHolmBonferroni`, which produces tables displaying the "league table", as shown in the examples of Figure 9, obtained with the Holm–Bonferroni test explained in Section 4.2;
- `TableAvgStdStat`, which produces tables displaying the average fitness value ± standard deviation and the results of the statistical analysis in terms of Wilcoxon rank-sum test, described in Section 4.1, or the ASA procedure, described in Section 4.3. Examples are given in Figure 10.

Of note, these four classes are all extensions of the abstract class `TableStatistics`, from which they inherit several miscellaneous methods, such as the `setErrorFlag()` method (used to indicate if tables should display average fitness or errors values), the `setRefenceAlgorithm()` method (used to indicate the reference algorithm) and the `execute()` method (which implements the specific statistical test), as well as attributes, e.g., variables to store the significance levels α, confidence levels $\delta = 1 - \alpha$, flags, etc. If customised tables are needed, this can be simply done by extending the `TableStatistics` superclass and making use of the auxiliary methods provided in such a class to implement the `execute()` method where new statistical tests and/or new table layouts can be implemented.

Let us now describe in detail the tables produced with the `TableHolmBonferroni` and `TableAvgStdStat` classes, shown respectively in the examples reported in Figures 9 and 10.

With reference to Figure 9, which shows the outcome of the Holm–Bonferroni test, it can be noted that SOS reports the rank of the reference algorithm in the caption (which is automatically generated). Regardless of the number of benchmark or real-world problems added in the experiment file, SOS performs the test as described in Section 4.2 and displays the comparison algorithms in descending order according to their rank. As can be seen by comparing the Rank and p_j columns, this corresponds to a descending order also for the p-values. This way, algorithms in the top positions are quite likely to behave similarly to the reference algorithm (i.e., the null-hypothesis is accepted), while those occupying lower positions behave worse than the reference algorithm (i.e., the null-hypothesis is rejected). Other relevant information as the z-statistic values (z_j) and the normalised significance/confidence levels (δ/j) are also reported in the table.

Table 25: Holm-Bonferroni procedure on non-rotated CEC2014 at 10, 50 and 100 dimension values (reference: DE/rand/1/exp, Rank = 8.35e + 00)

j	Optimizer	Rank	z_j	p_j	δ/j	Hypothesis
1	DE/rand/1/bin	7.68e+00	-1.76e+00	3.89e-02	5.00e-02	Rejected
2	eigen-DE/rand/1/exp	6.08e+00	-5.98e+00	1.09e-09	2.50e-02	Rejected
3	RIDE/rand/1/exp	5.11e+00	-8.57e+00	5.30e-18	1.67e-02	Rejected
4	MMCDE	4.76e+00	-9.48e+00	1.26e-21	1.25e-02	Rejected
5	RIDE/rand/1/bin	3.86e+00	-1.19e+01	8.04e-33	1.00e-02	Rejected
6	DE/rand/1/no-xo	3.79e+00	-1.21e+01	8.25e-34	8.33e-03	Rejected
7	eigen-DE/rand/1/bin	3.08e+00	-1.39e+01	2.34e-44	7.14e-03	Rejected
8	DE/current-to-rand/1	2.14e+00	-1.64e+01	8.12e-61	6.25e-03	Rejected

(a)

TABLE 10. Holm-Bonferroni procedure (reference: RI-(1 + 1)-CMA-ES, Rank = 7.16e+00)

j	Optimizer	Rank	z_j	p_j	δ/j	Hypothesis
1	μDEA	7.00e+00	-3.81e-01	3.52e-01	5.00e-02	Accepted
2	MDE-pBX	6.83e+00	-7.89e-01	2.15e-01	2.50e-02	Accepted
3	PMS	6.68e+00	-1.17e+00	1.21e-01	1.67e-02	Accepted
4	cDE-light	6.63e+00	-1.28e+00	1.00e-01	1.25e-02	Accepted
5	(1 + 1)-CMA-ES	6.12e+00	-2.53e+00	5.68e-03	1.00e-02	Rejected
6	Rosenbrock	5.19e+00	-4.82e+00	7.27e-07	8.33e-03	Rejected
7	ISPO	4.17e+00	-7.32e+00	1.23e-13	7.14e-03	Rejected
8	JADE	3.93e+00	-7.89e+00	1.48e-15	6.25e-03	Rejected
9	SPSA	1.27e+00	-1.44e+01	1.81e-47	5.56e-03	Rejected

(b)

Figure 9. Two examples of `TableHolmBonferroni` tables. In (**a**), some results from the study in [4] obtained over the non-rotated functions of the R-CEC14 benchmark suite presented in Section 3. More examples are available in the extended results files stored in the repository [99]. In (**b**), some results from the study in [73] obtained over the functions of the CEC 2014 benchmark suite [2]. Extended results for this study are available in the repository [98].

Figure 10 shows instead two examples of comparative tables obtained with the class TableAvgStdStat. At first glance, it can be noticed that the best performance (in terms of average error) is highlighted in boldface. This is obtained by setting the `useBold` flag equal to `true`, as in the example of Figure 7. Generally, the boldface option is useful as it allows for spotting the "winner" algorithm in a facilitated way. However, it can be turned off by simply setting `useBold` to `false`. To strengthen the validity of the displayed results, the outcome of a statistical test is also added to the table. With reference to Figure 7, if the `useAdvancedStatistic` flag is activated (i.e., it is equal to `true`), the ASA test presented in Section 4.3 is performed. Otherwise, the Wilcoxon rank-sum test, described in Section 4.1, is performed. Regardless of the employed statistical test, the same compact notation is adopted to report its outcome in the table:

- if the reference algorithm A (i.e., the first one in the table) is statistically equivalent to a comparison algorithm B (i.e., $H_0 : A = B$ cannot be rejected), the symbol "=" is placed in the corresponding column next to the average $\pm$ std. dev. fitness/error values of B;
- if A statistically outperforms B, the symbol "+" is placed in the corresponding column next to the average $\pm$ std. dev. fitness/error values of B;
- if A is statistically outperformed by B, the symbol "−" is placed in the corresponding column next to the average $\pm$ std. dev. fitness/error values of B;

TABLE 1. Average error ± standard deviation and Wilcoxon Rank-Sum Test (reference: RI-(1 + 1)-CMA-ES) for RI-(1 + 1)-CMA-ES against (1 + 1)-CMA-ES, SPSA and Rosenbrock on CEC2014[4] in 10 dimensions.

	RI-(1 + 1)-CMA-ES	(1 + 1)-CMA-ES		SPSA		Rosenbrock	
f_1	1.80e + 00 ± 5.70e + 00	**0.00e + 00 ± 0.00e + 00**	-	5.40e + 07 ± 8.85e + 07	+	2.57e + 05 ± 6.33e + 05	+
f_2	0.00e + 00 ± 1.80e − 14	**0.00e + 00 ± 0.00e + 00**	=	8.02e + 06 ± 9.32e + 06	+	5.03e + 03 ± 3.80e + 03	+
f_3	**0.00e + 00 ± 0.00e + 00**	0.00e + 00 ± 0.00e + 00	=	6.33e + 04 ± 2.54e + 04	+	5.73e + 03 ± 6.03e + 03	+
f_4	**9.72e + 00 ± 1.52e + 01**	2.49e + 01 ± 1.51e + 01	+	2.87e + 02 ± 2.37e + 02	+	1.66e + 01 ± 1.92e + 01	+
f_5	2.00e + 01 ± 3.72e − 04	**2.00e + 01 ± 2.87e − 03**	-	2.06e + 01 ± 2.48e − 01	+	2.00e + 01 ± 5.72e − 03	+
f_6	**1.26e + 01 ± 2.26e + 00**	1.55e + 01 ± 2.29e + 00	+	1.89e + 01 ± 1.69e + 00	+	1.84e + 01 ± 2.65e + 00	+
f_7	**6.03e − 02 ± 3.45e − 02**	1.52e − 01 ± 1.29e − 01	+	6.21e + 02 ± 2.28e + 02	+	9.70e − 01 ± 3.09e + 00	+
f_8	**7.99e + 01 ± 2.67e + 01**	1.36e + 02 ± 4.46e + 01	+	1.47e + 02 ± 5.35e + 01	+	1.43e + 02 ± 3.68e + 01	+
f_9	**9.28e + 01 ± 3.43e + 01**	1.54e + 02 ± 5.14e + 01	+	1.55e + 02 ± 5.06e + 01	+	1.63e + 02 ± 6.68e + 01	+
f_{10}	**1.09e + 03 ± 3.13e + 02**	1.66e + 03 ± 3.20e + 02	+	1.56e + 03 ± 6.75e + 02	+	1.66e + 03 ± 4.85e + 02	+
f_{11}	**1.18e + 03 ± 2.08e + 02**	1.77e + 03 ± 3.88e + 02	+	1.62e + 03 ± 6.24e + 02	+	1.82e + 03 ± 4.56e + 02	+
f_{12}	**4.34e − 01 ± 2.70e − 01**	1.19e + 00 ± 1.19e + 00	+	3.95e + 00 ± 2.14e + 00	+	3.58e + 00 ± 2.56e + 00	+
f_{13}	**2.57e − 01 ± 8.39e − 02**	5.34e − 01 ± 1.85e − 01	+	1.34e + 01 ± 3.58e + 00	+	3.36e − 01 ± 1.22e − 01	+
f_{14}	**2.27e − 01 ± 6.07e − 02**	4.66e − 01 ± 3.08e − 01	+	1.80e + 02 ± 6.73e + 01	+	4.01e − 01 ± 2.27e − 01	+
f_{15}	**1.76e + 00 ± 7.83e − 01**	3.53e + 00 ± 2.31e + 00	+	2.87e + 02 ± 8.62e + 02	+	6.52e + 00 ± 6.38e + 00	+
f_{16}	**4.35e + 00 ± 3.27e − 01**	4.63e + 00 ± 3.04e − 01	+	4.85e + 00 ± 2.05e − 01	+	4.65e + 00 ± 2.78e − 01	+
f_{17}	**2.71e + 02 ± 1.28e + 02**	5.43e + 02 ± 2.98e + 02	+	2.08e + 06 ± 1.75e + 06	+	3.37e + 04 ± 8.08e + 04	+
f_{18}	**3.28e + 01 ± 1.60e + 01**	3.88e + 01 ± 2.47e + 01	=	2.34e + 05 ± 1.05e + 06	+	1.66e + 04 ± 1.15e + 04	+
f_{19}	**4.31e + 00 ± 9.56e − 01**	7.40e + 00 ± 2.60e + 00	+	1.27e + 02 ± 1.13e + 02	+	5.00e + 01 ± 4.24e + 01	+
f_{20}	**5.94e + 01 ± 2.61e + 01**	1.16e + 02 ± 7.45e + 01	+	1.64e + 06 ± 2.84e + 06	+	8.30e + 03 ± 8.82e + 03	+
f_{21}	**1.70e + 02 ± 1.35e + 02**	3.59e + 02 ± 2.10e + 02	+	1.48e + 06 ± 1.44e + 06	+	1.19e + 04 ± 1.13e + 04	+
f_{22}	**1.06e + 02 ± 8.99e + 01**	2.65e + 02 ± 1.48e + 02	+	4.92e + 02 ± 1.97e + 02	+	4.99e + 02 ± 1.88e + 02	+
f_{23}	3.22e + 02 ± 4.11e + 01	3.18e + 02 ± 5.91e + 01	=	5.67e + 02 ± 1.39e + 02	+	**3.07e + 02 ± 8.21e + 01**	-
f_{24}	**1.92e + 02 ± 2.11e + 01**	3.26e + 02 ± 1.78e + 02	+	3.38e + 02 ± 1.52e + 02	+	2.94e + 02 ± 1.55e + 02	+
f_{25}	**1.91e + 02 ± 1.39e + 01**	1.99e + 02 ± 1.14e + 01	+	3.32e + 02 ± 9.62e + 01	+	2.00e + 02 ± 1.09e + 01	+
f_{26}	**1.04e + 02 ± 1.79e + 01**	1.56e + 02 ± 9.08e + 01	+	3.00e + 02 ± 1.08e + 02	+	1.76e + 02 ± 1.05e + 02	+
f_{27}	**3.62e + 02 ± 1.43e + 02**	5.01e + 02 ± 1.43e + 02	+	7.17e + 02 ± 1.59e + 02	+	5.38e + 02 ± 1.57e + 02	+
f_{28}	**1.09e + 03 ± 3.69e + 02**	2.83e + 03 ± 1.65e + 03	+	1.78e + 03 ± 6.40e + 02	+	2.78e + 03 ± 1.77e + 03	+
f_{29}	**2.69e + 02 ± 2.84e + 01**	1.19e + 05 ± 4.43e + 05	+	3.48e + 06 ± 7.26e + 06	+	4.75e + 05 ± 7.87e + 05	+
f_{30}	**1.10e + 03 ± 2.31e + 02**	1.38e + 03 ± 4.85e + 02	+	9.40e + 04 ± 1.37e + 05	+	1.68e + 03 ± 4.82e + 02	+

(a) Wilcoxon rank-sum test

Table 21: Average fitness ± standard deviation and statistic comparison (reference: *hyperSPAM-MT*) for *hyperSPAM-MT* against *CMAES MDE-pBX*, and *CCPSO2* on CEC2011[1] in 6, 30, and 20 dimensions.

	hyperSPAM-MT	*CMAES*		*MDE-pBX*		*CCPSO2*	
$Prob_1(6D)$	1.87e + 01 ± 1.21e + 01	3.37e + 01 ± 1.28e + 01	+	8.16e + 00 ± 6.77e + 00	-	**6.77e + 00 ± 3.79e + 00**	-
$Prob_2(30D)$	−1.86e + 01 ± 4.68e + 00	−2.53e + 01 ± 2.74e + 00	-	−2.20e + 01 ± 4.32e + 00	-	**−2.67e + 01 ± 1.74e + 00**	-
$Prob_7(20D)$	7.29e − 01 ± 1.47e − 01	**5.82e − 01 ± 8.31e − 02**	-	1.08e + 00 ± 1.77e − 01	+	1.16e + 00 ± 1.25e − 01	+

(b) ASA test

Figure 10. Two examples of `TableAvgStdStat` tables based on the Wilcoxon rank-sum test (**a**) and the ASA test (**b**), respectively. In (**a**), some results from the study in [73] obtained over the functions of the CEC 2014 benchmark suite [2]. In (**b**), some from the study in [24] obtained over three real-world problems from the CEC 2011 benchmark suite [42]. Extended results for the studies in (**a**,**b**) are available in the repository [98].

It is important to report the outcome of the statistical tests next to the average fitness/error value. For example, with reference to Figure 10a, it can be noticed that despite (1+1)-CMA-ES displaying the best average error value for f_2, this is quite likely to be an isolated event as the "=" symbol next to its value suggests that its general behaviour is actually statistically equivalent to the one of the reference algorithm, i.e., RI-(1+1)-CMA-ES. Indeed, there is a very small difference between the two average error values, i.e., of about 10^{-14}, which is mathematically in favour of the comparison algorithm, but practically, it is not sufficient to state that the comparison algorithm outperforms the reference algorithm.

Online sources with further numerical tables and graphs produced with SOS are inidcated in the Supplementary Materials section of this manuscript.

6. Conclusions

This paper highlighted the need for appropriate software platforms and procedures for rigorously evaluating, comparing, tuning and studying the behaviour of stochastic optimisation algorithms. Contrary to the current research trends, which led to an inflation of "novel" nature-inspired approaches whose algorithmic behaviours are often difficult to comprehend, we argued that regardless of their inspiring metaphor, modern optimisation algorithms should be considered simply as stochastic processes, due to their randomised nature, and therefore studied accordingly. The proposed SOS platform facilitates the design of stochastic optimisation algorithms by: (1) proposing a general approach for their implementation, which stresses the fact the metaheuristic algorithms are an implementation of the same concept (i.e., convergence to near-optimal solutions by alternating exploratory and exploitative phases); (2) providing mathematical tools to compare algorithms statistically regardless of their inspiring metaphor; (3) providing tools for studying the internal dynamics of population-based algorithms. The examples reported in this article illustrated how SOS can be used to address the aforementioned points and display several successful studies where SOS played a key role in fine-tuning algorithm parameters and studying the applicability of algorithmic components on specific classes of optimisation problems. In this light, we can conclude that SOS is not only a useful software tool for designing new algorithms, but most importantly, it is also ideal for studying the algorithmic behaviour of established optimisation frameworks.

Supplementary Materials: Further material can be obtained from the following open-access repositories: Raw data and extended galleries of tables and fitness trend graphs, generated with SOS for some of the studies mentioned in this article, are available at www.doi.org/10.17632/6st7grtxfr.2 and http://www.doi.org/10.17632/psp65d2nbc.1; Extensive galleries of images obtained with the joint use of SOS, for generating data, and IOHprofiler, for plotting them, are available at www.doi.org/10.17632/zdh2phb3b4.2 and www.doi.org/10.17632/cjjw6hpv9b.1.

Author Contributions: Writing, original draft: F.C.; Conceptualisation: F.C.; Investigation: F.C.; Data Curation: F.C.; Writing, review and editing: F.C. and G.I.; Software Development: F.C. and G.I. All authors read and agreed to the published version of the manuscript.

Funding: This research received no external funding.

Conflicts of Interest: The authors declare no conflict of interest.

Terminology

The following terminology is used in this manuscript:

Computational budget	Maximum allowed number of fitness evaluations for an optimisation process
Fitness function	The objective function (from the EA jargon; usually, it refers to a scalar function)
Fitness	The value returned by the fitness function
Fitness landscape	Refers to the topology of the fitness function co-domain
Basin of attraction	A set of solutions from which the search moves to a particular attractor
Run	An optimisation process during which one algorithm optimises one problem
Initial guess	Initial (random/passed) solution of an algorithm
Population-based	A metaheuristic algorithm requiring a set of candidate solutions to function
Individual	In the EA jargon, a candidate solution to a given problem
Population	A set of individuals (i.e., candidate solutions)
Population size	Number of solutions processed by a population-based algorithm
Variation operators	Perturbation operators typical of EAs, e.g., recombination and mutation
Recombination	Operator producing a new solution from two or more individuals
Mutation	Operator perturbing a single solution
Parent selection	Method to select candidate solutions to perform recombination
Survivor selection	Method to form a new population from existing individuals
Local searcher	Operator suitable for refining a solution rather than exploring the search space
Non-parametric test	A statistical test that does not require any assumption on how data are distributed
Null-hypothesis	In statistics, it is the hypothesis of a lack of significant difference between two distributions

Abbreviations

The following abbreviations are used in this manuscript:

CEC	Congress on Evolutionary Computation
CMA	Covariance matrix adaptation
DE	Differential evolution
EA	Evolutionary algorithm
EC	Evolutionary computation
ES	Evolution strategy
FWER	Family-wise error rate
GA	Genetic algorithm
IEEE	Institute of Electrical and Electronics Engineers
NFLT	No free lunch theorem
PSO	Particle swarm optimisation
SI	Swarm intelligence
SOS	Stochastic optimisation software

Appendix A. Producing the Fitness Trend

To save a plottable file containing the average fitness trend of an algorithm over a specific problem, SOS first fills an array (of a size equal to the computational budget, by default $5000 \times n$ with n being the problem dimensionality) with the function evaluations' values contained in the `FTrend` object returned by the algorithm. This is done for each performed run. It should be noted that the `FTrend` object contains a sequence of values that is monotonically decreasing (or increasing, depending on the problem), i.e., each element of the array contains the best fitness value so far. These values reproduce the optimisation trend of a single run and can be averaged with those obtained from other runs in order to get an average fitness trend. The latter is stored in another array of equal dimension, with each element computed as the average of the corresponding elements from each single run. Subsequently, these arrays (the single run arrays and the average fitness array) are downsampled to have 500 equispaced (in terms of fitness evaluation counter) elements. This results in better quality graphs (as those shown in Figure 8), which are at the same time easier to handle and read. However, the number of points plotted in the fitness trend can be changed, thus allowing for a higher or lower number of fitness values to be included in the graphs to adapt to specific needs.

References

1. Caraffini, F.; Kononova, A.V.; Corne, D. Infeasibility and structural bias in differential evolution. *Inf. Sci.* **2019**, *496*, 161–179. [CrossRef]
2. Liang, J.J.; Qu, B.Y.; Suganthan, P.N. *Problem Definitions and Evaluation Criteria for the CEC 2014 Special Session and Competition on Single Objective Real-Parameter Numerical Optimization*; Technical Report; Computational Intelligence Laboratory, Zhengzhou University: Zhengzhou, China; Nanyang Technological University: Singapore, 2013.
3. Caraffini, F.; Neri, F.; Picinali, L. An analysis on separability for Memetic Computing automatic design. *Inf. Sci.* **2014**, *265*, 1–22. [CrossRef]
4. Caraffini, F.; Neri, F. A study on rotation invariance in differential evolution. *Swarm Evol. Comput.* **2019**, *50*, 100436. [CrossRef]
5. Mittelmann, H.D.; Spellucci, P. Decision Tree for Optimization Software. 2005 Available online: http://plato.asu.edu/guide.html (accessed on 31 March 2020).
6. Hansen, N.; Auger, A.; Mersmann, O.; Tusar, T.; Brockhoff, D. COCO: A Platform for Comparing Continuous Optimizers in a Black-Box Setting. *arXiv* **2016**, arXiv:1603.08785.
7. Doerr, C.; Ye, F.; Horesh, N.; Wang, H.; Shir, O.M.; Bäck, T. Benchmarking discrete optimization heuristics with IOHprofiler. *Appl. Soft Comput.* **2020**, *88*, 106027. [CrossRef]
8. Durillo, J.J.; Nebro, A.J. jMetal: A Java framework for multi-objective optimization. *Adv. Eng. Softw.* **2011**, *42*, 760–771. [CrossRef]

9. Tian, Y.; Cheng, R.; Zhang, X.; Jin, Y. PlatEMO: A MATLAB platform for evolutionary multi-objective optimization. *IEEE Comput. Intell. Mag.* **2017**, *12*, 73–87. [CrossRef]
10. Reed, D.H.; Frankham, R. Correlation between fitness and genetic diversity. *Conserv. Biol.* **2003**, *17*, 230–237. [CrossRef]
11. Yaman, A.; Iacca, G.; Caraffini, F. A comparison of three differential evolution strategies in terms of early convergence with different population sizes. *AIP Conf. Proc.* **2019**, *2070*, 020002. [CrossRef]
12. Kononova, A.; Corne, D.; De Wilde, P.; Shneer, V.; Caraffini, F. Structural bias in population-based algorithms. *Inf. Sci.* **2015**, *298*. [CrossRef]
13. García, S.; Fernández, A.; Luengo, J.; Herrera, F. A study of statistical techniques and performance measures for genetics-based machine learning: Accuracy and interpretability. *Soft Comput.* **2009**, *13*, 959–977. [CrossRef]
14. Del Ser, J.; Osaba, E.; Molina, D.; Yang, X.S.; Salcedo-Sanz, S.; Camacho, D.; Das, S.; Suganthan, P.N.; Coello Coello, C.A.; Herrera, F. Bio-inspired computation: Where we stand and what's next. *Swarm Evol. Comput.* **2019**, *48*, 220–250. [CrossRef]
15. Coello, C.A.C. Theoretical and numerical constraint-handling techniques used with evolutionary algorithms: A survey of the state of the art. *Comput. Methods Appl. Mech. Eng.* **2002**, *191*, 1245–1287. [CrossRef]
16. Kononova, A.V.; Caraffini, F.; Wang, H.; Bäck, T. Can Single Solution Optimisation Methods Be Structurally Biased? *Preprints* **2020**. [CrossRef]
17. Caraffini, F. The Stochastic Optimisation Software (SOS) platform. *Zenodo* **2019**. [CrossRef]
18. Caraffini, F. *Algorithmic Issues in Computational Intelligence Optimization: From Design to Implementation, from Implementation to Design*; Number 243 in Jyväskylä studies in computing; University of Jyväskylä: Jyväskylä, Finland, 2016.
19. Eiben, A.; Smith, J. *Introduction to Evolutionary Computing*; Natural Computing Series, Springer: Berlin/Heidelberg, Germany, 2015; Volume 53. [CrossRef]
20. Kennedy, J.; Shi, Y.; Eberhart, R.C. *Swarm Intelligence*; Elsevier: Amsterdam, The Netherlands, 2001; p. 512.
21. Burke, E.K.; Hyde, M.; Kendall, G.; Ochoa, G.; Özcan, E.; Woodward, J.R., A Classification of Hyper-heuristic Approaches. In *Handbook of Metaheuristics*; Springer: Boston, MA, USA, 2010; pp. 449–468._15. [CrossRef]
22. Burke, E.K.; Gendreau, M.; Hyde, M.; Kendall, G.; Ochoa, G.; Özcan, E.; Qu, R. Hyper-heuristics: A survey of the state of the art. *J. Oper. Res. Soc.* **2013**, *64*, 1695–1724. [CrossRef]
23. Caraffini, F.; Neri, F.; Iacca, G.; Mol, A. Parallel memetic structures. *Inf. Sci.* **2013**, *227*, 60–82. [CrossRef]
24. Caraffini, F.; Neri, F.; Epitropakis, M. HyperSPAM: A study on hyper-heuristic coordination strategies in the continuous domain. *Inf. Sci.* **2019**, *477*, 186–202. [CrossRef]
25. Kenneth, V.P.; Rainer, M.S.; Jouni, A.L. *Differential Evolution*; Natural Computing Series; Springer: Berlin/Heidelberg, Germany, 2005. [CrossRef]
26. Xinchao, Z. Simulated annealing algorithm with adaptive neighborhood. *Appl. Soft Comput.* **2011**, *11*, 1827–1836. [CrossRef]
27. Kennedy, J.; Eberhart, R. Particle swarm optimization. In Proceedings of the ICNN'95—International Conference on Neural Networks, Perth, Australia, 27 November–1 December 1995; Volume 4, pp. 1942–1948. [CrossRef]
28. Auger, A. Benchmarking the (1+1) evolution strategy with one-fifth success rule on the BBOB-2009 function testbed. In *Proceedings of the 11th Annual Conference Companion on Genetic and Evolutionary Computation Conference—GECCO '09*; ACM Press: New York, NY, USA, 2009; p. 2447. [CrossRef]
29. Hansen, N.; Ostermeier, A. Completely Derandomized Self-Adaptation in Evolution Strategies. *Evol. Comput.* **2001**, *9*, 159–195. [CrossRef]
30. Hansen, N., The CMA Evolution Strategy: A Comparing Review. In *Towards a New Evolutionary Computation: Advances in the Estimation of Distribution Algorithms*; Springer: Berlin/Heidelberg, Germany, 2006; pp. 75–102._4. [CrossRef]
31. Iacca, G.; Caraffini, F.; Neri, F.; Mininno, E. Single particle algorithms for continuous optimization. In Proceedings of the 2013 IEEE Congress on Evolutionary Computation, Cancun, Mexico, 20–23 June 2013; IEEE: Piscataway, NJ, USA, 2013; pp. 1610–1617.
32. Liang, J.J.; Qin, A.K.; Suganthan, P.N.; Baskar, S. Comprehensive learning particle swarm optimizer for global optimization of multimodal functions. *IEEE Trans. Evol. Comput.* **2006**, *10*, 281–295. [CrossRef]
33. Li, X.; Yao, X. Cooperatively Coevolving Particle Swarms for Large Scale Optimization. *Evol. Comput. IEEE Trans.* **2012**, *16*, 210–224. [CrossRef]

34. Brest, J.; Greiner, S.; Boskovic, B.; Mernik, M.; Zumer, V. Self-Adapting Control Parameters in Differential Evolution: A Comparative Study on Numerical Benchmark Problems. *IEEE Trans. Evol. Comput.* **2006**, *10*, 646–657. [CrossRef]
35. Brest, J.; Maucec, M.S. Self-adaptive differential evolution algorithm using population size reduction and three strategies. *Soft Comput.* **2011**, *15*, 2157–2174. [CrossRef]
36. Alic, A.; Berkovic, K.; Boskovic, B.; Brest, J. Population Size in Differential Evolution. In *Swarm, Evolutionary, and Memetic Computing and Fuzzy and Neural Computing, Proceedings of the 7th International Conference, SEMCCO 2019, and 5th International Conference, FANCCO 2019, Maribor, Slovenia, July 10–12, 2019*; Revised Selected Papers; Communications in Computer and Information Science Book Series; Zamuda, A., Das, S., Suganthan, P.N., Panigrahi, B.K., Eds.; Springer: Berlin/Heidelberg, Germany, 2019, Volume 1092, pp. 21–30._3. [CrossRef]
37. Zhang, J.; Sanderson, A. JADE: Adaptive Differential Evolution With Optional External Archive. *Evol. Comput. IEEE Trans.* **2009**, *13*, 945–958. [CrossRef]
38. Islam, S.M.; Das, S.; Ghosh, S.; Roy, S.; Suganthan, P.N. An Adaptive Differential Evolution Algorithm With Novel Mutation and Crossover Strategies for Global Numerical Optimization. *IEEE Trans. Syst. Man, Cybern. Part B (Cybernetics)* **2012**, *42*, 482–500. [CrossRef]
39. Iacca, G.; Neri, F.; Mininno, E.; Ong, Y.S.; Lim, M.H. Ockham's razor in memetic computing: Three stage optimal memetic exploration. *Inf. Sci.* **2012**, *188*, 17–43. [CrossRef]
40. Molina, D.; Lozano, M.; Herrera, F. MA-SW-Chains: Memetic algorithm based on local search chains for large scale continuous global optimization. In Proceedings of the IEEE Congress on Evolutionary Computation, Barcelona, Spain, 18–23 July 2010; pp. 1–8. [CrossRef]
41. Epitropakis, M.G.; Caraffini, F.; Neri, F.; Burke, E.K. A Separability Prototype for Automatic Memes with Adaptive Operator Selection. In *IEEE SSCI 2014—2014 IEEE Symposium Series on Computational Intelligence, Proceedings of the FOCI 2014: 2014 IEEE Symposium on Foundations of Computational Intelligence, Orlando, FL, USA, 9–12 December 2014*; IEEE: Piscataway, NJ, USA; pp. 70–77. [CrossRef]
42. Das, S.; Suganthan, P.N. *Problem Definitions and Evaluation Criteria for CEC 2011 Competition on Testing Evolutionary Algorithms on Real World Optimization Problems*; Technical Report; Jadavpur University: Kolkata, India; Nanyang Technological University: Singapore, 2010.
43. Caraffini, F.; Kononova, A.V. Structural bias in differential evolution: A preliminary study. *AIP Conf. Proc.* **2019**, *2070*, 020005. [CrossRef]
44. Kononova, A.V.; Caraffini, F.; Bäck, T. Differential evolution outside the box. *arXiv* **2020**, arXiv:2004.10489.
45. Structural Bias in Optimisation Algorithms: Extended Results. *Mendeley Data* **2020**, doi:10.17632/zdh2phb3b4.2. [CrossRef]
46. Caraffini, F.; Kononova, A.V. Differential evolution outside the box—Extended results. *Mendeley Data* **2020**, doi:10.17632/cjjw6hpv9b.1. [CrossRef]
47. Wolpert, D.; Macready, W. No free lunch theorems for optimization. *IEEE Trans. Evol. Comput.* **1997**, *1*, 67–82. [CrossRef]
48. Ho, Y.C.; Pepyne, D.L. Simple Explanation of the No Free Lunch Theorem of Optimization. *Cybern. Syst. Anal.* **2002**, *38*, 292–298.:1021251113462. [CrossRef]
49. Mason, K.; Duggan, J.; Howley, E. A meta optimisation analysis of particle swarm optimisation velocity update equations for watershed management learning. *Appl. Soft Comput.* **2018**, *62*, 148–161. [CrossRef]
50. Neumüller, C.; Wagner, S.; Kronberger, G.; Affenzeller, M. Parameter Meta-optimization of Metaheuristic Optimization Algorithms. In *Computer Aided Systems Theory—EUROCAST 2011*; Moreno-Díaz, R., Pichler, F., Quesada-Arencibia, A., Eds.; Springer: Berlin/Heidelberg, Germany, 2012; pp. 367–374._47. [CrossRef]
51. Andre, J.; Siarry, P.; Dognon, T. An improvement of the standard genetic algorithm fighting premature convergence in continuous optimization. *Adv. Eng. Softw.* **2001**, *32*, 49–60. [CrossRef]
52. Lampinen, J.; Zelinka, I. On stagnation of the differential evolution algorithm. In Proceedings of the MENDEL, Brno, Czech Republic, 7–9 June 2000; pp. 76–83.
53. Caraffini, F.; Neri, F. Rotation Invariance and Rotated Problems: An Experimental Study on Differential Evolution. In *Applications of Evolutionary Computation*; Sim, K., Kaufmann, P., Eds.; Springer International Publishing: Cham, Switzerland, 2018; pp. 597–614._41. [CrossRef]
54. Sang, H.Y.; Pan, Q.K.; Duan, P.y. Self-adaptive fruit fly optimizer for global optimization. *Nat. Comput.* **2019**, *18*, 785–813. [CrossRef]

55. Mirjalili, S.; Mirjalili, S.M.; Lewis, A. Grey Wolf Optimizer. *Adv. Eng. Softw.* **2014**, *69*, 46–61. [CrossRef]
56. Chu, S.C.; Tsai, P.W.; Pan, J.S. Cat Swarm Optimization. In *PRICAI 2006: Trends in Artificial Intelligence*; Yang, Q., Webb, G., Eds.; Springer: Berlin/Heidelberg, Germany, 2006; pp. 854–858._94. [CrossRef]
57. Meng, X.B.; Gao, X.; Lu, L.; Liu, Y.; Zhang, H. A new bio-inspired optimisation algorithm: Bird Swarm Algorithm. *J. Exp. Theor. Artif. Intell.* **2016**, *28*, 673–687. [CrossRef]
58. Eusuff, M.; Lansey, K.; Pasha, F. Shuffled frog-leaping algorithm: a memetic meta-heuristic for discrete optimization. *Eng. Opt.* **2006**, *38:2*, 129–154. [CrossRef]
59. Mirjalili, S. Moth-flame optimization algorithm: A novel nature-inspired heuristic paradigm. *Knowl.-Based Syst.* **2015**, *89*, 228–249. [CrossRef]
60. Wang, Y.; Du, T. An Improved Squirrel Search Algorithm for Global Function Optimization. *Algorithms* **2019**, *12*, 80. [CrossRef]
61. Sulaiman, M.H.; Mustaffa, Z.; Saari, M.M.; Daniyal, H. Barnacles Mating Optimizer: A new bio-inspired algorithm for solving engineering optimization problems. *Eng. Appl. Artif. Intell.* **2020**, *87*, 103330. [CrossRef]
62. Geem, Z.W.; Kim, J.H.; Loganathan, G.V. A new heuristic optimization algorithm: Harmony search. *Simulation* **2001**, *76*, 60–68. [CrossRef]
63. Atashpaz-Gargari, E.; Lucas, C. Imperialist competitive algorithm: An algorithm for optimization inspired by imperialistic competition. In Proceedings of the 2007 IEEE Congress on Evolutionary Computation, Singapore, 25–28 September 2007; pp. 4661–4667. [CrossRef]
64. Rao, R.; Savsani, V.; Vakharia, D. Teaching-learning-based optimization: A novel method for constrained mechanical design optimization problems. *Comput.-Aided Des.* **2011**, *43*, 303–315. [CrossRef]
65. Bidar, M.; Kanan, H.R.; Mouhoub, M.; Sadaoui, S. Mushroom Reproduction Optimization (MRO): A Novel Nature-Inspired Evolutionary Algorithm. In Proceedings of the 2018 IEEE Congress on Evolutionary Computation (CEC), Rio de Janeiro, Brazil, 8–13 July 2018; pp. 1–10. [CrossRef]
66. Hatamlou, A. Black hole: A new heuristic optimization approach for data clustering. *Inf. Sci.* **2013**, *222*, 175–184. Including Special Section on New Trends in Ambient Intelligence and Bio-inspired Systems, doi:10.1016/j.ins.2012.08.023. [CrossRef]
67. Taradeh, M.; Mafarja, M.; Heidari, A.A.; Faris, H.; Aljarah, I.; Mirjalili, S.; Fujita, H. An evolutionary gravitational search-based feature selection. *Inf. Sci.* **2019**, *497*, 219–239. [CrossRef]
68. Iacca, G.; Caraffini, F.; Neri, F. Compact Differential Evolution Light: High Performance Despite Limited Memory Requirement and Modest Computational Overhead. *J. Comput. Sci. Technol.* **2012**, *27*, 1056–1076. [CrossRef]
69. Neri, F.; Mininno, E.; Iacca, G. Compact Particle Swarm Optimization. *Inf. Sci.* **2013**, *239*, 96–121. [CrossRef]
70. Iacca, G.; Caraffini, F. Compact Optimization Algorithms with Re-Sampled Inheritance. In *Applications of Evolutionary Computation*; Kaufmann, P., Castillo, P.A., Eds.; Springer International Publishing: Cham, Switzerland, 2019; pp. 523–534._35. [CrossRef]
71. Takahama, T.; Sakai, S. Solving nonlinear optimization problems by Differential Evolution with a rotation-invariant crossover operation using Gram-Schmidt process. In Proceedings of the 2010 Second World Congress on Nature and Biologically Inspired Computing (NaBIC), Kitakyushu, Japan, 15–17 December 2010; pp. 526–533. [CrossRef]
72. Guo, S.; Yang, C. Enhancing Differential Evolution Utilizing Eigenvector-Based Crossover Operator. *IEEE Trans. Evol. Comput.* **2015**, *19*, 31–49. [CrossRef]
73. Caraffini, F.; Iacca, G.; Yaman, A. Improving (1+1) covariance matrix adaptation evolution strategy: A simple yet efficient approach. *AIP Conf. Proc.* **2019**, *2070*, 020004. [CrossRef]
74. Igel, C.; Suttorp, T.; Hansen, N. A Computational Efficient Covariance Matrix Update and a (1+1)-CMA for Evolution Strategies. In Proceedings of the 8th Annual Conference on Genetic and Evolutionary Computation, GECCO '06, Seattle, WA, USA, 8–12 July 2006; Association for Computing Machinery: New York, NY, USA, 2006; pp. 453–460. [CrossRef]
75. Yao, X.; Liu, Y.; Lin, G. Evolutionary programming made faster. *IEEE Trans. Evol. Comput.* **1999**, *3*, 82–102. [CrossRef]
76. Suganthan, P.N.; Hansen, N.; Liang, J.J.; Deb, K.; Chen, Y.P.; Auger, A.; Tiwari, S. *Problem Definitions and Evaluation Criteria for the CEC 2005 Special Session on Real-Parameter Optimization*; Technical Report 201212; Zhengzhou University: Zhengzhou, China; Nanyang Technological University: Singapore, 2005.

77. Tang, K.; Yao, X.; Suganthan, P.N.; MacNish, C.; Chen, Y.P.; Chen, C.M.; Yang, Z. *Benchmark Functions for the CEC 2008 Special Session and Competition on Large Scale Global Optimization.* Technical Report; Nature Inspired Computation and Applications Laboratory, {USTC}: Hefei, China, 2007.
78. Tang, K.; Li, X.; Suganthan, P.N.; Yang, Z.; Weise, T. *Benchmark Functions for the CEC 2010 Special Session and Competition on Large-Scale Global Optimization*; Technical Report 1, Technical Report; University of Science and Technology of China: Hefei, China, 2010.
79. Liang, J.J.; Qu, B.Y.; Suganthan, P.N.; Hernández-Díaz, A.G. *Problem Definitions and Evaluation Criteria for the CEC 2013 Special Session on Real-Parameter Optimization*; Technical Report 201212; Zhengzhou University: Zhengzhou, China; Nanyang Technological University: Singapore, 2013.
80. Li, X.; Tang, K.; Omidvar, M.N.; Yang, Z.; Qin, K. *Benchmark Functions for the CEC'2013 Special Session and Competition on Large-Scale Global Optimization*; Technical Report; RMIT University: Melbourne, Australia; University of Science and Technology of China: Hefei, China; National University of Defense Technology: Changsha, China, 2013.
81. Li, X.; Tang, K.; Omidvar, M.N.; Yang, Z.; Qin, K. *Benchmark Functions for the CEC'2015 Special Session and Competition on Large Scale Global Optimization*; Technical Report; University of Science and Technology of China: Hefei, China, 2015.
82. Herrera, F.; Lozano, M.; Molina, D. *Test Suite for the Special Issue of Soft Computing on Scalability of Evolutionary Algorithms and other Metaheuristics for Large Scale Continuous Optimization Problems*; Technical Report; University of Granada: Granada, Spain, 2010.
83. Hansen, N.; Auger, A.; Finck, S.; Ros, R.; Hansen, N.; Auger, A.; Finck, S.; Optimiza, R.R.R.p.B.b. *Real-Parameter Black-Box Optimization Benchmarking 2010 : Experimental Setup*; Technical Report; INRIA: Paris, France, 2010.
84. Holm, S. A simple sequentially rejective multiple test approach. *Scand. J. Stat.* **1979**, *6*, 65–70.
85. Rey, D.; Neuhäuser, M. Wilcoxon-Signed-Rank Test. In *International Encyclopedia of Statistical Science*; Springer: Berlin/Heidelberg, Germany, 2011; pp. 1658–1659._616. [CrossRef]
86. Iacca, G.; Caraffini, F.; Neri, F. Multi-Strategy Coevolving Aging Particle Optimization. *Int. J. Neural Syst.* **2013**, *24*, 1450008. [CrossRef] [PubMed]
87. WILCOXON, F. Individual comparisons of grouped data by ranking methods. *J. Econ. Entomol.* **1946**, *39*, 269. [CrossRef] [PubMed]
88. Mann, H.B.; Whitney, D.R. On a Test of Whether one of Two Random Variables is Stochastically Larger than the Other. *Ann. Math. Stat.* **1947**, *18*, 50–60. [CrossRef]
89. Middleton, D.; Rice, J.A. Mathematical Statistics and Data Analysis. *Math. Gaz.* **1988**, *72*, 330. [CrossRef]
90. Tango, T. 100 Statistical Tests. *Stat. Med.* **2000**, *19*, 3018.:21<3018::AID-SIM594>3.0.CO;2-U. [CrossRef]
91. Frey, B.B. Holm's Sequential Bonferroni Procedure. In *The SAGE Encyclopedia of Educational Research, Measurement, and Evaluation*; SAGE Publications, Inc.: Thousand Oaks, CA, USA, 2018; pp. 1–8. [CrossRef]
92. Mudholkar, G.S.; Srivastava, D.K.; Thomas Lin, C. Some p-variate adaptations of the shapiro-wilk test of normality. *Commun. Stat. Theory Methods* **1995**, *24*, 953–985. [CrossRef]
93. Cacoullos, T. The F-test of homoscedasticity for correlated normal variables. *Stat. Probab. Lett.* **2001**, *54*, 1–3. [CrossRef]
94. Kim, T.K. T test as a parametric statistic. *Korean J. Anesthesiol.* **2015**, *68*, 540. [CrossRef]
95. Cressie, N.A.C.; Whitford, H.J. How to Use the Two Samplet-Test. *Biom. J.* **1986**, *28*, 131–148. [CrossRef]
96. Zimmerman, D.W.; Zumbo, B.D. Rank transformations and the power of the Student t test and Welch t' test for non-normal populations with unequal variances. *Can. J. Exp. Psychol. Can. Psychol. Exp.* **1993**, *47*, 523–539. [CrossRef]
97. Zimmerman, D.W. A note on preliminary tests of equality of variances. *Br. J. Math. Stat. Psychol.* **2004**, *57*, 173–181. [CrossRef] [PubMed]
98. Caraffini, F. Novel Memetic Structures (raw data & extended results). *Mendeley Data* **2020**, doi:10.17632/6st7grtxfr.2. [CrossRef]
99. Caraffini, F.; Neri, F. Raw data & extended results for: A study on rotation invariance in differential evolution. *Mendeley Data* **2019**, doi:10.17632/psp65d2nbc.1. [CrossRef]

Article

EvoPreprocess—Data Preprocessing Framework with Nature-Inspired Optimization Algorithms

Sašo Karakatič

Faculty of Electrical Engineering and Computer Science, University of Maribor, Maribor 2000, Slovenia; saso.karakatic@um.si

Received: 29 April 2020; Accepted: 27 May 2020; Published: 2 June 2020

Abstract: The quality of machine learning models can suffer when inappropriate data is used, which is especially prevalent in high-dimensional and imbalanced data sets. Data preparation and preprocessing can mitigate some problems and can thus result in better models. The use of meta-heuristic and nature-inspired methods for data preprocessing has become common, but these approaches are still not readily available to practitioners with a simple and extendable application programming interface (API). In this paper the EvoPreprocess open-source Python framework, that preprocesses data with the use of evolutionary and nature-inspired optimization algorithms, is presented. The main problems addressed by the framework are *data sampling* (simultaneous over- and under-sampling data instances), *feature selection* and *data weighting* for supervised machine learning problems. EvoPreprocess framework provides a simple object-oriented and parallelized API of the preprocessing tasks and can be used with scikit-learn and imbalanced-learn Python machine learning libraries. The framework uses self-adaptive well-known nature-inspired meta-heuristic algorithms and can easily be extended with custom optimization and evaluation strategies. The paper presents the architecture of the framework, its use, experiment results and comparison to other common preprocessing approaches.

Keywords: data sampling; feature selection; instance weighting; nature-inspired algorithms; meta-heuristic algorithms

---

1. Introduction

Data preprocessing is one of the standard procedures in data mining, which can greatly improve the performance of machine learning models or statistical analysis [1]. Three common data preprocessing tasks that are addressed by the presented EvoPreprocess framework are feature selection, data sampling, and data weighting. This paper presents the EvoPreprocess framework, which addresses the listed preprocessing tasks with the use of supervised machine learning based evaluation. All three tasks deal with inappropriate and high-dimensional data, which can result in either over-fitted and non-generalizable machine learning models [2,3].

Many different techniques have been proposed and applied to these of preprocessing data tasks [1,4]; from various feature selection methods based on statistics (information gain, covariance, Gini index, χ^2 etc.) [5,6]; under-sampling data with neighborhood cleaning [7], prototype selection [8], over-sampling data with SMOTE [9], and other SMOTE data over-sampling variants [10]. These methods are mainly deterministic and have limited variability in the resulting solutions. Due to this, researchers also extensively focused on preprocessing with meta-heuristic optimization methods [11,12].

Meta-heuristic optimization methods provide sufficiently good solutions to NP-hard problems while not guaranteeing that the solutions are globally optimal. Since feature selection, data sampling and data weighting are NP-hard problems [6], meta-heuristics present a valid approach, which is also supported by a wide body of research presented in the following sections. Applications of nature-inspired algorithms in data preprocessing have already been used. Genetic algorithms were used for feature selection in high-dimensional datasets to select six biomarker genes that linked with colon cancer [13,14]. Evolution strategy was used for data sampling for the in the early software defect detection [15]. Also, particle swarm optimization was used for the data sampling for classification of hyperspectral images [16].

While research on the topic is wide and prolific, the standard libraries, packages and frameworks are sparse. There are some well maintained and well-documented data preprocessing libraries [17–22], but none of them provide the ability to use nature-inspired approaches in Python programming language. The EvoPreprocess framework aims to fill in the gap between the data preprocessing and nature-inspired meta-heuristics and provide easy to use and extend Python API that can be used by practitioners in their data mining pipelines, or by researchers developing novel nature-inspired methods on the problem of data preprocessing.

This paper presents the implementation details and examples of use of the EvoPreprocess framework, which offers API for solving the three mentioned data preprocessing tasks with nature-inspired meta-heuristic optimization methods. While the data preprocessing approaches for classification tasks are already available, there is a lack of resources for data preprocessing on regression tasks. The presented framework works with both tasks of supervised learning (i.e., classification and regression) and therefore fills this gap.

The novelty of the framework is the following:

1. The framework provides simple Python object-oriented and parallel implementation of three common data preprocessing tasks with nature-inspired optimization methods.
2. The framework is compatible with the well-established Python machine learning and data analysis libraries, scikit-learn, imbalanced-learn, pandas and NumPy. It is also compatible with the nature-inspired optimization framework NiaPy.
3. The framework provides an easily extendable and customizable API, that can be customized with any scikit-learn compatible decision model or NiaPy compatible optimization method.
4. The framework provides data preprocessing for regression supervised learning problems.

The implementation of preprocessing tasks in the provided framework is on-par or better in comparison to other available approaches. Thus, the framework can be used as-is, but its main strength is that it provides a framework on which others can build upon to provide various specialized preprocessing approaches. The framework handles the parallelization and the evaluation of the optimization process and addresses the data leakage problem without any additional input needed.

The rest of the paper is organized as follows. The next section contains the problem formulation of the three preprocessing tasks and the literature overview of data preprocessing with the nature-inspired method. Section 3 contains the implementation details of the presented framework, which is followed by the fourth section with examples of use. Finally, concluding remarks are provided in the fifth section of the paper.

2. Problem Formulation

Let $X \in \mathbb{R}^{n \times m}$ be a data matrix (data set) where n denotes the number of instances (samples) in the data set, and m is the number of features. The data set X consists of instances $x_1, x_2, \ldots, x_n$, where $x_i \in \mathbb{R}^m$ and is written as $X = [x_1, x_2, \ldots, x_n]$. Also, the data set X consists of features $f_1, f_2, \ldots f_m$, where $f_i \in \mathbb{R}^n$ and is denoted as $X = [f_1, f_2, \ldots, f_m]$. As we are dealing with a supervised data mining problem, there is also Y, which is a vector of target values which are to be predicted. If the problem is in a type of classification, the values in Y are nominal $Y \in \{c_1, c_2, \ldots, c_k\}$, where k is the number of predefined discreet categories or classes. On the contrary, if we are dealing with regressing, the target values in Y are continuous $Y \in \mathbb{R}^n$. The goal of supervised learning is to construct the model

M, which can map X to $\hat{Y}$ while minimizing the difference between the predicted target values $\hat{Y}$ and true target values Y.

Feature selection handles the curse of dimensionality when machine learning models tend to over-fit on data with a too large set of features [23]. It is a technique where most relevant features are selected, while the redundant, noisy or irrelevant features are discarded. The result of using feature selection is improved learning performance, increased computational efficiency, decreased memory storage space needed and more generalizable models [5]. In mathematical terms, the feature selection transforms the original data X to the new X_{FS} (Equation (1)), with a potentially smaller set of features f_i' than in the original set.

$$\begin{aligned} X &= [f_1, f_2, \ldots, f_m] \\ X_{FS} &= [f_1', f_2', \ldots, f_l'] \\ l &\leq m \\ l &> 0 \end{aligned} . \quad (1)$$

Feature selection has already been addressed with meta-heuristic and nature-inspired optimization methods, as has been demonstrated in review papers [24,25]. Lately, the research topic has gained extensive focus from the nature-inspired optimization research community, with the application of every type of nature-inspired method to the given problem—whale optimization algorithm [26], dragonfly [27] and chaotic dragonfly algorithm [28], grasshopper algorithm [29], grey wolf optimizer [30,31], differential evolution and artificial bee colony [32], crow search [33], swarm optimization [34], genetic algorithm [35] and many others.

On the other hand, **data sampling** and **data weighting** (sometimes *instance weighting*) address the problem of improper ratios of instances in the learning data set [36]. The problem is two-fold—some types of instances can be over-represented (the majority), and other instances can be under-represented (the minority).

Imbalanced data sets can form for various reasons. Either there is a natural imbalance in real-life cases, or certain instances are more difficult to collect. For example, some diseases are not common and therefore patients with similar symptoms without the rare disease are much more common than the patients with symptoms that have the rare condition [37,38]. In other cases, the balanced and representative collection of data that reflects the population is sometimes problematic or even impossible. One major cause of this problem in social domains is the well-documented self-selection bias, where only a non-representative group of individuals select themselves into the group [39]. Convenience sampling is the next reason for over-representation of some and under-representation of other samples [40]. For example, the cost-effectiveness of data collection can also contribute to the emergence of majority and minority cases, when minorities are expensive (be it time- or financial cost-wise) to obtain.

Furthermore, machine learning models require an appropriate representation of all types of instances to be able to extract the signal and not confuse it with noise [41]. Data sampling [42–44] and cost-sensitive instance weighting [45] have already been tackled with meta-heuristic methods, evolutionary algorithms, and nature-inspired methods.

In the **data sampling** task, we transform the original data set X to X_S, with a potentially different distribution of instances. Note, that we can (1) *under-sample* the data set–only select the most relevant instances, (2) *over-sample* the data set–introduce copies or new instances to the original set X, or (3) *simultaneously under- and over-sample* the data set. The latter removes redundant instances and introduces new ones in the new data set X_S. With the EvoPreprocess framework, the simultaneous under- and over-sampling is used, where instances can be removed, and copies of existing instances can be introduced. The size n' of the new set X_S can be different or equal to the size n of the original set X, but the distribution of the instances in the X_S should be different from the X. This is shown in Equation (2).

$$\begin{aligned} n &= |X| \\ n' &= |X_S| \\ \\ n' &> 0 \\ X_S &\neq X \\ X_S \cap X &\neq \emptyset. \end{aligned} \tag{2}$$

On the other hand, **data weighting** does not alter the original data set X, but introduces the importance factor of instances in the X, called weights. The greater the importance of the instance, the bigger the weight for that instance, and vice versa—the lesser the importance, the smaller the instance weight. Fitting the machine learning model on weighted instances is called *cost-sensitive* learning, and can be utilized in several different machine learning models [46]. Let us denote the vector of weights as W, which consists of individual weights $w_1, w_2, \ldots, w_n$, where $w_i \in \mathbb{R}^+$ as is presented in Equation (3).

$$\begin{aligned} W &= [w_1, w_2, \ldots, w_n] \\ |W| &= |X| \end{aligned}. \tag{3}$$

While there are some rudimentary feature selection, data weighting, and data sampling methods in the Python machine learning framework scikit-learn, again, there is a lack of evolutionary and nature-inspired methods either included in this framework or, as independent open-source libraries, compatible with it. EvoPreprocess intends to fill this gap by providing a scikit-learn compatible toolkit in Python, which can be extended easily with the custom nature-inspired algorithms.

2.1. Nature-Inspired Preprocessing Optimization

Nature-inspired optimization algorithms is a broad term for meta-heuristic optimization techniques inspired by nature, more specifically by biological systems, swarm intelligence, physical and chemical systems [47,48]. In essence, these algorithms look for good enough solutions to any optimization problem with the formulation [47,49] in Equation (4).

$$\begin{aligned} S_t &= [s_1, s_2, \ldots, s_k] \\ S_{t+1} &= A\left\{S_t; (p_1, p_2, \ldots, p_h); (w_1, w_2, \ldots, w_g))\right\}. \end{aligned} \tag{4}$$

The set S_t is a set of solutions $[s_1, s_2, \ldots, s_k]$ in the iteration t. The next iteration of solutions S_{t+1} is generated using an algorithm A in accordance to the solution set in the previous iteration S_t, the parameters $(p_1, p_2, \ldots, p_h)$ of the algorithm A, and random variables $(w_1, w_2, \ldots, w_g)$.

These algorithms have already been successfully applied to the preprocessing tasks [11,12,25], proving the validity and the efficacy of these methods.

2.1.1. Solution Encoding for Preprocessing Tasks

The base of any optimization method is the solution s to the problem, which is encoded in a way that changing operators can be applied on it to minimize the distance between the solution s and the optimal solution. A broad set of nature-inspired algorithms work with the encoding of one solution in the form of an array of values. If the values in the array are continuous numbers, we are dealing with a continuous optimization problem. Alternatively, we also have a discrete optimization problem where values in the array are discrete. Even though some problems need a discrete solution, one can use continuous optimization techniques that can be mapped to the discrete search space [50]. Figure 1 shows the common solution encoding s for feature selection, data sampling, and data weighting and the transformation from a continuous search space to a discrete one, where this is applicable (data sampling and feature selection).

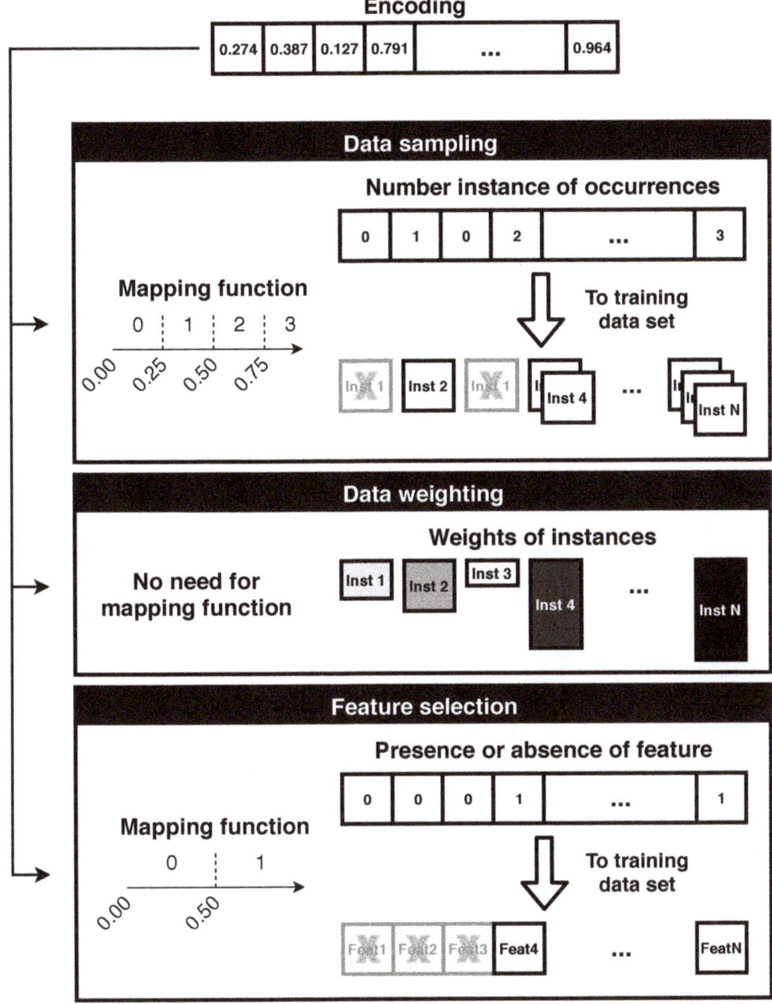

Figure 1. Encoding and decoding of solutions for data preprocessing tasks.

As Figure 1 shows, the encoding of *data sampling* with discrete values is straightforward—one value in the encoding array corresponds to one instance from the original data set X, and the scalar value represents the number of occurrences in the sampled data set X_S. When using continuous optimization, one can use mapping function m, which splits the continuous search space into bins, each with its corresponding discrete value. Note that discrete values of occurrences can take any of the non-negative integers, but solution encoding can be any non-negative real value. This is reflected in Equation (5).

$$
\begin{aligned}
s_S &= [e_1, e_2, \ldots, e_k] \\
e_i &\in \mathbb{R}_{\geq 0} \\
\\
G_S &= m(s_S) \\
m &: \mathbb{R}_{\geq 0} \longrightarrow \mathbb{N}_0 \\
\\
G_S &= [g_1, g_2, \ldots, g_k] \\
g_i &\in \mathbb{N}_0
\end{aligned}
\tag{5}
$$

The encoding for *data weighting* is even more straightforward. Again, each value in the array corresponds to one instance in the data set X, and the scalar values represent the actual weights W of the instances. There is no need for the mapping function, as long as we limit the interval of allowed values for scalars in solution e_i to $[0, max\_weight]$ as shown in Equation (6). Some implementations of the machine learning algorithms accept only weights up to 1, thus limiting the search space to $[0, 1]$; others have no such limit, broadening the search space to $[0, \infty)$.

$$
\begin{aligned}
s_W &= [e_1, e_2, \ldots, e_k] \\
e_i &\in \mathbb{R}_{\geq 0}.
\end{aligned}
\tag{6}
$$

Encoding for *feature selection* is in the form of an array of binary values—the feature is either present (value 1) or absent (value 0) from the changed data set X_{FS} (see Equation (7)). Using the continuous solution encoding, one should again use the mapping function m, which splits the search space into two bins (with arbitrary limits), one for the feature being present and one for the feature being absent.

$$
\begin{aligned}
s_{FS} &= [e_1, e_2, \ldots, e_k] \\
e_i &\in \mathbb{R}_{\geq 0} \\
\\
G_{FS} &= m(s_{FS}) \\
m &: \mathbb{R}_{\geq 0} \longrightarrow \{0, 1\} \\
\\
G_{FS} &= [g_1, g_2, \ldots, g_k] \\
g_i &\in \{0, 1\}.
\end{aligned}
\tag{7}
$$

2.1.2. Self-Adaptive Solutions

The presented solution encoding shows that there are different options for mapping the encoding to the actual data set. These mapping settings are the following:

- In *data weighting* the maximum weight can be set,
- in *data sampling* the mapping from the continuous value to the appearance count can be set, and
- in *feature selection* the mapping from the continuous value to presence or absence of the feature can be set.

In general, these values can all be set arbitrary, but could also be one of the objectives of the optimization process itself. The values of these parameters can seriously influence the quality of the results in preprocessing tasks [51–53], as it can guide the evolution of the optimization to the global optima, rather than to the local one. The implementation of the preprocessing tasks in EvoPreprocess uses this self-adaptive approach with the additional genes in the genotype [54].

In the *data weighting* task, there is one additional gene, which corresponds to the maximum weight that can be assigned. All of the genes corresponding to the weights are normalized to this maximum value while leaving the minimum at 0. When using the imbalanced data set it is preferred that bigger differences in weights are possible.

In *data sampling* the mapping of the interval $[0, 1]$ to the instance occurrence count is set. Here, n_{max} setting genes are added to the genotype, where n_{max} is the maximum number of occurrences on the individual instance after it is over-sampled. Each setting gene presents the size of the mapping interval of an individual occurrence count. Figure 2 presents the process of splitting the encoding genome to the sampling and the self-adaptation parts and mapping it to the solution. Note, that in Figure 2 the mapping intervals are just an example of one such self-adaptation and are not set as the final values used in all mappings. These values differ from solutions and are also data set dependent.

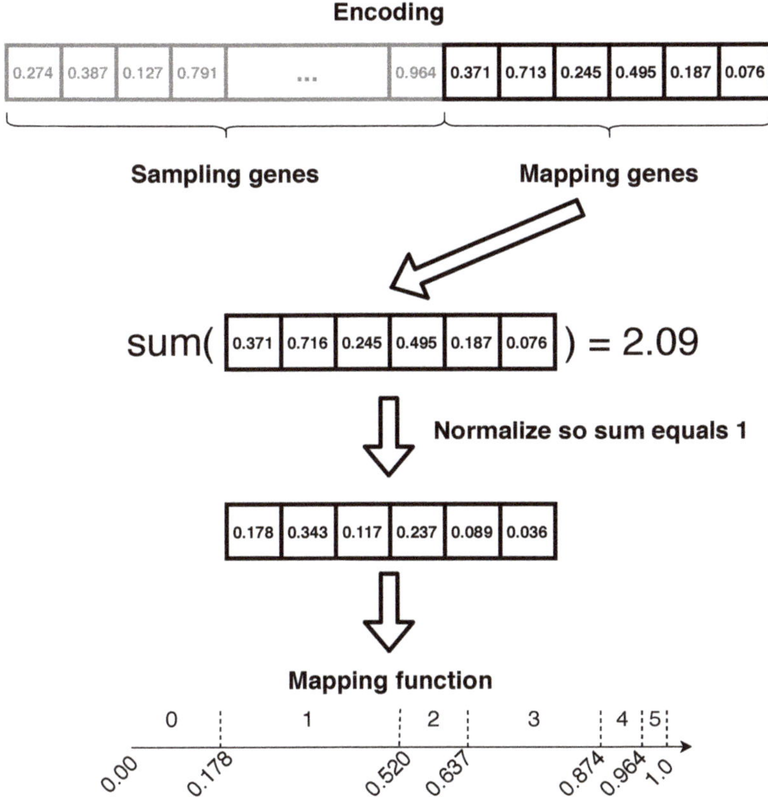

Figure 2. The self-adaptation with mapping genes in the encoding for data sampling task. The mapping values presented in this example are determined by the genotype and are not fixed, but adapt during the evolution process.

Some heavily imbalanced data sets could be better analysed if the emphasis is on under-sampling—the interval for 0 occurrences (the absence of the instance) would become bigger in the process. Other data sets could be more suitable if there are more minority instances, and therefore intervals for many occurrences become bigger within the optimization process.

With *feature selection* where is only one setting gene added to the genotype. This gene represents the size of the first interval, which maps other genes to the absence of the feature. Wider the data sets with more features could be analyses with better results if more features are removed from the data sets. Therefore, the larger the number in the setting gene is preferred, when the more emphasis is given to the shrinkage of the data set as more features are not selected. Again, the mapping interval in Figure 3 are just an example for that particular mapping gene in the encoded genotype.

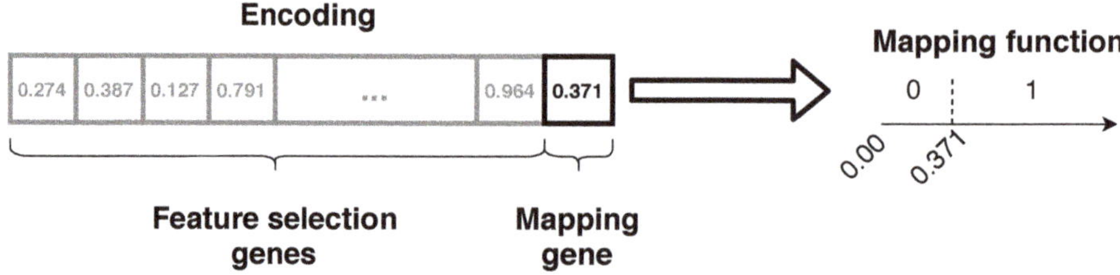

Figure 3. The self-adaptation with mapping gene in the encoding for feature selection task. The mapping value presented in this example are determined by the genotype and are not fixed, but adapt during the evolution process.

The self-adaptation is implemented in the framework but could be removed in the extension or customization of the individual optimization process. It is included in the framework as it is an integral part of the optimization process in recent state-of-the-art papers [53,55–57] on data preprocessing with nature-inspired methods.

2.1.3. Optimization Process for Preprocessing Tasks

All nature-inspired methods run in iterations during which the optimization process is applied. The broad overview of the nature-inspired optimization algorithm is shown in Figure 4 and is based on the rudimentary framework for nature-inspired optimization of feature selection by [25].

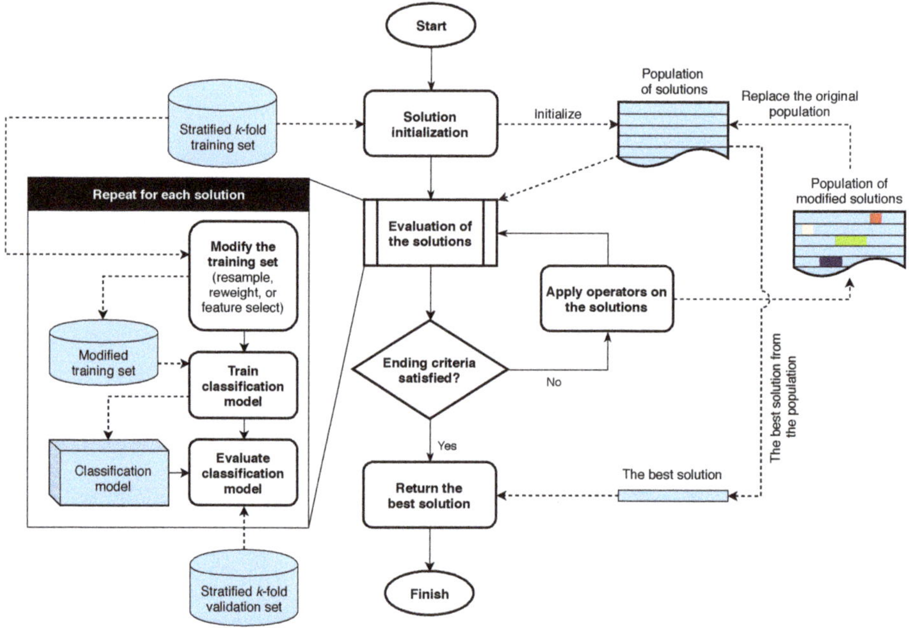

Figure 4. The optimization process of preprocessing data with nature-inspired optimization methods in EvoPreprocess.

The optimization process starts with the initialization of solutions, which is either random or with some heuristic methodology. After that, the iteration loop starts. First, every solution is evaluated, which will be discussed later in this section. Next, the optimization operators are applied to the solutions, which are specific to each the nature-inspired algorithm. The general goal of the operators is to select, repair, change, combine, improve, mutate, and so forth,. solutions in the direction of a perceived (either local or, hopefully, global) optimum. The iteration loop is stopped when a predefined limit of ending criteria is reached, be it the maximum number of iterations, the maximum number of iterations when the solutions stagnate, the quality that solutions reach, and others.

Nature-inspired algorithms strive to keep good solutions and discard the bad ones. Here, the method of evaluating solutions plays an important role. One should define one or more objectives that solutions should meet, and the evaluation function grades the quality of the solution based on these objectives. As solutions are evaluated every iteration, this is usually the most computationally time-consuming process of all steps in the optimization process.

When dealing with the optimization of the data set for the machine learning process, multiple objectives have been considered in the literature [25,58–61]. Usually, one of the most important objectives is the quality of the fitted model from the given data set. If the classification task is used the standard classification metrics could be used: accuracy, error rate, F-measure, G-measure, area under the ROC curve (AUC). If we are dealing with the regression task, the following regression tasks can be used—mean squared error, mean absolute error, or explained variance. By default, the EvoPreprocess framework uses a single-objective optimization to obtain either the error rate and F-score for the classification and the mean squared error for the regression tasks. As the later sections will show, the framework is easily extendable to be used for multiple objectives, be it the size of the data set or others.

It is important to note that not all researchers consider the problem of data leakage. *Data leakage* occurs when information from outside the training data set is used in the fitting of the machine learning models. This manifests in over-fitted models that perform exceptionally well on the training set, but poorly on the hold-out testing set. A common mistake that leads to data leakage is not using the validation set when optimizing the model fitting. Applying the same logic to nature-inspired preprocessing, data leakage occurs when the same data are used in the evaluation process and the final testing process. The EvoPreprocess framework automatically holds-out the separate validation sets and thus, prevents data leakage.

2.2. Nature-Inspired Algorithms

A large number of different nature-inspired algorithms were proposed in recent years. The recent meta-heuristic research of nature-inspired algorithm was reviewed by Lones [62] in 2020, where he concluded that most recent innovations are usually small variations of already existing optimization operators. Still, there are some novel algorithms worth further investigation: polar bear optimization algorithm [63], bison algorithm optimization [64], butterfly optimization [65], cheetah based optimization algorithm [66], coyote optimization algorithm [67] and squirrel search algorithm [68]. Due to the amount of nature-inspired algorithms, a broad overview is beyond the scope of this paper. Consequently, this section provides the formulation for the well-established and most used nature-inspired algorithms.

First, *Genetic algorithm* (GA) is one of the cornerstones of nature-inspired algorithms, presented by Sampson [69] in 1976, but research efforts on its the variations and novel applications are still numerous. The basic operators here are the selection of the solutions (called individuals) that are then used for the crossover (the mixing of genotype) which forms new individuals, which have a chance to go through a mutation procedure (random changing of genotype). The crossover is an exploitation operator where the solutions are varied to find their best variants. On the other hand, the mutation is a prime example of an exploration operator, which prevent the optimization to get stuck in the local optima. Each iteration is called a generation and can repeat until predefined criteria, be it in a form of maximum generations or stagnation limit. Most of the following optimization algorithms use a variation of the presented operators.

Next, *Differential evolution* (DE) [70] includes the same operators as GA, selection, crossover and mutation, but multiple solutions (called agents) can be used. In its basic form, the crossover of three agents is not done with the simple mixing of genes (like in GA), but the calculation from Equation (8) for each gene is used. Here $x_{parent1}$, $x_{parent2}$ and $x_{parent3}$ are gene values from three parents, x_{new} is the new value of the gene, and F is the differential weight parameter.

$$x_{new} = x_{parent1} + F * (x_{parent2} - x_{parent3}). \tag{8}$$

Evolution strategy (ES) [71] is an optimization algorithm, which has similar operators to GA, but the emphasis is given to the selection and mutation of the individuals. The most common variants are the following: $(\mu/\rho, \lambda)$, where only new solutions form the next generation and $(\mu/\rho + \lambda)$ where the old solutions compete with the new ones. The parameter μ presents the number of selected solutions for the crossover and mutation from where ρ are derived. Value λ denotes the size of the generation.

A variation of Evolution strategy is the *Harmony search* (HS) optimization algorithm [72] which mimics the improvisation of a musician. The main search operator is the pitch adjustment, which adds random noise to existing solutions (called harmony) or creates a new random harmony. Creation of new harmony is shown in Equation (9). where x_{old} denotes the old gene from the old genotype, x_{new} is the new gene solution, b_{range} denotes the range of maximal improvisation (change of solution) and ϵ is a random number in the interval $[-1, 1]$.

$$x_{new} = x_{old} + b_{range} * \epsilon. \tag{9}$$

Next group of nature-inspired algorithms mimic the behaviour of swarms and are called *swarm optimization* algorithms. *Particle swarm optimization* (PSO) [73] is a prime example of swarm algorithms. Here each solution (called particle) is supplemented with its velocity. This velocity represents the difference of change from its current position (genotype encoding) in the new iteration for this solution. After each iteration, the velocities are recalculated to direct the particle to the best position. The moving of particles in the search space represents both, the exploitation (moving around the best solution) and the exploration (moving towards the best solution) parts of optimization. The iterative moving of particles stops one of the following criteria is satisfied—the maximum number of iterations is reached, the stagnation limit is reached, or particles converge to one best position.

One variation of PSO is *Artificial bee colony* (ABC) algorithm [74] which imitates the foraging process of a honey bee swarm. This optimization algorithm consists of three operators which modify existing solutions. First, employed bees use local search to exploit already existing solutions. Next, the onlooker bees, which serve as a selection operator, search for new sources in their vicinity of existing ones (exploitation). And last, the scout bees are used for the exploration where they use a random search to find new food sources (new solutions).

Next, the *Bat algorithm* (BA) [75] which is a variant of PSO which imitates swarms of microbats, where every solution (called bat) still has a velocity, but also emits pulses with varying levels of loudness and frequency. The velocity of bats changes in consideration to the pulses from other bats and the pulses are determined by the quality of the solution. Equation (10) shows the procedure for updating the genes of individual bats. Here x_{old} and x_{new} denote the old and the new genes respectively, v_{old} and v_{new} are the old and the new velocities of the bats, and f_i, f_{min} and f_{max} are current, minimal and maximal frequencies. β is a random number in the interval $[0, 1]$.

$$\begin{aligned} f_i &= f_{min} + (f_{max} - f_{min}) * \beta \\ v_{new} &= v_{old} + (x_{old} - x_{best}) * f_i \\ x_{new} &= x_{old} + v_{new}. \end{aligned} \tag{10}$$

Finally, one of the widely used swarm algorithms is *Cuckoo search* (CS) [76], which imitates laying of eggs in the foreign nest by cuckoo birds. The solutions are eggs in the nests and those are repositioned to new nests every iteration (exploitation), while the worst ones are abandoned. The modification of the solutions is done with the optimization operator called the Lévy flight, which is in the form of long random flights (exploration) or short random flights (exploitation). The migration of the eggs with Lévy flight is shown in Equation (11), where α denotes the size of the maximal flight and $Levi(\lambda)$ represents a random number from Lévi distribution.

$$x_{new} = x_{old} + \alpha * Levi(\lambda). \tag{11}$$

2.3. Computation Complexity

In general, most nature-inspired algorithms have time complexity of $O(m * p * C_{operators} + p * C_{fitness_eval})$, where m is the size of the solution, p is the number of modified and evaluated solutions during the whole process, $C_{operators}$ is the complexity of the operators (i.e., crossover, recombination, mutation, selection, random jumps...), and $C_{fitness_eval}$ is the complexity of the evaluation.

The evaluation of solutions is the most computationally expensive part. One of the classification/regression algorithms must be used in order to (1) build the model, (2) make the predictions, and (3) evaluate the predictions. This part is heavily reliant on the chosen solution evaluator (the classification algorithm used). The evaluation of the prediction is a fixed $O(n)$ complexity, dependent on n samples in the data set. Training of the models and making predictions vary: decision tree with $O_{training}(n^2p)$ and $O_{prediction}(p)$, linear and logistic regression with $O_{training}(p^2n + p^3)$ and $O_{prediction}(p)$, and naive Bayes with $O_{training}(np)$ and $O_{prediction}(p)$. Some of the ensemble methods have and additional factor of the number of models in the ensembles (i.e., Random forest and AdaBoost), and neural networks depend on the net architecture. Usually, the more complex the training process, the more complex patterns can be extracted from the data and consequently, the predictions are better.

If the evaluator and its use are fixed (as is in the experimental part of the paper), optimization time varies in relation to the nature-inspired optimization algorithms used. Considering only the array style solution encoding, the basic exploration and exploitation type operators (i.e., crossover, recombination, mutation and random jumps) have $O(m * r * C_{operation})$ complexity, where m is the size of the encoded solution, r is the number of solutions used in the operator, and $C_{operation}$ is the complexity of the operation (i.e., linear combination, sum, distance calculation...). As the m size of the solution is usually fixed, researchers must optimize other aspects of the algorithms to get shorter computation times. Thus, the fastest optimization algorithms are the ones with the fewest operators, the operators with the fewest solutions participating in the optimization, or the lest complex operators. For example, the genetic algorithm has multiple relatively simple operators: crossover, mutation, elitism, selection (i.e., tournament where the fitness values are compared) and thus is one of the more time-consuming ones.

Furthermore, algorithms, where operators can be vectorized, can take advantage of computational speed-ups when low-level massive computation calls can be used. This is especially prevalent in swarm algorithms, where calculating distances or velocities and then moving the solutions can be done with matrix multiplications for all of the solution set at once, instead of individually for every solution. If the speed of the data preprocessing is of the essence, this should be taken into account in the implementation of the optimization algorithms. EvoPreprocess framework already provides parallel runs of the optimization process data set folds, but its speed is still reliant on the nature-inspired algorithms and evaluators used in the process.

3. EvoPreprocess Framework

The present section provides a detailed architecture description of the EvoPreprocess framework, which is meant to serve as a basis for further third party additions and extensions.

The EvoPreprocess framework includes three main modules:

- `data_sampling`
- `data_weighting`
- `feature_selection`

Each of the modules contains two files:

1. Main **task class** is to be used for running tasks data sampling, feature selection or data weighting;
2. Standard **benchmark class**, which is a default class used in the evaluation of the task, and can be replaced or extended by custom evaluation class.

Figure 5 shows a UML class diagram for the EvoPreprocess framework and its relation to the Python packages scikit-learn, imbalanced-learn and NiaPy. The custom implementation of benchmark classes are in the classes `CustomSamplingBenchmark`, `CustomWeightingBenchmark` and `CustomFeatureSelectionBenchmark`, which are denoted with dark background color.

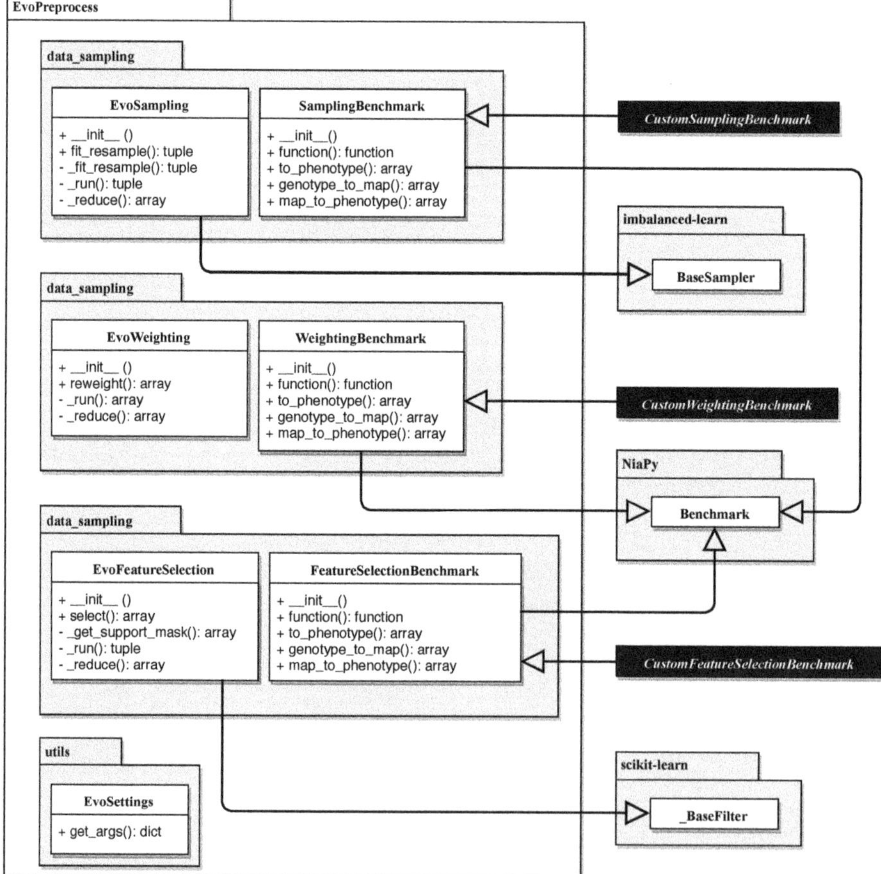

Figure 5. UML class diagram of the EvoPreprocess framework.

3.1. Task Classes

The task class for data sampling is `EvoSampling`, which extends the imbalanced-learn class for sampling data `BaseOverSampler`. The task class for feature selection, `EvoFeatureSelection`, extends `_BaseFilter` class from scikit-learn. The main task class for data weighting does not use any parent class. All three task classes are initialized with the following parameters.

- `random_seed` ensures reproducibility. The default value is the current system time in milliseconds.
- `evaluator` is the machine learning supervised approach used for the evaluation of preprocessed data. Here, scikit-learn compatible classifier or regressor should be used. See description of the **benchmark** classes for more details.
- `optimizer` is the optimization method used to get the preprocessed data. Here, the NiaPy compatible optimization method is expected, with the function `run` and the usage of the evaluation benchmark function. The default optimization method is the genetic algorithm.
- `n_folds` is the number of folds for the cross-validation split into the training and the validation sets. To prevent data leakage, the evaluations of optimized data samplings should be done on the hold-out validation sets. The default number of folds is set to 2 folds.
- `n_runs` is the number of independent runs of the optimizer on each fold. If the optimizer used is deterministic, just one run should be sufficient, otherwise more runs are suggested. The default number of runs is set to 10.
- `benchmark` is the evaluation class which contains the function that returns the quality of data sampling. The custom benchmark classes should be used here if the data preprocessing objective is different from the singular objective of optimizing error rate and F-score (for classification) or mean squared error (for regression).
- `n_jobs` is the number of optimizers to be run in parallel. The default number of jobs is set to the number of CPU cores.

The base optimization procedure pseudo-code is demonstrated in Algorithm 1, where the train-validation split (lines 3 and 4) and multiple parallel runs (for loop in lines 6, 7 and 8) are shown. Note that the X_T data set given to the procedure should already be the training set and not the whole data set X. The further splitting of X_T into training and validation data sets ensures that data leakage does not happen. To ensure that random splitting into training and validation sets does not produce split dependent results, the stratified k-fold splitting is applied (into n_folds folds). As the nature of nature-inspired optimization methods is non-deterministic, optimization is done multiple times (*n_runs* parameter) and is run in parallel—every optimization on the separate CPU core. The reduction (aggregation) of the results to one final results is done in different ways, dependent on the preprocessing task.

- In *data sampling* the best performing instance occurrences in every fold are aggregated with mode.
- In *data weighting* the best weights in every fold are averaged.
- In *feature selection* the best performing selected features in every fold are aggregated with mode.

The preprocessing procedure from Algorithm 1 is called in different way in every task, which is the consequence of different inheritance for every task: data sampling inherits from scikit-learn `_BaseFilter`, feature selection inherits from imbalanced-learn `BaseSampler` and data weighting does not inherit from any class.

All three task classes contain the following functions.

- `_run` is a private static function to create and run the `optimizer` with the provided evaluation `benchmark` function. Multiple calls of this function can be run in parallel.
- `_reduce` is a private static function used to aggregate (reduce) the results of individual runs on multiple folds in one final sampling.

Algorithm 1: Base procedure for running preprocessing optimization.

```
input  : X_T training data set
         n_folds number of folds for k-fold validation
         n_runs number of individual parallel runs for each fold
         optimizer nature-inspired optimization algorithm
         evaluator classifier/regressor for evaluation processed data set
         benchmark function to evaluate preprocessed data with evaluator
output: X_processed processed data set

results ← ∅
for i ← 1 to n_folds do
    training_set ← getTrainingData(X_T)
    validation_set ← getValidationData(X_T)
    split_results ← ∅
    // This for loop is run in parallel
    for r ← 1 to n_runs do
        X′ ← _run(training_set, validation_set, optimizer, evaluator, benchmark)
        split_results ← split_results ∪ X′
    end
    best_result ← best (split_results)
    results ← results ∪ best_result
end
X_processed ←_reduce(results)
```

In addition to those functions, the class `EvoSampling` also contains the function which is used for sampling data.

- `_fit_resample` private function which gets the data to be sampled `X` and the corresponding target values y. This is an implementation of the abstract function from imbalanced-learn package, is called from the public `fit_resample` function from the `BaseOverSampler` class, and provides the possibility to be included in imbalanced-learn pipelines. It contains the main logic of the sampling process. The function returns a tuple with two values: `X_S` which is a sampled `X` and `y_S` which is a sampled y.

  The class `EvoWeighting` also contains the following function.

- `reweight`, which gets the data to be reweighted `X` and the corresponding target values y. This function returns the array `weights` with weights for every instance in `X`.

  The class `EvoFeatureSelection` contains the following functions.

- `_get_support_mask` private function, which checks if features are already selected. It overrides the function from the `_BaseFilter` class from scikit-learn.
- `select` is a score function that gets the data `X` and the corresponding target values y and selects the features from `X`. This function is provided as the scoring function for the `_BaseFilter` class so it can be used in scikit-learn pipelines as the feature selection function. This function returns the `X_FS`, which is derived from `X` with potentially some features removed.

3.2. Evaluation of Solutions with Benchmark Classes

The second part of classes in all the modules are **benchmark** classes, which are helper classes meant to be used in the evaluation of the samplings, weighting or feature selection tasks. The implementation of custom fitness evaluation function should be done by replacing or extending of these classes. The benchmark classes are initialized with the following parameters.

- `X` data to be preprocessed.
- y target values for each data instance from `X`.
- `train_indices` array of indices of which instances from `X` should be used to train the evaluator.
- `valid_indices` array of indices of which instances from `X` should be used for validation.
- `random_seed` is the random seed for the evaluator and ensures reproducibility. The default value is 1234 to prevent different results from evaluators initialized at different times.
- `evaluator` is the machine learning supervised approach used for the evaluation of the results of the task. Here, scikit-learn compatible classifier or regressor should be used: evaluator's function `fit` is used to construct the model, and function `predict` is used to get the predictions. If the target of the data set is numerical (regression task) the regression method should be provided, otherwise the classification method is needed. The default evaluator is `None`, which sets the evaluator to either linear regression if the data set target is a number, or Gaussian naive Bayes classifier if the target is nominal.

The benchmark classes all must provide one function—`function` which returns the evaluation function of the tasks. This architecture is in accordance with NiaPy benchmark classes for their optimization methods and was used in EvoPreprocess for compatibility reasons.

Three benchmark classes are provided, each for its task to be evaluated:

- `SamplingBenchmark` for data sampling,
- `WeightingBenchmark` for data weighting, and
- `FeatureSelectionBenchmark` for feature selection.

All provided benchmarks classes evaluate the task in the same way: evaluator is trained on the training set selected with `train_indices` and evaluated on the validation set selected with `valid_indices`. If the evaluator provided is a classifier, the sampled, weighted or feature selected data are evaluated with *error rate*, which is defined in the Equation (12) and presents the ratio of misclassified instances [77].

$$Error\ rate = \frac{1}{n}\sum_{n}^{i-1} z \begin{cases} 0 & \text{if } y_i = \widehat{y_i} \\ 1 & \text{if } y_i \neq \widehat{y_i} \end{cases} \tag{12}$$

The *F-score* can be used to evaluate the classification quality, where the balance between precision and recall is considered [77]. As the optimization strives to minimize the solution values, 1 − F-score is used then. If the evaluator is a regressor, the new data set is evaluated using *mean squared error* presented in Equation 13 and calculates the average of absolute error in predicting the outcome variable [77].

$$MSE = \frac{1}{n}\sum_{n}^{i-1} (y_i - \widehat{y_i})^2 . \tag{13}$$

4. Examples of Use

This section presents examples of using EvoPreprocess in all three supported data preprocessing tasks: first, the problem of data sampling, next, we cover data weighting, and finally, the feature selection problem. Before using the framework, the following requirements must be met.

- Python 3.6 must be installed,
- The NumPy package,
- The scikit-learn package, at least version 0.19.0,
- The imbalanced-learn package, at least version 0.3.1,
- The NiaPy package, at least version 2.0.0*rc*5, and
- The EvoPreprocess package can be accessed at https://github.com/karakatic/EvoPreprocess.

4.1. Data Sampling

One of the problems addressed by the EvoPreprocess is data sampling; or more specifically, simultaneous under- and over-sampling of data with class `EvoSampling` in the module `data_sampling` and its function `fit_resample()`. This function returns a two arrays; a two-dimensional array of sampled instances (rows) and features (columns), and a one-dimensional array of sampled target values from the corresponding instances from the first array. The code below shows the basic use of the framework for resampling data for the classification problem, with default parameter values.

```
>>> from sklearn.datasets import load_breast_cancer
>>> from EvoPreprocess.data_sampling import EvoSampling
>>>
>>> dataset = load_breast_cancer()
>>> print(dataset.data.shape, len(dataset.target))
[569,30] 569

>>> X_resampled, y_resampled = EvoSampling().fit_resample(dataset.data, dataset.target)
>>> print(X_resampled.shape, len(y_resampled))
[341,30] 341
```

The results of the run show that there were 569 instances in the data set before sampling and there are 341 instances in the data set after the sampling, which shows that more samples were removed from the data set than there were added.

The following code shows the usage of data sampling for the data set for the regression problem. The code also demonstrates the setting of parameter values: a non-default optimizer method of evolution strategy, 5 folds for validation, 5 numbers of individual runs on each fold split and 4 parallel executions of individual runs.

```
>>> from sklearn.datasets import load_boston
>>> from EvoPreprocess.data_sampling import EvoSampling, SamplingBenchmark
>>>
>>> dataset = load_boston()
>>> print(dataset.data.shape, len(dataset.target))
(506, 13) 506

>>> X_resampled, y_resampled = EvoSampling(
evaluator=DecisionTreeRegressor(),
optimizer=nia.Evolution strategy,
n_folds=5,
n_runs=5,
n_jobs=4,
benchmark=SamplingBenchmark
).fit_resample(dataset.data, dataset.target)
>>> print(X_resampled.shape, len(y_resampled))
(703, 13) 703
```

Note that, in this case, the optimized resampled set is bigger (703 instances) than the original non-resampled data set (506 instances). This is a clear example of when more instances are added than removed from the set. If only under-sampling is preferred instead of simultaneous under- and over-sampling, the appropriate under-sampling benchmark class would be provided as the value for the `benchmark` parameter. In this example, a CART regression decision tree is used as the evaluator for the resampled data sets (`DecisionTreeRegressor`), but any scikit-learn regressor could take its place.

4.2. Data Weighting

Some scikit-learn models can handle weighted instances. Class `EvoWeighting` from the module `data_weighting` optimizes weights of individual instances with the call of `reweight()` function. This function serves to find instance weights that can lead to better classification or regression results—it returns the array of the real numbers, which are weights of instances in the order of those instances in the given data set. The following code shows the basic example of reweighting and the resulting array of weights.

```
>>> from sklearn.datasets import load_breast_cancer
>>> from EvoPreprocess.data_weighting import EvoWeighting
>>>
>>> dataset = load_breast_cancer()
>>> instance_weights = EvoWeighting().reweight(dataset.data,
dataset.target)
>>> print(instance_weights)
[1.568983893273244 1.2899430717992133 ... 0.7248390003761751]
```

The following code shows the example of combining the data weight optimization with the scikit-learn classifier of the decision tree, which supports weighted instances. The accuracies of both classifiers, the one fitted with unweighted data and the one built with weighted data, are outputted.

```
>>> from sklearn.datasets import load_breast_cancer
>>> from sklearn.model_selection import train_test_split
>>> from sklearn.tree import DecisionTreeClassifier
>>> from EvoPreprocess.data_weighting import EvoWeighting
>>>
>>> random_seed = 1234
>>> dataset = load_breast_cancer()
>>> X_train, X_test, y_train, y_test = train_test_split(
dataset.data, dataset.target,
test_size=0.33,
random_state=random_seed)
>>> cls = DecisionTreeClassifier(random_state=random_seed)
>>> cls.fit(X_train, y_train)
>>>
>>> print(X_train.shape,
accuracy_score(y_test, cls.predict(X_test)),
sep=':')
(381, 30): 0.8936170212765957

>>> instance_weights = EvoWeighting(random_seed=random_seed).reweight(X_train,
y_train)
>>> cls.fit(X_train, y_train, sample_weight=instance_weights)
>>> print(X_train.shape,
accuracy_score(y_test, cls.predict(X_test)),
sep=':')
(381, 30): 0.9042553191489362
```

The example shows that the number of instances stays 381 and the number of feature stays at 30, but the accuracy rises from 89.36% with unweighted data to 90.43% with weighted instances.

4.3. Feature Selection

Another task performed by the EvoPreprocess framework is feature selection. This is done with nature-inspired algorithms from the NiaPy framework, and can be used independently, or as one of the steps in the scikit-learn pipeline. The feature selection is used with the construction of the `EvoFeatureSelection` class from the module `feature_selection` and calling the function `fit_transform()`, which returns the new data set with only selected features.

```
>>> from sklearn.datasets import load_breast_cancer
>>> from EvoPreprocess.feature_selection import EvoFeatureSelection
>>>
>>> dataset = load_breast_cancer()
>>> print(dataset.data.shape)
(569, 30)

>>> X_new = EvoFeatureSelection().fit_transform(dataset.data,
dataset.target)
>>> print(X_new.shape)
(569, 17)
```

The results of the example demonstrate, that the original data set contains 30 features, and the modified data set after the feature selection contains only 17 features.

Again, numerous settings can be changed: the nature-inspired algorithm used for the optimization process, the classifier/regressor used for the evaluation, the number of folds for validation, the number of repeated runs for each fold, the number of parallel runs and the random seed. The following code shows the example of combining `EvoFeatureSelection` with the regressor from scikit-learn, and the evaluation of the quality of both approaches – the regressor built with the original data set, and the regressor built with the modified data set with only some features selected.

```
>>> from sklearn.datasets import load_boston
>>> from sklearn.metrics import mean_squared_error
>>> from sklearn.model_selection import train_test_split
>>> from sklearn.tree import DecisionTreeRegressor
>>> from EvoPreprocess.feature_selection import EvoFeatureSelection
>>>
>>> random_seed = 654
>>> dataset = load_boston()
>>> X_train, X_test, y_train, y_test = train_test_split(
dataset.data,
dataset.target,
test_size=0.33,
random_state=random_seed)
>>> model = DecisionTreeRegressor(random_state=random_seed)
>>> model.fit(X_train, y_train)
>>> print(X_train.shape,
mean_squared_error(y_test, model.predict(X_test)),
sep=':')
(339, 13): 24.475748502994012

>>> evo = EvoFeatureSelection(evaluator=model, random_seed=random_seed)
>>> X_train_new = evo.fit_transform(X_train, y_train)
>>>
>>> model.fit(X_train_new, y_train)
>>> X_test_new = evo.transform(X_test)
>>> print(X_train_new.shape,
mean_squared_error(y_test, model.predict(X_test_new)),
sep=':')
(339, 6): 18.03443113772455
```

The results show that, using the new data set with only 6 features selected during the fitting of the decision tree regressor, outperforms the decision tree regressor built with the original data set with all of the 13 features—MSE of 18.03 for the regressor with feature selected data set vs. MSE of 24.48 for the regressor with the original data set.

4.4. Compatibility and Extendability

The compatibility with existing well-established data analysis machine learning libraries is one of the main features of EvoPreprocess. For this reason, all the modules accept the data in the form of extensively used NumPy array [78] and the pandas `DataFrame` [79]. The following examples demonstrate the further compatibility capacity of the EvoPreprocess with scikit-learn and imbalanced-learn Python machine learning packages.

The scikit-learn already includes various feature selection methods and supports their usage with the provided machine learning pipelines. `EvoFeatureSelection` extends scikit-learn's feature selection base class, and so it can be included in the pipeline, as the following code demonstrates.

```
>>> from sklearn.linear_model import LinearRegression
>>> from sklearn.pipeline import Pipeline
>>> from sklearn.datasets import load_boston
>>> from sklearn.metrics import mean_squared_error
>>> from sklearn.model_selection import train_test_split
>>> from sklearn.tree import DecisionTreeRegressor
>>> from EvoPreprocess.feature_selection import EvoFeatureSelection
>>>
>>> random_seed = 987
>>> dataset = load_boston()
>>>
>>> X_train, X_test, y_train, y_test = train_test_split(
dataset.data,
dataset.target,
test_size=0.33,
random_state=random_seed)

>>> model = DecisionTreeRegressor(random_state=random_seed)
>>> model.fit(X_train, y_train)

>>> print(mean_squared_error(y_test, cls.predict(X_test)))
20.227544910179642

>>> pipeline = Pipeline(steps=[
('feature_selection', EvoFeatureSelection(
evaluator=LinearRegression(),
n_folds=4,
n_runs=8,
random_seed=random_seed)),
('regressor', DecisionTreeRegressor(random_state=random_seed))
])
>>> pipeline.fit(X_train, y_train)

>>> print(mean_squared_error(y_test, pipeline.predict(X_test)))
19.073532934131734
```

As the example demonstrates, the pipeline with `EvoFeatureSelection` builds a better regressor than the model without the feature selection, as the MSE is 19.07 vs. the original regressor's MSE of 20.23. The example also shows that one can choose different evaluators in any of the preprocessing tasks from the final classifier or regressor. Here, the linear regression is chosen as the evaluator, but the decision tree regressor is used in the final model fitting. Using a different, lightweight, evaluation model can be useful when fitting the decision models that can be computationally expensive.

All three EvoPreprocess tasks are compatible and can be used in imbalanced-learn pipelines [17], as is demonstrated in the following code. Note that both tasks in the example are parallelized and the number of simultaneous runs on different CPU cores is set with the `n_jobs` parameter (default value of `None` utilizes all cores available). The reproducibility of the results is guaranteed with setting the `random_seed` parameter.

```
>>> from sklearn.datasets import load_breast_cancer
>>> from sklearn.tree import DecisionTreeClassifier
>>> from imblearn.pipeline import Pipeline
>>> from EvoPreprocess.feature_selection import EvoFeatureSelection
>>> from EvoPreprocess.data_sampling import EvoSampling
>>> random_seed = 1111
>>> dataset = load_breast_cancer()
>>> X_train, X_test, y_train, y_test = train_test_split(
dataset.data,
dataset.target,
test_size=0.33,
random_state=random_seed)
>>>
>>> cls = DecisionTreeClassifier(random_state=random_seed)
>>> cls.fit(X_train, y_train)
>>> print(accuracy_score(y_test, cls.predict(X_test)))
0.8829787234042553

>>> pipeline = Pipeline(steps=[
('feature_selection', EvoFeatureSelection(n_folds=10,
random_seed=random_seed)),
('data_sampling', EvoSampling(n_folds=10,
random_seed=random_seed)),
('classifier', DecisionTreeClassifier(random_state=random_seed))])
])
>>> pipeline.fit(X_train, y_train)
>>> print(accuracy_score(y_test, pipeline.predict(X_test)))
0.9148936170212766
```

The results show that the pipeline with both feature selection and data sampling results in a superior classifier than the one without these preprocessing steps, as the accuracy of the pipeline is 91.49 vs. 88.30 for the classifier without preprocessing steps.

Using any of the EvoPreprocess tasks can be further optimized with the custom settings of the optimizer (the nature-inspired algorithm used to optimize the task). This can be done with the parameter `optimizer_settings`, which is in the form of a Python dictionary. The code listing below shows one such example, where the bat algorithm [80] is used for the optimization process, and samples the new data set with only 335 instances instead of the original 569. One should refer to the NiaPy package for the available nature-inspired optimization methods and their settings.

```
>>> from sklearn.datasets import load_breast_cancer
>>> from EvoPreprocess.data_sampling import EvoSampling
>>> import NiaPy.algorithms.basic as nia
>>>
>>> dataset = load_breast_cancer()
>>> print(dataset.data.shape, len(dataset.target))
[569,30] 569

>>> settings = {'NP': 1000, 'A': 0.5, 'r': 0.5, 'Qmin': 0.0, 'Qmax': 2.0}
>>> X_resampled, y_resampled = EvoSampling(optimizer=nia.Bat algorithm,
optimizer_settings=settings
).fit_resample(dataset.data,
dataset.target)
>>> print(X_resampled.shape, len(y_resampled))
(335, 30) 335
```

As EvoPreprocess uses any NiaPy compatible continuous optimization algorithm, one can implement their own, or customize and extend the existing ones. The customization is also possible on the evaluation functions (i.e., favoring or limiting to under-sampling with more emphasis on a smaller data set than on the quality of the model). This can be done with the extension or replacement of benchmark classes (`SamplingBenchmark`, `FeatureSelectionBenchmark` and `WeightingBenchmark`). The following code listing shows the usage of the custom optimizer (random search) and custom benchmark function for data sampling (a mixture of both model quality and size of the data set).

```
>>> import numpy as np
>>> from NiaPy.algorithms import Algorithm
>>> from numpy import apply_along_axis, math
>>> from sklearn.datasets import load_breast_cancer
>>> from sklearn.utils import safe_indexing
>>> from EvoPreprocess.data_sampling import EvoSampling
>>> from EvoPreprocess.data_sampling.SamplingBenchmark import SamplingBenchmark
>>>
>>> class RandomSearch(Algorithm):
Name = ['RandomSearch', 'RS']

def runIteration(self, task, pop, fpop, xb, fxb, **dparams):
pop = task.Lower + self.Rand.rand(self.NP, task.D) * task.bRange
fpop = apply_along_axis(task.eval, 1, pop)
return pop, fpop, {}
>>>
>>> class CustomSamplingBenchmark(SamplingBenchmark):
# ________________0___1_____2______3_______4___
mapping = np.array([0.5, 0.75, 0.875, 0.9375, 1])

def function(self):
def evaluate(D, sol):
phenotype = SamplingBenchmark.map_to_phenotype(
CustomSamplingBenchmark.to_phenotype(sol))
X_sampled = safe_indexing(self.X_train, phenotype)
y_sampled = safe_indexing(self.y_train, phenotype)

if X_sampled.shape[0] > 0:
cls = self.evaluator.fit(X_sampled, y_sampled)
y_predicted = cls.predict(self.X_valid)
quality = accuracy_score(self.y_valid, y_predicted)
size_percentage = len(y_sampled) / len(sol)

return (1 - quality) * size_percentage
else:
return math.inf

return evaluate

@staticmethod
def to_phenotype(genotype):
return np.digitize(genotype[:-5], CustomSamplingBenchmark.mapping)
>>>
>>> dataset = load_breast_cancer()
>>> print(dataset.data.shape, len(dataset.target))
(569, 30) 569

>>> X_resampled, y_resampled = EvoSampling(optimizer=RandomSearch,
benchmark=CustomSamplingBenchmark
).fit_resample(dataset.data, dataset.target)
>>> print(X_resampled.shape, len(y_resampled))
(311, 30) 311
```

5. Experiments

In this section, the results of the experiments of all three data preprocessing tasks with EvoPreprocess framework are presented. The results of the EvoPreprocess framework are compared to other publicly available preprocessing approaches in Python programming language. The classification data set *Libras Movement* [81] was selected, as is the prototype of heavily imbalanced data set with many features. Thus, data sampling, feature selection and data weighting could all be used. The data set contains 360 instances of two classes with imbalance ration 14:1 (334 vs. 24 instances) and 90 continuous features. All tests were run with five-fold cross-validation. The classifier used was scikit-learn implementation of CART decision tree `DecisionTreeClassifier` with default setting. The settings for all EvoPreprocess experiments were the following:

- 2 folds for internal cross-validation,
- 10 independent runs of meta-heuristic algorithm on each fold,
- `Genetic algorithm` meta-heuristic optimizer from NiaPy with the population size of 200, with 0.8 chance of crossover, 0.5 chance of mutation and 20,000 total evaluations,
- `DecisionTreeClassifier` for the evaluation of the solutions (denoted as CART),
- random seed was set to 1111 for all algorithms, and
- all other settings were left at their default values.

The results of weighting the data set instances are shown in Table 1. Classification of the weighted data set (denoted as EvoPreprocess) produced better results, in terms of overall accuracy and F-score, than the classification of the original non-weighted set. This was evident by looking at the average and median score as the average rank (the smaller ranks are preferred). There are no other weighting approaches available in either scikit-learn or imbalanced-learn packages, so the results of EvoPreprocess weighting was compared to non-weighted instances.

Table 1. Accuracy and F-score results of weighting. The best values are bolded.

	CART		EvoPreprocess	
Fold	**Acc**	**Fsc**	**Acc**	**Fsc**
1	83.56	33.33	**87.67**	**47.06**
2	90.28	36.36	**95.83**	**57.14**
3	88.89	20.00	**91.67**	**57.14**
4	**91.67**	25.00	88.89	**42.86**
5	85.92	0.00	**88.73**	**0.00**
Average	88.06	22.94	**90.56**	**40.84**
Median	**88.89**	25.00	**88.89**	**47.06**
Average rank	1.8	1.8	**1.2**	**1.0**

Next, the feature selection EvoPreprocess method was experimented with. All of the scikit-learn feature selection methods were included in this comparison: `chi2`, `ANOVA F` and `Mutual information`. Also, one additional implementation of feature election with genetic algorithm was included in the experiment—`sklearn-genetic` [82]. This implementation was included in the experiment as it is freely available for use when applying the genetic algorithm to feature selection problems and it is compatible with scikit-learn framework (aligning it with of EvoPreprocess). The settings of `sklearn-genetic` were left to their default values, as with EvoPreprocess, to prevent over-fitting to the provided data set.

The results in Table 2 show that, again, in both classification metrics, the classification results were superior when feature selection with EvoPreprocess was used in comparisons to other available methods. Final classification with features selected by EvoPreprocess resulted in the best accuracy and F-score (according to average, median and average rank) Even when compared to the already existing implementation of feature selection with genetic algorithm `sklearn-genetic`, it is evident, that EvoPreprocess framework results in data sets which tend to perform better in classification. Even if this superiority of EvoPreprocess over `sklearn-genetic` is as a result of chance, due to No Free Lunch Theorem, `sklearn-genetic` does not provide self-adaptation and parallelization as is readily available with EvoPreprocess. Furthermore, `sklearn-genetic` provides feature selection only with genetic algorithms, while EvoPreprocess provides preprocessing approaches with many more nature-inspired algorithms and is easily extendable with any future new methods.

Table 2. Accuracy and F-score results of feature selection. The best values are bolded.

	CART		chi2		ANOVA F		Mutual Information		Sklearn Genetic		EvoPreprocess	
Fold	**Acc**	**Fsc**	**Acc**	**Fsc**	**Acc**	**Fsc**	**Acc**	**Fsc**	**Acc**	**Fsc**	**Acc**	**Fsc**
1	84.93	35.29	97.26	80.00	**98.63**	**88.89**	90.41	53.33	82.19	31.58	97.26	80.00
2	83.33	33.33	94.44	33.33	94.44	33.33	**95.83**	**57.14**	94.44	60.00	**95.83**	**57.14**
3	93.06	28.57	91.67	0.00	87.50	0.00	86.11	16.67	91.67	25.00	**95.83**	**57.14**
4	86.11	0.00	93.06	**61.54**	91.67	40.00	93.06	54.55	91.67	50.00	**94.44**	50.00
5	81.69	0.00	83.10	14.29	**90.14**	0.00	84.51	**35.29**	87.32	0.00	**90.14**	0.00
Average	85.82	19.44	91.91	37.83	92.48	32.44	89.98	43.40	89.46	33.32	**94.70**	**48.86**
Median	84.93	28.57	93.06	33.33	91.67	33.33	90.41	53.33	90.56	32.45	**95.83**	**57.14**
Avg. rank	5.0	4.0	3.0	2.8	2.8	3.6	3.4	2.6	3.8	3.2	**1.2**	**2.2**

Lastly, the experiment with data sampling was conducted. Here, several imbalanced-learn implementation for under- and over-sampling were included: `under-sampling with Tomek Links`, `over-sampling with SMOTE`, and `over- and under-sampling with SMOTE and Tomek Links`. The results in Table 3 show the mixed performance of EvoPreprocess for data sampling. In general, both over-sampling methods (SMOTE and SMOTE with Tomek Links) performed better than EvoPreprocess. SMOTE over-sampling does not over-sample instance with duplication but constructs synthetic new instances. Regular over-sampling with duplication can only go so far, while the creation of new knowledge with new instances can further diversify the data set and prevent over-fitting of the model. Also, the implementation of data sampling is such that the search space is positively correlated with the number of instances in the original data set. Here, the data set contains 558 instances, which means, the length of the genotype is at least as large.

Table 3. Accuracy and F-score results of data sampling (under- and over- sampling). The best values are bolded.

	CART		Under-Sampling Tomek Links		Over-Sampling SMOTE		Over-under- Sampling SMOTE with Tomek		EvoPreprocess	
Fold	**Acc**	**Fsc**	**Acc**	**Fsc**	**Acc**	**Fsc**	**Acc**	**Fsc**	**Acc**	**Fsc**
1	84.93	35.29	84.93	35.29	**94.52**	**66.67**	**94.52**	**66.67**	84.93	0.00
2	83.33	33.33	83.33	33.33	**97.22**	**75.00**	**97.22**	**75.00**	94.44	33.33
3	93.06	28.57	93.06	28.57	88.89	**33.33**	88.89	**33.33**	**94.44**	**33.33**
4	86.11	0.00	86.11	0.00	**93.06**	**54.55**	**93.06**	**54.55**	87.50	40.00
5	81.69	0.00	81.69	0.00	92.96	**61.54**	92.96	**61.54**	**94.37**	0.00
Average	85.82	19.44	85.82	19.44	**93.33**	**58.22**	**93.33**	**58.22**	91.14	21.33
Median	84.93	28.57	84.93	28.57	93.06	**61.54**	93.06	**61.54**	**94.37**	33.33
Avg. rank	3.4	3.4	3.4	3.4	**1.8**	**1.0**	**1.8**	**1.0**	2.2	3

It was shown [83] that larger search spaces tend to be harder to solve for meta-heuristic algorithms, making them stuck in the local optima. Numerous techniques tackle large problem solving for meta-heuristic and nature-inspired methods [84–86], but the scope of this paper was to demonstrate the usability of the EvoPreprocess framework in general. The framework could easily be extended and modified with any large-scale mitigation approaches. On the other hand, the data sampling with EvoPreprocess still outperforms the included under-sampling with Tomek Links by a great margin.

As the results of the experiment demonstrate, the EvoPreprocess framework is a viable and competitive alternative to already existing data preprocessing approaches. Even though it did not provide the best results in all cases on the one data set used in the experiment, its effectiveness is not questionable. Because of the No Free Lunch Theorem, there is no perfect setting for the used methods that would perform the best in all of the tasks and different data sets. Naturally, with the right parameter setting, all of the methods used in the experiment would probably perform in better. Tuning the EvoPreprocess setting parameters would change the results, but the optimization to the one used data set is not the goal of this paper. This experiment serves to demonstrate the straightforward use and general effectiveness of the framework.

Comparison of Nature-Inspired Algorithms

In this section, the results of different nature-inspired algorithms in their ability to preprocess data with feature selection, data sampling and data weighting, are compared. A synthetic classification data set was constructed, which contained 400 instances, 10 features and 2 overlapping classes. Out of all 10 features, 3 of them were informative, 3 were redundant of informative features, 2 were transformed copies of informative and redundant features and 2 were random noise. The experiment was done with 5-fold cross-validation and the final results were averaged across all folds. The overall accuracy and F-score were measured, as well as computation time of nature-inspired algorithms. All three preprocessing tasks were applied individually before the preprocessed data was used in training of the CART classification decision tree. The experiment was done on a computer with Intel i5-6500 CPU @ 3.2 GHz with four cores, 32GB of RAM and Windows 10.

The nature-inspired algorithms used in this experiment were the following: *Artificial cee colony*, *Bat algorithm*, *Cuckoo search*, *Differential evolution*, *Evolution strategy* ($\mu + \lambda$), *Genetic algorithm*, *Harmony search*, and *Particle swarm optimization*. The settings of the optimization algorithms were left at their default values, only the size of the solution set (individual solutions in one generation/iteration) was set to 100, and the maximal number of fitness evaluations was set to 100,000.

Table 4. Comparison of different nature-inspired algorithms on three data preprocessing task. Classification accuracy, F-score and computation time (in seconds) are presented. The best results are bolded.

Task	Algorithm	Acc	Fsc	Computation Time (s) with Four Parallel Runs	Computation Time (s) with no Parallel Runs
Feature selection	Artificial bee colony	86.24	66.02	15.0062	42.1964
	Bat algorithm	85.24	64.12	15.2828	47.6856
	Cuckoo search	84.49	60.80	**13.8600**	**35.9927**
	Differential evolution	86.52	66.98	16.2162	61.5314
	Evolution strategy	**87.01**	**68.72**	15.2992	50.4924
	Genetic algorithm	86.51	67.50	15.0095	45.0393
	Harmony search	85.76	65.99	15.9159	50.2762
	Particle swarm optimization	85.25	63.66	14.8623	52.1369
Data sampling	Artificial bee colony	85.50	64.27	19.4990	61.4075
	Bat algorithm	**87.01**	**67.85**	19.0993	58.1645
	Cuckoo search	85.50	63.95	18.9420	53.1541
	Differential evolution	86.25	65.33	23.4585	67.7465
	Evolution strategy	84.00	61.56	19.2668	60.3277
	Genetic algorithm	85.02	62.89	24.8603	73.1325
	Harmony search	86.01	65.59	39.1134	105.1702
	Particle swarm optimization	85.25	64.30	**18.7993**	**52.8433**
Data weighting	Artificial bee colony	87.76	**69.98**	16.7115	43.9391
	Bat algorithm	86.25	66.47	**14.7816**	**37.3560**
	Cuckoo search	86.74	67.55	15.9384	42.0680
	Differential evolution	85.52	65.64	18.8684	51.4843
	Evolution strategy	86.00	67.16	15.6373	72.5965
	Genetic algorithm	86.24	64.35	22.0804	62.2260
	Harmony search	87.01	67.04	37.2622	109.2218
	Particle swarm optimization	**88.00**	68.76	15.5451	40.0671

The results of classification metrics and computation times for all preprocessing tasks are in Table 4 and show no clear winner. Evolution strategy is the best algorithm in feature selection (acc = 87.01%, F-score = 68.72%), and Bat algorithm is the best in data sampling (acc = 87.01%,

F-score = 67.85%). Artificial bee colony (acc = 87.76%, F-score = 69.98%) and Particle swarm optimization (acc = 88%, F-score = 68.76%) perform best in data weighting. Again, this is due to No Free Lunch Theorem so there is no setting or algorithm that could perform the best in all situations (preprocessing tasks or data sets).

Computation times are more consistent for different algorithms. Harmony search is clearly among the slowest optimization algorithms in all three tasks, while Particle swarm optimization, Cuckoo search and Bat algorithm rank among the fastest one in all three tasks. Table 4 also shows the computation times of the same algorithms when no parallelization is used. The computation time advantage of the parallelized implementation of the framework is evident in every nature-inspired algorithm in every data preprocessing tasks. The parallelization is done by the framework and is optimization algorithm independent, which makes custom implementations of optimization algorithms straightforward. The speed differences are due to the differences in the operators (see Section 2.3) and implementation differences (i.e., vectorized operators tend to compute faster, than non-vectorized ones).

6. Conclusions

This paper presents the EvoPreprocess framework for preprocessing data with nature-inspired optimization algorithms for three different tasks: data sampling, data weighting, and feature selection. The architectural structure and the documentation of the application programming interface of the framework are overviewed in detail. The EvoPreprocess framework offers a simple object-oriented implementation, which is easily extendable with custom metrics and custom optimization methods. The presented framework is compatible with the established Python data mining libraries pandas, scikit-learn and imbalanced-learn.

The framework effectiveness is demonstrated in the provided results of the experiment, where it was compared to the well-established data preprocessing packages. As the results suggest, the EvoPreprocess framework already provides competitive results in all preprocessing tasks (in some more than in others) in its current form. However, its true potential is in the customization ability with other variants of data preprocessing tasks.

In the future, unit tests are planned to be included in the framework modules, which could make customization of the provided tasks and classes easier. There are numerous preprocessing tasks which could still benefit from the easy to use package (feature extraction, missing value imputation, data normalization, etc.). Ultimately, the framework may serve the community by providing a simple and customizable tool for use in practical applications and future research endeavors.

Funding: This research was funded by the Slovenian Research Agency (research core funding No. P2-0057).

Conflicts of Interest: The author declares no conflict of interest.

References

1. García, S.; Luengo, J.; Herrera, F. *Data Preprocessing in Data Mining*; Springer: Berlin/Heidelberg, Germany, 2015.
2. Japkowicz, N.; Stephen, S. The class imbalance problem: A systematic study. *Intell. Data Anal.* **2002**, *6*, 429–449. [CrossRef]
3. García, V.; Sánchez, J.S.; Mollineda, R.A. On the effectiveness of preprocessing methods when dealing with different levels of class imbalance. *Knowl.-Based Syst.* **2012**, *25*, 13–21. [CrossRef]
4. Kotsiantis, S.; Kanellopoulos, D.; Pintelas, P. Data preprocessing for supervised leaning. *Int. J. Comput. Sci.* **2006**, *1*, 111–117.
5. Li, J.; Cheng, K.; Wang, S.; Morstatter, F.; Trevino, R.P.; Tang, J.; Liu, H. Feature selection: A data perspective. *ACM Comput. Surv. (CSUR)* **2018**, *50*, 94. [CrossRef]
6. Guyon, I.; Elisseeff, A. An introduction to variable and feature selection. *J. Mach. Learn. Res.* **2003**, *3*, 1157–1182.

7. Laurikkala, J. Improving Identification of Difficult Small Classes by Balancing Class Distribution. In Proceedings of the 8th Conference on AI in Medicine in Europe: Artificial Intelligence Medicine, Cascais, Portugal, 1–4 July 2001; Springer: London, UK, 2001; pp. 63–66.
8. Liu, H.; Motoda, H. *Instance Selection and Construction for Data Mining*; Springer Science & Business Media: Berlin/Heidelberg, Germany, 2013; Volume 608.
9. Chawla, N.V.; Bowyer, K.W.; Hall, L.O.; Kegelmeyer, W.P. SMOTE: Synthetic minority over-sampling technique. *J. Artif. Intell. Res.* **2002**, *16*, 321–357. [CrossRef]
10. Fernández, A.; Garcia, S.; Herrera, F.; Chawla, N.V. SMOTE for learning from imbalanced data: Progress and challenges, marking the 15-year anniversary. *J. Artif. Intell. Res.* **2018**, *61*, 863–905. [CrossRef]
11. Diao, R.; Shen, Q. Nature inspired feature selection meta-heuristics. *Artif. Intell. Rev.* **2015**, *44*, 311–340. [CrossRef]
12. Galar, M.; Fernández, A.; Barrenechea, E.; Herrera, F. EUSBoost: Enhancing ensembles for highly imbalanced data-sets by evolutionary undersampling. *Pattern Recognit.* **2013**, *46*, 3460–3471. [CrossRef]
13. Sayed, S.; Nassef, M.; Badr, A.; Farag, I. A nested genetic algorithm for feature selection in high-dimensional cancer microarray datasets. *Expert Syst. Appl.* **2019**, *121*, 233–243. [CrossRef]
14. Ghosh, M.; Adhikary, S.; Ghosh, K.K.; Sardar, A.; Begum, S.; Sarkar, R. Genetic algorithm based cancerous gene identification from microarray data using ensemble of filter methods. *Med. Biol. Eng. Comput.* **2019**, *57*, 159–176. [CrossRef] [PubMed]
15. Rao, K.N.; Reddy, C.S. A novel under sampling strategy for efficient software defect analysis of skewed distributed data. *Evol. Syst.* **2019**, *11*, 119–131. [CrossRef]
16. Subudhi, S.; Patro, R.N.; Biswal, P.K. Pso-based synthetic minority oversampling technique for classification of reduced hyperspectral image. In *Soft Computing for Problem Solving*; Springer: Berlin/Heidelberg, Germany, 2019; pp. 617–625.
17. Lemaître, G.; Nogueira, F.; Aridas, C.K. Imbalanced-learn: A python toolbox to tackle the curse of imbalanced datasets in machine learning. *J. Mach. Learn. Res.* **2017**, *18*, 559–563.
18. Kursa, M.B.; Rudnicki, W.R. Feature selection with the Boruta package. *J. Stat. Softw.* **2010**, *36*, 1–13. [CrossRef]
19. Lagani, V.; Athineou, G.; Farcomeni, A.; Tsagris, M.; Tsamardinos, I. Feature selection with the r package mxm: Discovering statistically-equivalent feature subsets. *arXiv* **2016**, arXiv:1611.03227.
20. Scrucca, L.; Raftery, A.E. clustvarsel: A Package Implementing Variable Selection for Gaussian Model-based Clustering in R. *J. Stat. Softw.* **2018**, *84*. [CrossRef]
21. Dramiński, M.; Koronacki, J. rmcfs: An R Package for Monte Carlo Feature Selection and Interdependency Discovery. *J. Stat. Softw.* **2018**, *85*, 1–28. [CrossRef]
22. Lunardon, N.; Menardi, G.; Torelli, N. ROSE: A Package for Binary Imbalanced Learning. *R J.* **2014**, *6*, 79–89. [CrossRef]
23. Liu, H.; Motoda, H. *Computational Methods of Feature Selection*; CRC Press: Boca Raton, FL, USA, 2007.
24. Xue, B.; Zhang, M.; Browne, W.N.; Yao, X. A survey on evolutionary computation approaches to feature selection. *IEEE Trans. Evol. Comput.* **2015**, *20*, 606–626. [CrossRef]
25. Brezočnik, L.; Fister, I.; Podgorelec, V. Swarm intelligence algorithms for feature selection: A review. *Appl. Sci.* **2018**, *8*, 1521. [CrossRef]
26. Mafarja, M.M.; Mirjalili, S. Hybrid whale optimization algorithm with simulated annealing for feature selection. *Neurocomputing* **2017**, *260*, 302–312. [CrossRef]
27. Mafarja, M.; Aljarah, I.; Heidari, A.A.; Faris, H.; Fournier-Viger, P.; Li, X.; Mirjalili, S. Binary dragonfly optimization for feature selection using time-varying transfer functions. *Knowl.-Based Syst.* **2018**, *161*, 185–204. [CrossRef]
28. Sayed, G.I.; Tharwat, A.; Hassanien, A.E. Chaotic dragonfly algorithm: An improved metaheuristic algorithm for feature selection. *Appl. Intell.* **2019**, *49*, 188–205. [CrossRef]
29. Aljarah, I.; Ala'M, A.Z.; Faris, H.; Hassonah, M.A.; Mirjalili, S.; Saadeh, H. Simultaneous feature selection and support vector machine optimization using the grasshopper optimization algorithm. *Cogn. Comput.* **2018**, *10*, 478–495. [CrossRef]
30. Abdel-Basset, M.; El-Shahat, D.; El-henawy, I.; de Albuquerque, V.H.C.; Mirjalili, S. A new fusion of grey wolf optimizer algorithm with a two-phase mutation for feature selection. *Expert Syst. Appl.* **2020**, *139*, 112824. [CrossRef]

31. Al-Tashi, Q.; Kadir, S.J.A.; Rais, H.M.; Mirjalili, S.; Alhussian, H. Binary Optimization Using Hybrid Grey Wolf Optimization for Feature Selection. *IEEE Access* **2019**, *7*, 39496–39508. [CrossRef]
32. Zorarpacı, E.; Özel, S.A. A hybrid approach of differential evolution and artificial bee colony for feature selection. *Expert Syst. Appl.* **2016**, *62*, 91–103. [CrossRef]
33. Sayed, G.I.; Hassanien, A.E.; Azar, A.T. Feature selection via a novel chaotic crow search algorithm. *Neural Comput. Appl.* **2019**, *31*, 171–188. [CrossRef]
34. Gu, S.; Cheng, R.; Jin, Y. Feature selection for high-dimensional classification using a competitive swarm optimizer. *Soft Comput.* **2018**, *22*, 811–822. [CrossRef]
35. Dong, H.; Li, T.; Ding, R.; Sun, J. A novel hybrid genetic algorithm with granular information for feature selection and optimization. *Appl. Soft Comput.* **2018**, *65*, 33–46. [CrossRef]
36. Ali, A.; Shamsuddin, S.M.; Ralescu, A.L. Classification with class imbalance problem: A review. *Int. J. Adv. Soft. Comput. Appl.* **2015**, *7*, 176–204.
37. Dragusin, R.; Petcu, P.; Lioma, C.; Larsen, B.; Jørgensen, H.; Winther, O. Rare disease diagnosis as an information retrieval task. In Proceedings of the Conference on the Theory of Information Retrieval, Bertinoro, Italy, 12–14 September 2011; pp. 356–359.
38. Griggs, R.C.; Batshaw, M.; Dunkle, M.; Gopal-Srivastava, R.; Kaye, E.; Krischer, J.; Nguyen, T.; Paulus, K.; Merkel, P.A. Clinical research for rare disease: Opportunities, challenges, and solutions. *Mol. Genet. Metab.* **2009**, *96*, 20–26. [CrossRef]
39. Weigold, A.; Weigold, I.K.; Russell, E.J. Examination of the equivalence of self-report survey-based paper-and-pencil and internet data collection methods. *Psychol. Methods* **2013**, *18*, 53. [CrossRef]
40. Etikan, I.; Musa, S.A.; Alkassim, R.S. Comparison of convenience sampling and purposive sampling. *Am. J. Theor. Appl. Stat.* **2016**, *5*, 1–4. [CrossRef]
41. Haixiang, G.; Yijing, L.; Shang, J.; Mingyun, G.; Yuanyue, H.; Bing, G. Learning from class-imbalanced data: Review of methods and applications. *Expert Syst. Appl.* **2017**, *73*, 220–239. [CrossRef]
42. Triguero, I.; Galar, M.; Vluymans, S.; Cornelis, C.; Bustince, H.; Herrera, F.; Saeys, Y. Evolutionary undersampling for imbalanced big data classification. In Proceedings of the 2015 IEEE Congress on Evolutionary Computation (CEC), Sendai, Japan, 25–28 May 2015; pp. 715–722.
43. Fernandes, E.; de Leon Ferreira, A.C.P.; Carvalho, D.; Yao, X. Ensemble of Classifiers based on MultiObjective Genetic Sampling for Imbalanced Data. *IEEE Trans. Knowl. Data Eng.* **2019**, *32*, 1104–1115. [CrossRef]
44. Ha, J.; Lee, J.S. A new under-sampling method using genetic algorithm for imbalanced data classification. In Proceedings of the 10th International Conference on Ubiquitous Information Management and Communication, DaNang, Vietnam, 4–6 January 2016; p. 95.
45. Zhang, L.; Zhang, D. Evolutionary cost-sensitive extreme learning machine. *IEEE Trans. Neural Netw. Learn. Syst.* **2016**, *28*, 3045–3060. [CrossRef]
46. Elkan, C. The foundations of cost-sensitive learning. In Proceedings of the International Joint Conference on Artificial Intelligence; Seattle, WA, USA, 4–10 August 2001; Volume 17, pp. 973–978.
47. Yang, X.S. *Nature-Inspired Optimization Algorithms*; Elsevier: Amsterdam, The Netherlands, 2014.
48. Fister, I., Jr.; Yang, X.S.; Fister, I.; Brest, J.; Fister, D. A brief review of nature-inspired algorithms for optimization. *arXiv* **2013**,arXiv:1307.4186.
49. Yang, X.S.; Cui, Z.; Xiao, R.; Gandomi, A.H.; Karamanoglu, M. *Swarm Intelligence and Bio-Inspired Computation: Theory and Applications*; Newnes: Newton, MA, USA, 2013.
50. Pardalos, P.M.; Prokopyev, O.A.; Busygin, S. Continuous approaches for solving discrete optimization problems. In *Handbook on Modelling for Discrete Optimization*; Springer: Berlin/Heidelberg, Germany, 2006; pp. 39–60.
51. Fister, D.; Fister, I.; Jagrič, T.; Brest, J. Wrapper-Based Feature Selection Using Self-adaptive Differential Evolution. In *Swarm, Evolutionary, and Memetic Computing and Fuzzy and Neural Computing*; Springer: Berlin/Heidelberg, Germany, 2019; pp. 135–154.
52. Ghosh, A.; Datta, A.; Ghosh, S. Self-adaptive differential evolution for feature selection in hyperspectral image data. *Appl. Soft Comput.* **2013**, *13*, 1969–1977. [CrossRef]
53. Tao, X.; Li, Q.; Guo, W.; Ren, C.; Li, C.; Liu, R.; Zou, J. Self-adaptive cost weights-based support vector machine cost-sensitive ensemble for imbalanced data classification. *Inf. Sci.* **2019**, *487*, 31–56. [CrossRef]

54. Brest, J.; Greiner, S.; Boskovic, B.; Mernik, M.; Zumer, V. Self-adapting control parameters in differential evolution: A comparative study on numerical benchmark problems. *IEEE Trans. Evol. Comput.* **2006**, *10*, 646–657. [CrossRef]
55. Zainudin, M.; Sulaiman, M.; Mustapha, N.; Perumal, T.; Nazri, A.; Mohamed, R.; Manaf, S. Feature selection optimization using hybrid relief-f with self-adaptive differential evolution. *Int. J. Intell. Eng. Syst.* **2017**, *10*, 21–29. [CrossRef]
56. Xue, Y.; Xue, B.; Zhang, M. Self-adaptive particle swarm optimization for large-scale feature selection in classification. *ACM Trans. Knowl. Discov. Data (TKDD)* **2019**, *13*, 1–27. [CrossRef]
57. Fister, D.; Fister, I.; Jagrič, T.; Brest, J. A novel self-adaptive differential evolution for feature selection using threshold mechanism. In Proceedings of the 2018 IEEE Symposium Series on Computational Intelligence (SSCI), Bangalore, India, 18–21 November 2018; pp. 17–24.
58. Mafarja, M.; Mirjalili, S. Whale optimization approaches for wrapper feature selection. *Appl. Soft Comput.* **2018**, *62*, 441–453. [CrossRef]
59. Soufan, O.; Kleftogiannis, D.; Kalnis, P.; Bajic, V.B. DWFS: A wrapper feature selection tool based on a parallel genetic algorithm. *PLoS ONE* **2015**, *10*, e0117988. [CrossRef]
60. Mafarja, M.; Eleyan, D.; Abdullah, S.; Mirjalili, S. S-shaped vs. V-shaped transfer functions for ant lion optimization algorithm in feature selection problem. In Proceedings of the international conference on future networks and distributed systems, Cambridge, UK, 19–20 July 2017; p. 21.
61. Ghareb, A.S.; Bakar, A.A.; Hamdan, A.R. Hybrid feature selection based on enhanced genetic algorithm for text categorization. *Expert Syst. Appl.* **2016**, *49*, 31–47. [CrossRef]
62. Lones, M.A. Mitigating metaphors: A comprehensible guide to recent nature-inspired algorithms. *SN Comput. Sci.* **2020**, *1*, 49. [CrossRef]
63. Połap, D. Polar bear optimization algorithm: Meta-heuristic with fast population movement and dynamic birth and death mechanism. *Symmetry* **2017**, *9*, 203. [CrossRef]
64. Kazikova A.; Pluhacek M.; Senkerik R.; Viktorin, A. Proposal of a New Swarm Optimization Method Inspired in Bison Behavior. *Recent Adv. Soft. Comput.* **2019**, 146–156. [CrossRef]
65. Arora, S.; Singh, S. Butterfly optimization algorithm: A novel approach for global optimization. *Soft. Comput.* **2019**, *23*, 715–734. [CrossRef]
66. Klein, C.E.; Mariani, V.C.; dos Santos Coelho, L. *Cheetah Based Optimization Algorithm: A Novel Swarm Intelligence Paradigm*; ESANN: Bruges, Belgium, 2018.
67. Pierezan, J.; Coelho, L.D.S. Coyote optimization algorithm: A new metaheuristic for global optimization problems. In Proceedings of the 2018 IEEE Congress on Evolutionary Computation (CEC), Rio de Janeiro, Brazil, 8–13 July 2018; pp. 1–8.
68. Jain, M.; Singh, V.; Rani, A. A novel nature-inspired algorithm for optimization: Squirrel search algorithm. *Swarm Evol. Comput.* **2019**, *44*, 148–175. [CrossRef]
69. Sampson, J.R. *Adaptation in Natural and Artificial Systems*; Holland, J.H., Ed.; The MIT Press: Cambridge, MA, USA, 1976.
70. Storn, R.; Price, K. Differential evolution–a simple and efficient heuristic for global optimization over continuous spaces. *J. Glob. Optim.* **1997**, *11*, 341–359. [CrossRef]
71. Beyer, H.G.; Schwefel, H.P. Evolution strategies—A comprehensive introduction. *Nat. Comput.* **2002**, *1*, 3–52. [CrossRef]
72. Yang, X.S. Harmony search as a metaheuristic algorithm. In *Music-Inspired Harmony Search Algorithm*; Springer: Berlin/Heidelberg, Germany, 2009; pp. 1–14.
73. Kennedy, J.; Eberhart, R. Particle swarm optimization. In Proceedings of ICNN'95-International Conference on Neural Networks, Perth, Australia, 27 November–1 December 11995; Volume 4, pp. 1942–1948.
74. Karaboga, D.; Basturk, B. A powerful and efficient algorithm for numerical function optimization: Artificial bee colony (ABC) algorithm. *J. Glob. Optim.* **2007**, *39*, 459–471. [CrossRef]
75. Yang, X.S.; Gandomi, A.H. Bat algorithm: A novel approach for global engineering optimization. *Eng. Comput.* **2012**. [CrossRef]
76. Yang, X.S.; Deb, S. Cuckoo search via Lévy flights. In Proceedings of the 2009 World congress on nature & biologically inspired computing (NaBIC), Coimbatore, India, 9–11 December 2009; pp. 210–214.
77. Friedman, J.; Hastie, T.; Tibshirani, R. *The Elements of Statistical Learning*; Springer: New York, NY, USA, 2001; Volume 1.

78. Oliphant, T. *NumPy: A guide to NumPy*; Trelgol Publishing: Spanish Fork, UT, USA, 2006.
79. McKinney, W. Data Structures for Statistical Computing in Python. In Proceedings of the 9th Python in Science Conference, Austin, TX, USA, 28 June–3 July 2010; pp. 51–56.
80. Yang, X.S. A new metaheuristic bat-inspired algorithm. In *Nature inspired cooperative strategies for optimization (NICSO 2010)*; Springer: Berlin/Heidelberg, Germany, 2010; pp. 65–74.
81. Dias, D.B.; Madeo, R.C.; Rocha, T.; Bíscaro, H.H.; Peres, S.M. Hand movement recognition for brazilian sign language: A study using distance-based neural networks. In Proceedings of the 2009 International Joint Conference on Neural Networks, Atlanta, GA, USA, 14–19 June 2009; pp. 697–704.
82. Calzolari, M. *Manuel-Calzolari/Sklearn-Genetic: Sklearn-Genetic 0.2*; Zenodo: Geneva, Switzerland, 2019. [CrossRef]
83. Reeves, C.R. Landscapes, operators and heuristic search. *Ann. Oper. Res.* **1999**, *86*, 473–490. [CrossRef]
84. Yang, Z.; Tang, K.; Yao, X. Large scale evolutionary optimization using cooperative coevolution. *Inf. Sci.* **2008**, *178*, 2985–2999. [CrossRef]
85. Zhang, H.; Ishikawa, M. An extended hybrid genetic algorithm for exploring a large search space. In Proceedings of the 2nd International Conference on Autonomous Robots and Agents, Kyoto, Japan, 10–11 December 2004; pp. 244–248.
86. Siedlecki, W.; Sklansky, J. A note on genetic algorithms for large-scale feature selection. In *Handbook of Pattern Recognition and Computer Vision*; World Scientific: London, UK, 1993; pp. 88–107.

MDPI
St. Alban-Anlage 66
4052 Basel
Switzerland
Tel. +41 61 683 77 34
Fax +41 61 302 89 18
www.mdpi.com

Mathematics Editorial Office
E-mail: mathematics@mdpi.com
www.mdpi.com/journal/mathematics

www.ingramcontent.com/pod-product-compliance
Lightning Source LLC
LaVergne TN
LVHW070158170726
843515LV00007B/1948

* 9 7 8 3 0 3 9 4 3 4 5 4 1 *